Numerical Methods for Modeling Metasurface with GSTC

Numerical Methods for Modeling Metasurface with GSTC

Qiang Ren
School of Electronic and Information Engineering
Beihang University, Beijing, China

Kaiming Wu
School of Electronic and Information Engineering
Beihang University, Beijing, China

Na Liu
School of Electronic Science and Engineering
Xiamen University, Xiamen, China

Guoxiong Cai
School of Electronic Science and Engineering
Xiamen University, Xiamen, China

IEEE Press
WILEY

Published by John Wiley & Sons, Inc., Hoboken, New Jersey.

For general information on our other products and services or for technical support, please contact our Customer Care Department within the United States at (800) 762-2974, outside the United States at (317) 572-3993 or fax (317) 572-4002.

Wiley also publishes its books in a variety of electronic formats. Some content that appears in print may not be available in electronic formats. For more information about Wiley products, visit our web site at www.wiley.com.

Library of Congress Cataloging-in-Publication Data

Names: Ren, Qiang (Professor) author
Title: Numerical methods for modeling metasurface with GSTC / Qiang Ren [and three others].
Description: Hoboken, New Jersey : Wiley-IEEE Press, [2026] | Includes index.
Identifiers: LCCN 2026006204 (print) | LCCN 2026006205 (ebook) | ISBN 9781394244294 cloth | ISBN 9781394244317 adobe pdf | ISBN 9781394244300 epub
Subjects: LCSH: Metasurfaces | Metasurfaces–Computer simulation | Electromagnetic waves
Classification: LCC TK7871.15.M48 R47 2026 (print) | LCC TK7871.15.M48 (ebook)
LC record available at https://lccn.loc.gov/2026006204
LC ebook record available at https://lccn.loc.gov/2026006205

Cover Design: Wiley
Cover Image: © juanma hache/Getty Images

Set in 9.5/12.5pt STIXTwoText by Straive, Chennai, India

This book is dedicated to our families, mentors, present and former students, and the pioneers who paved the way in this field.

Qiang Ren: to my wife, Prof. Zhu, and to my daughter, Muxi (Yvonne) Ren.

Contents

About the Authors

Qiang Ren received the BS degree in electrical engineering from Beihang University, Beijing, China, in 2008, the MS degree in electrical engineering from the Institute of Acoustics, Chinese Academy of Sciences, Beijing, in 2011, and the PhD degree in electrical engineering from Duke University, Durham, NC, USA, in 2015. From 2016 to 2017, he was a Post-Doctoral Researcher with the Computational Electromagnetics and Antennas Research Laboratory (CEARL), The Pennsylvania State University, University Park, PA, USA. In September 2017, he joined the School of Electronics and Information Engineering, Beihang University, where he is currently a "Blue Sky" Full Professor. He has authored 3 books and over 170 articles in peer-reviewed journals and conference proceedings. His research interests include numerical modeling for complex media, multiscale and multiphysics problems, electromagnetic compatibility, and deep learning. Dr. Ren has received several prestigious honors, including the ACES Early Career Award (2021), the Xiaomi Young Scholar (2022), and the Young Scientist Awards from PIERS 2024, APEMC 2022, and the 2018 ACES-China Symposium. He serves as an Associate Editor for *IEEE Journal on Multiscale and Multiphysics Computational Techniques* (JMMCT) *and Microwave and Optical Technology Letters* (MOTL).

Kaiming Wu received the BS and PhD degrees in electronic information engineering from Beihang University, Beijing, China, in 2020 and 2025. He is now with College of Photonic and Electronic Engineering from Fujian Normal University, Fuzhou, China, as a lecturer.

Na Liu received the BS degree in information and computation science from Henan University, Kaifeng, China, in 2008, and the PhD degree in computational mathematics from the University of Chinese Academy of Sciences, Beijing, China, in 2013. From 2013 to 2017, she was a postdoctoral researcher at the Institute of Electromagnetics and Acoustics, Xiamen University. Since March 2017, she has been an Associate Professor with the Institute of Electromagnetics and Acoustics, Xiamen University. Her research interests include computational electromagnetics, especially fast and efficient methods for complex media and their applications in multiphysics simulations.

Guoxiong Cai received the BS and MS degrees and the PhD degree in electrical engineering from Xiamen University, Xiamen, China, in 2005, 2008, and 2013, respectively. He was a Visiting Scholar with the Department of Electrical and Computer Engineering, Duke University, Durham, NC, USA, from 2010 to 2012. He conducts extensive yet intensive research on nanophotonics and optoelectronic techniques, especially applications in sensing, imaging, optical communication, and on-chip optical interconnection. Since July 2013, he has been with the Institute of Electromagnetics and Acoustics, Xiamen University, where he is currently an Associate Professor.

Preface

The past two decades have witnessed an explosive growth in the research and application of metasurfaces – artificial two-dimensional structures that redefine the manipulation of electromagnetic waves with unprecedented flexibility. From flat lenses and polarization converters to radar cross-section reducers and orbital angular momentum generators, metasurfaces have transcended the limitations of natural materials, opening new frontiers in optics, communications, radar, and antennas. However, this revolutionary potential is accompanied by a critical challenge: the numerical simulation of metasurfaces. As a multiscale system composed of electrically small meta-atoms arranged in electrically large arrays, direct full-wave simulation of metasurfaces demands enormous computational resources, often becoming a bottleneck for rapid design, optimization, and innovation.

To address this bottleneck, the generalized sheet transition conditions (GSTC) have emerged as a powerful homogenization tool. By replacing complex subwavelength structures with equivalent surface susceptibility tensors, GSTC enables efficient macroscale analysis while preserving the essential electromagnetic response of metasurfaces. This book aims to provide a systematic, in-depth, and practical guide to integrating GSTC with mainstream computational electromagnetic (CEM) methods, bridging the gap between theoretical principles and engineering practice.

The book is structured to guide readers from foundational concepts to advanced applications through four cohesive chapters. Chapter 1 serves as an introduction, outlining the development, key functionalities, and application scenarios of metasurfaces, followed by an overview of core CEM methods classified into frequency-domain and time-domain categories. It establishes the motivation for GSTC-based fast simulation by highlighting the multiscale challenge of traditional metasurface modeling.

Chapter 2 delves into the theoretical foundation of GSTC. Starting with asymptotic analysis of Maxwell's equations and homogenization theory, it elaborates on electromagnetic material properties, dispersion characteristics, and traditional boundary conditions, before rigorously deriving GSTC through both distribution theory and the pill-box method. The chapter also details practical methods for determining surface susceptibilities, including retrieval from scattering parameters, transformation from bulk material parameters, and direct synthesis for targeted functionalities, supplemented by illustrative examples to reinforce key concepts.

Chapters 3 and 4 form the core of the book, focusing on the integration of GSTC with frequency-domain and time-domain CEM methods, respectively. Chapter 3 covers four frequency-domain techniques: finite-difference frequency-domain (FDFD), finite element method (FEM), spectral element method (SEM), and integral equation (IE) method. For each method, we present strategies for incorporating GSTC into its framework, followed by numerical examples – such as beam deflectors, polarization beam splitters, and electromagnetic illusions – that demonstrate accuracy, efficiency, and applicability. Chapter 4 extends the scope to time-domain methods, including the finite-difference time-domain (FDTD) method and discontinuous Galerkin time-domain (DGTD) method. It addresses unique time-domain challenges such as dispersion modeling, stability criteria, and auxiliary differential equations (ADEs), with examples ranging from wideband refraction metasurfaces to conformal curved arrays, showcasing the versatility of GSTC in handling broadband and complex geometric scenarios.

What sets this book apart is its emphasis on practicality and comprehensiveness. We avoid excessive mathematical abstraction while ensuring theoretical rigor, providing detailed implementation strategies, simulation scripts, and comparative analyses with commercial software results. Whether you are a graduate student entering the field of metasurfaces, a researcher seeking efficient simulation tools, or an engineer engaged in metasurface-based product development, this book offers valuable insights and actionable techniques.

As metasurface technology continues to evolve, with emerging trends such as spatially dispersive metasurfaces, time-varying metasurfaces, and multi-physics integration, the GSTC-based simulation framework presented in this book will remain a foundational tool. We hope this book inspires further innovations in both metasurface design and computational electromagnetics, advancing the translation of metasurface research into real-world applications.

February, 2026

Qiang Ren
Kaiming Wu
Na Liu
Guoxiong Cai

Acknowledgments

We would like to express our sincere gratitude to everyone who helped during the preparation and publication of this book. Special thanks go to our advisors, Professor Qing Huo Liu and Professor Douglas Werner, our collaborator Professor Yuanguo Zhou, as well as our former and current students – Mr. Shaowen Tian, Mr. Kaier Shuai, Dr. Heng Cao, Dr. Chao Li, and Ms. Lujie Wang – for their valuable guidance, collaboration, and contributions to this book.

We would like to thank our Managing Editor, Nandhini Karuppiah, for her professional and efficient coordination throughout the publication process. We also thank the reviewers for their constructive comments and suggestions, which have helped improve the quality of this book. Last but not least, we are deeply grateful to our families for their unwavering support.

1

Introduction

When electromagnetic waves pass through an interface between two media, the reflection angle is equal to the incidence angle, and the transmission angle follows Snell's law. So, what if one wants electromagnetic waves to have richer reflection and transmission characteristics at the interface? The answer is the artificial surface, i.e., metasurface, which is generally defined as an "artificial" two-dimensional material composed of an array of electrically small scatterers, with electromagnetic properties that cannot be found in natural media. Metasurface has been a research hotspot in many fields, such as optics, communications, radar, and antennas, in the past decade.

Figure 1.1 shows the number of publications with "metasurface" as a keyword in the core database of the Web of Science since 2000 [1]. As a comparison, we show the number of the keyword "FDTD" (finite-difference time-domain method), which is a classic and widely used computational electromagnetic algorithm and is also one of the algorithms employed in this book. It can be seen that the research on metasurface has increased rapidly in the past 10 years. Figure 1.2 shows the word frequency statistics of the titles of the accepted papers at AP-S/URSI 2024 in Italy [2], in which metasurface ranks fourth among all the keywords with a rate of 7.3%.

Before fabrication, electromagnetic simulation of the metasurface is a necessary step, which is an imitative representation of the possible response of metasurfaces in the real world. Different from the classical electromagnetic full-wave simulation problems such as waveguide transmission, antenna design, characteristic mode analysis, and radar cross-section calculation, the simulation of the metasurface is a complex electrical multiscale problem caused by the combination of the electrically small structure of meta-atoms and the electrically large array, which needs a lot of computational resources and hence limits the development of metasurface.

Therefore, this book introduces a fast simulation method for metasurfaces, i.e., the generalized sheet transition conditions (GSTC), which have become popular

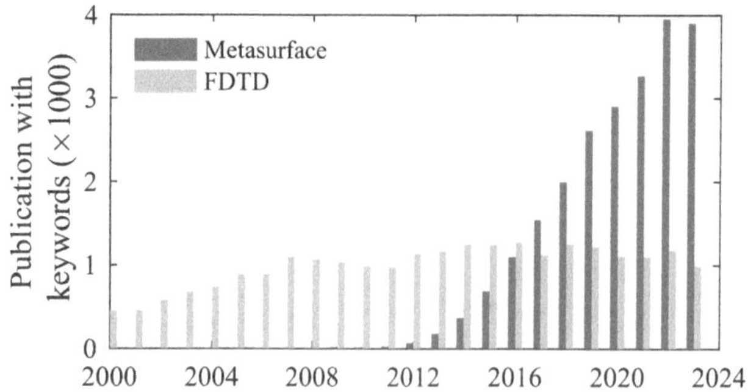

Figure 1.1 Number of publications with keywords "Metasurface" and "FDTD" in Web of Science.

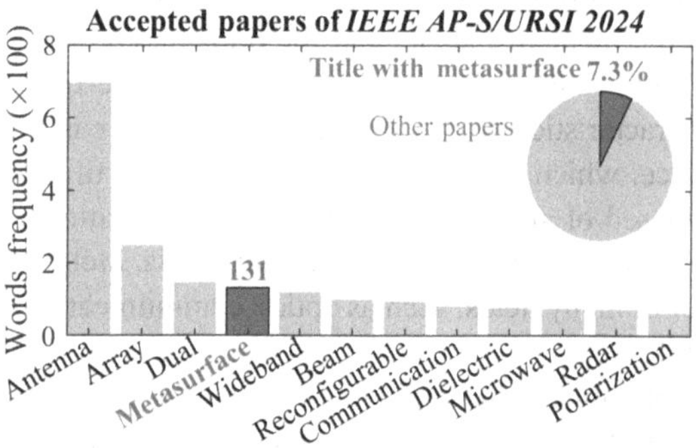

Figure 1.2 The word frequency of the titles of the accepted papers in the 2024 IEEE International Symposium on Antennas and Propagation and INC-USNC-URSI Radio Science Meeting (AP-S/URSI).

and well studied recently. Based on the homogenization theory, it uses surface parameters to replace the subwavelength structure equivalently in the full-wave simulations, avoiding the excessive computational resources of brute-force direct solving. This book provides a systematic and detailed description of how to combine GSTC with the six mainstream computational electromagnetic methods, with numerical examples. Since there is no commercial software including GSTC yet, this book is valuable in paving a new way for the fast analysis of metasurfaces.

1.1 Overview of Metasurfaces

The concept of metasurfaces emerged from the field of metamaterials, which are engineered to have properties not found in nature. Early work on metamaterials focused on three-dimensional (3-D) structures designed to manipulate electromagnetic waves, leading to the concept of negative-index materials, first theoretically proposed by Veselago in 1967 [3] and then experimentally realized by Smith in 2000 [4]. The invisibility cloak, which sounds like a magic tool from science fiction, has been successfully realized by metamaterials [5]. Since then, metamaterials have captured the interest of both researchers and the public and have found great applications in improved imaging systems, advanced telecommunications, and innovative medical devices [6].

However, metamaterials are limited by their thickness in the direction of propagation, which leads to excessive loss, and researchers need to use several times or even dozens of times the excitation power in order to obtain electromagnetic waves with the required properties. Besides, the complexity of large-scale 3-D manufacturing is another key point that hinders the wide industrial applications of metamaterials. In 2002, two groups [7, 8] proposed two-dimensional metamaterials based on the transmission line concept to decrease the loss significantly, which is the prototype of the metasurface.

Metasurfaces not only achieve similar or even superior abilities to manipulate electromagnetic waves at a surface rather than throughout a volume, but also have the advantages of easier fabrication, lower losses, and smaller physical dimensions. Over the years, metasurfaces have been applied across a large spectrum range, from radio frequency to optical regimes, enabling advanced applications such as flat lenses for focus and imaging, communication systems, and cloaking and stealth technology. Figure 1.3 depicts eight real applications with metasurfaces, including (a) metasurface-based wearable sensor, (b) antenna decoupling, (c) multicolor and 3-D holography generator, (d) orbital angular momentum (OAM) wave generator, (e) Bessel beam former, (f) polarization selector, (g) RCS reduction, and (h) conformal antenna. Functionalities have been greatly enriched to control one or more aspects of electromagnetic waves, and three typical ones are described as follows.

Amplitude Control: When electromagnetic waves impinge on a surface between two media, reflection and refraction happen, and the ratio of them depends on the refractive index contrast. Metasurface provides the possibility to manipulate the amplitudes of the reflected and transmitted waves at will. High-impedance metasurfaces not only enable high reflection but also reflect the incident plane wave in phase and suppress surface waves [17]. These unique properties are useful for applications such as low-profile directed antennas [18, 19], scattering reduction [20], and antenna decoupling [10, 21].

The concept of controlling amplitude by surface impedance also leads to the perfect absorbers and the frequency-selective surfaces [22–24]. By carefully

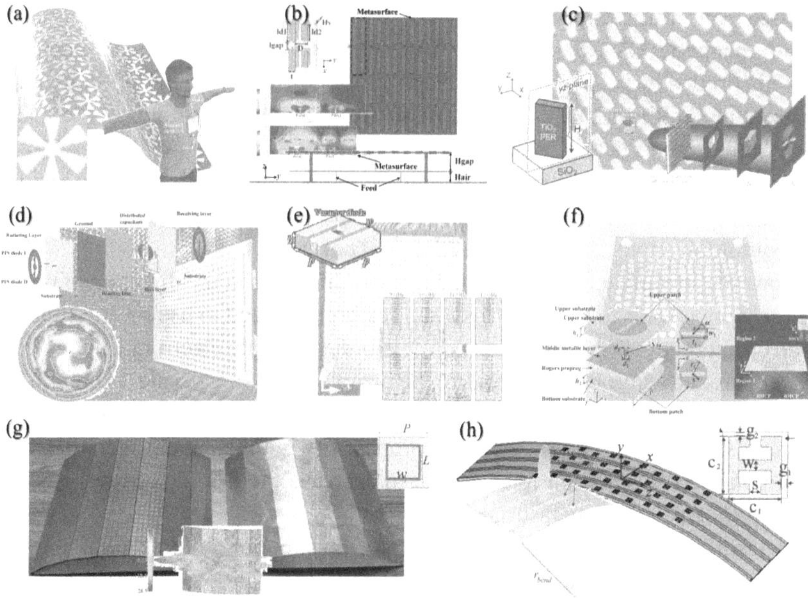

Figure 1.3 Some typical applications of metasurfaces. The figures with the gray filter are the metasurface array, the blue structures are the meta-atoms, while the orange sketches are the function of the metasurfaces. (a) Metasurface-based wearable sensor. *Source:* Chen et al. [9] / with permission of John Wiley & Sons, (b) antenna decoupling. *Source:* Liu et al. [10] / with permission of IEEE, (c) multicolor and 3-D holography generator. *Source:* Adapted from So et al. [11], (d) OAM generator. *Source:* Adapted from Bai et al. [12], (e) Bessel beam former. *Source:* Adapted from Feng et al. [13], (f) polarization selector. *Source:* Adapted from Yang et al. [14], (g) RCS reduction. *Source:* Adapted from Wang et al. [15] / with permission of IEEE, and (h) conformal antenna. *Source:* Adapted from Yoo and Smith [16].

designing the electric and/or magnetic impedances of the structured surfaces, they can perform functions such as energy harvesting [25, 26], sensing [9, 27], and spectral and spatial filtering [28].

Wavefront Control: Wavefront control is typically achieved through phase manipulation. In traditional surfaces, this is severely limited by the inherent properties of the material. Fortunately, metasurfaces offer the capability to introduce phase discontinuities by specifically designing and arranging meta-atoms, allowing for flexible control of electromagnetic wavefront [29]. Various phase strategies are available, including the propagation phase, geometric phase, and resonance phase [30]. By employing these design techniques, meta-atoms with 2π phase coverage can be engineered, and metasurfaces with controllable phase profiles can be constructed to achieve a range of advanced functionalities, like beam shaping and holographic projection [11, 31]. Consequently, metasurfaces have found widespread applications from microwave to optical wavelengths. In the microwave domain, metasurfaces are frequently used for beam focusing to enhance antenna gain and transmission efficiency [32]. In the terahertz and optical bands, in addition to beam focusing, metasurfaces are also engineered to generate specialized beams, such as vortex beams [33] and Bessel beams [13].

Polarization Control: Traditional polarization controllers rely on the anisotropy of natural materials, inherently giving rise to the drawbacks of bulky size, narrow-band operation, and limited functionality. Metasurfaces precisely control the polarization of electromagnetic waves on ultrathin thicknesses by anisotropic meta-atoms, e.g., T-, C-, and L-shaped structures and their different spatial positions and orientations [34].

For instance, by designing metal H-shaped structures or the metal cut wires as top resonators of the metal-insulator-metal (MIM) configuration, the polarization states of reflected waves can be manipulated and experimentally implemented in the microwave [35], terahertz [36], and optical frequency bands [37], with their polarization conversion ratios (PCR) approaching 100%. Due to the low conversion efficiencies, controls on transmitted waves would be a little bit difficult. Despite this, the T-shaped meta-atoms made a metasurface to perform as a quarter-wave plate (QWP) with a ~50% efficiency and a broad bandwidth across the visible to near-infrared range [38], while chiral metasurfaces provide the ability to improve the PCR and the working efficiency [39, 40]. The latter can even enhance chirality and chiral near-fields to stimulate biosensing for chiral molecules. Also recently, bifunctional or multifunctional metasurfaces involving polarization manipulation have promoted themselves to real-world applications in wireless communication and detection [14].

As research develops on the theory and mechanism, design, numerical simulation, physical implementation, and experimental measurement, the field of the metasurface is growing prosperously to profoundly revolutionize electromagnetic science [41, 42].

1.2 Computational Electromagnetic Methods

Some solutions can be obtained analytically, such as the plane wave solution in vacuum based on the rectangular coordinate system, the Bessel function solution in vacuum based on a cylindrical coordinate system, and the transverse electric (TE) and transverse magnetic (TM) mode solutions in a waveguide.

These are the analytical solutions of Maxwell's equations under corresponding boundary conditions. However, most practical engineering problems have no analytical solutions because of their irregular geometries. Hence, researchers use numerical methods to find approximate solutions to these problems. Based on numerical analysis and computer technology, the discipline of researching numerical solutions is called computational electromagnetics (CEM). Since its development in the early 1960s, CEM has been widely used in radar, antenna, microwave, circuit, electronic and nanodevices, biomedicine, electromagnetic imaging, and so on [43, 44].

There are two main types of CEM methods: high-frequency approximation methods and full-wave methods. The high-frequency approximation method is based on the optical description of high-frequency electromagnetic waves, and the size of the problem is usually much larger than the working wavelength. This category includes the basic geometrical optics (GO) method and physical optics (PO) method, and the methods incorporating diffraction theory, such as geometrical theory of diffraction (GTD) method and physical theory of diffraction (PTD) method. Subsequent developments include the unified theory of diffraction (UTD) and the shooting and bouncing ray (SBR) method. Since these high-frequency approximation methods are rarely used in metasurface simulation, they will not be covered in this book. The full-wave methods, which constitute the main content of this book, are based on strict numerical solutions of Maxwell's equations, Helmholtz equations, or various integral equations in the frequency domain or time domain. According to the different mathematical properties of the governing equations, they can be subdivided into two types of numerical full-wave methods, i.e., the partial differential equations (PDE)-based method and the integral equations (IE)-based method. Full-wave models extensively make use of symmetry, infinity, and periodicity, and solve for reduced dimensionality from 3-D problems to two-dimensional (2-D) ones or even one-dimensional (1-D) ones, which will be often seen in this book. The six numerical CEM methods introduced in the following chapters are

1) Finite-difference frequency-domain (FDFD) method [45]
2) Finite element method (FEM) [46]
3) Spectral element method (SEM) [47]
4) Boundary element method (BEM) [48]
5) Finite-difference time-domain (FDTD) method [49]
6) Discontinuous Galerkin time-domain (DGTD) method [50]

where methods 1, 2, 3, 5, and 6 are PDE methods, and method 4 is the IE method. The control equations of PDE methods, i.e., Maxwell's equations, are written as

$$\nabla \times \mathbf{E} = -\frac{\partial \mathbf{B}}{\partial t} \tag{1.1.a}$$

$$\nabla \times \mathbf{H} = \frac{\partial \mathbf{D}}{\partial t} + \mathbf{J} \tag{1.1.b}$$

$$\nabla \cdot \mathbf{D} = \rho \tag{1.1.c}$$

$$\nabla \cdot \mathbf{B} = 0 \tag{1.1.d}$$

where the field quantities **E**, **H**, **D**, and **B** are electric field intensity, magnetic field intensity, electric flux density, and magnetic flux density. The source terms **J** and ρ are volume current density and free volume charge density. The IE method is based on the equivalence principle, which requires integrating the equivalent electric current $\mathbf{J}_s$ and magnetic current $\mathbf{K}_s$ on the metasurface to solve the field response. The control equations are written as

$$\mathbf{E} = -jkZ \int \left[\mathbf{J}_s + \frac{1}{k^2}\nabla(\nabla\cdot\mathbf{J}_s)\right] Gdr_s + \int \mathbf{K}_s \times \nabla Gdr_s \tag{1.2.a}$$

$$\mathbf{H} = -j\frac{k}{Z} \int \left[\mathbf{K}_s + \frac{1}{k^2}\nabla(\nabla\cdot\mathbf{K}_s)\right] Gdr_s - \int \mathbf{J}_s \times \nabla Gdr_s \tag{1.2.b}$$

where Z, k, and G denote the wave impedance, wave number, and the Green's function. In metasurface simulations, when we want to know the macroscopic response of a metasurface under excitations such as linearly polarized plane waves, we need to identify a region around the geometry of this metasurface and discretize this region with grids, and this region is called the computational domain, and the discretization is called meshing. The PDE-based method also imposes a truncation condition on the boundary of the computational domain to mimic an infinite open space outside the computational region. Methods 1 and 5 use regular square grids, called structured mesh, while methods 2, 3, 4, and 6 support unstructured grids, such as tetrahedrons and hexahedrons.[1] The unstructured grids are geometrically flexible and suitable for metasurfaces with complex geometric structures and curved arrays. In each grid, whether to use a known base, such as a polynomial series, to approximate the solution is also one of the differences between methods 1, 5 and 2, 3, 4, 6. This base is called the basis function, and the highest power series that the basis function can represent is its order. The advantage of the basis function is that the precision can be improved by setting the order of basis functions, which provides a method to improve the precision in addition to enhancing the meshing density. The number of the basis functions is referred to as the degrees of freedom (DoFs). Finally, according to whether the fields are time-dependent or time-harmonic, the methods can be divided into frequency-domain methods and time-domain methods, which is also the reason for the division of Chapters 3 and 4 of this

1 Of course, the FD method also has related research using unstructured grids, but in this book, we only discuss the basic form of each algorithm.

Table 1.1 Simple comparison of the six numerical CEM methods introduced in this book.

	FDFD	FEM (SEM)	BEM	FDTD	DGTD
Controlling equation	PDE	PDE	IE	PDE	PDE
Grid type	Structured	Unstructured	Unstructured	Structured	Unstructured
Basis function	×	√	√	×	√
Working domain	Frequency	Frequency	Frequency	Time	Time

book. The frequency-domain methods 1, 2, 3, and 4 obtain the steady-state solution under harmonic illumination, which is suitable for single-frequency analysis, while the time-domain methods 5 and 6 obtain the transient solution, which is suitable for broadband analysis. To summarize, Table 1.1 gives a simple comparison of the six methods used in this book.

In addition to the above full-wave methods, there are other full-wave methods for metasurface analysis, such as the Cagniard-De Hoop technique [51], but until the publication of this book, there is no published work combining them with GSTC, so they will not be covered in this book.

1.3 Layout of This Book

Chapter 2 introduces the GSTC theory. Beginning with the asymptotic analysis of Maxwell's equations, we introduce the concept of using homogenization to replace electrically small structures and the related basic concepts of electromagnetism, and we also introduce some classical boundary conditions. Then, we derive the expression of GSTC rigorously from mathematics and electromagnetism and compare it with the traditional boundary conditions. Finally, we give the methods to determine the equivalent parameters of GSTC, i.e., the surface susceptibility tensor.

Chapter 3 introduces the FDFD, FEM, SEM, and IE (BEM) methods. All these methods are developed in the following order, i.e., basic theories (especially the parts related to GSTC), the strategies for integrating GSTC in their basic framework, and simulation examples of metasurface using the incorporated methods.

Chapter 4 introduces the FDTD and DGTD methods and follows the same way as Chapter 3.

Chapter 5 presents an overview of this book and discusses the state of the art in metasurface fast simulation, and finally closes with some remarks on future challenges.

References

1 Clarivate, Journal Citation Reports, 2024. https://jcr.clarivate.com/ Dec 10, 2024.

2 APS-URSI 2024, List of Accepted Papers, 2024. https://2024.apsursi.org/papers/accepted_papers.php Dec 10, 2024.

3 V. G. Veselago, "The electrodynamics of substances with simultaneously negative values of ε and μ," *Sov. Phys. Uspekhi*, vol. 10, no. 4, p. 509, 1968.

4 D. R. Smith, W. J. Padilla, D. Vier, S. C. Nemat-Nasser, and S. Schultz, "Composite medium with simultaneously negative permeability and permittivity," *Phys. Rev. Lett.*, vol. 84, no. 18, p. 4184, 2000.

5 D. Schurig *et al.*, "Metamaterial electromagnetic cloak at microwave frequencies," *Science*, vol. 314, no. 5801, pp. 977–980, 2006.

6 J. B. Pendry, D. Schurig, and D. R. Smith, "Controlling electromagnetic fields," *Science*, vol. 312, no. 5781, pp. 1780–1782, 2006.

7 A. Grbic and G. V. Eleftheriades, "Experimental verification of backward-wave radiation from a negative refractive index metamaterial," *J. Appl. Phys.*, vol. 92, no. 10, pp. 5930–5935, 2002.

8 C. Caloz, H. Okabe, T. Iwai, and T. Itoh, "Transmission line approach of left-handed (LH) materials," in *Proc. USNC/URSI National Radio Science Meeting*, vol. 1. San Antonio, TX, June 2002, p. 39.

9 L. Chen *et al.*, "Flexible and robust metasurface-based wearable sensor for intelligent human monitoring," *Adv. Mater.*, e14150, Dec. 2025.

10 F. Liu, J. Guo, L. Zhao, G.-L. Huang, Y. Li, and Y. Yin, "Dual-band metasurface-based decoupling method for two closely packed dual-band antennas," *IEEE Trans. Antennas Propag.*, vol. 68, no. 1, pp. 552–557, Jan. 2020, doi: 10.1109/TAP.2019.2940316.

11 S. So *et al.*, "Multicolor and 3D holography generated by inverse-designed single-cell metasurfaces," *Adv. Mater.*, vol. 35, no. 17, p. 2208520, 2023.

12 X. Bai *et al.*, "High-efficiency transmissive programmable metasurface for multimode OAM generation," *Adv. Opt. Mater.*, vol. 8, no. 17, p. 2000570, 2020.

13 R. Feng *et al.*, "Flexible manipulation of bessel-like beams with a reconfigurable metasurface," *Adv. Opt. Mater.*, vol. 8, no. 23, p. 2001084, 2020.

14 W. Yang *et al.*, "Polarization-selective bifunctional metasurface for high-efficiency millimeter-wave folded transmitarray antenna with circular polarization," *IEEE Trans. Antennas Propag.*, vol. 70, no. 9, pp. 8184–8194, 2022.

15 Y. Wang, J. Su, Z. Li, Q. Guo, and J. Song, "A prismatic conformal metasurface for radar cross-sectional reduction," *IEEE Antennas Wirel. Propag. Lett.*, vol. 19, no. 4, pp. 631–635, Apr. 2020, doi: 10.1109/LAWP.2020.2974018.

16 I. Yoo and D. R. Smith, "Design of conformal array of rectangular waveguide-fed metasurfaces," *IEEE Trans. Antennas Propag.*, vol. 70, no. 7, pp. 6060–6065, Jul. 2022, doi: 10.1109/TAP.2022.3140822.
17 D. Sievenpiper, L. Zhang, R. F. J. Broas, N. G. Alexopolous, and E. Yablonovitch, "High-impedance electromagnetic surfaces with a forbidden frequency band," *IEEE Trans. Microwave Theory Tech.*, vol. 47, no. 11, pp. 2059–2074, 1999.
18 J. Kim and Y. Rahmat-Samii, "Low-profile loop antenna above EBG structure," in *2005 IEEE APS Int. Symp.*, 2005, vol. 2A. IEEE, pp. 570–573.
19 A. P. Feresidis, G. Goussetis, S. Wang, and J. C. Vardaxoglou, "Artificial magnetic conductor surfaces and their application to low-profile high-gain planar antennas," *IEEE Trans. Antennas Propag.*, vol. 53, no. 1, pp. 209–215, 2005.
20 D. F. Sievenpiper, "Nonlinear grounded metasurfaces for suppression of high-power pulsed RF currents," *IEEE Antennas Wirel. Propag. Lett.*, vol. 10, pp. 1516–1519, 2011.
21 R. F. J. Broas, D. F. Sievenpiper, and E. Yablonovitch, "An application of high-impedance ground planes to phased array antennas," *IEEE Trans. Antennas Propag.*, vol. 53, no. 4, pp. 1377–1381, 2005.
22 Y. Ra'di, C. R. Simovski, and S. A. Tretyakov, "Thin perfect absorbers for electromagnetic waves: Theory, design, and realizations," *Phys. Rev. Appl.*, vol. 3, no. 3, 037001, 2015.
23 B. A. Munk, *Frequency Selective Surfaces: Theory and Design*. John Wiley & Sons, 2005.
24 N. Engheta, "Thin absorbing screens using metamaterial surfaces," in *IEEE APS Int. Symp. (IEEE Cat. No. 02CH37313)*, 2002, vol. 2. IEEE, pp. 392–395.
25 E. Cortés *et al.*, "Optical metasurfaces for energy conversion," *Chem. Rev.*, vol. 122, no. 19, pp. 15082–15176, 2022.
26 B. Alavikia, T. S. Almoneef, and O. M. Ramahi, "Electromagnetic energy harvesting using complementary split-ring resonators," *Appl. Phys. Lett.*, vol. 104, no. 16, 163903, 2014.
27 N. Liu, M. Mesch, T. Weiss, M. Hentschel, and H. Giessen, "Infrared perfect absorber and its application as plasmonic sensor," *Nano Lett.*, vol. 10, no. 7, pp. 2342–2348, 2010.
28 K. Sarabandi and N. Behdad, "A frequency selective surface with miniaturized elements," *IEEE Trans. Antennas Propag.*, vol. 55, no. 5, pp. 1239–1245, 2007.
29 N. Yu and F. Capasso, "Flat optics with designer metasurfaces," *Nat. Mater.*, vol. 13, no. 2, pp. 139–150, 2014.
30 S. Chen, Z. Li, Y. Zhang, H. Cheng, and J. Tian, "Phase manipulation of electromagnetic waves with metasurfaces and its applications in nanophotonics," *Adv. Opt. Mater.*, vol. 6, no. 13, p. 1800104, 2018.

31 H. Sroor *et al.*, "High-purity orbital angular momentum states from a visible metasurface laser," *Nat. Photonics*, vol. 14, no. 8, pp. 498–503, 2020.

32 Z. N. Chen *et al.*, "Microwave metalens antennas," *Proc. IEEE*, vol. 111, no. 8, pp. 978–1010, 2023.

33 Q. Zhou *et al.*, "Generation of perfect vortex beams by dielectric geometric metasurface for visible light," *Laser Photonics Rev.*, vol. 15, no. 12, p. 2100390, 2021.

34 A. Arbabi, Y. Horie, M. Bagheri, and A. Faraon, "Dielectric metasurfaces for complete control of phase and polarization with subwavelength spatial resolution and high transmission," *Nat. Nanotechnol.*, vol. 10, no. 11, pp. 937–943, 2015.

35 J. Hao *et al.*, "Manipulating electromagnetic wave polarizations by anisotropic metamaterials," *Phys. Rev. Lett.*, vol. 99, no. 6, 063908, 2007.

36 N. K. Grady *et al.*, "Terahertz metamaterials for linear polarization conversion and anomalous refraction," *Science*, vol. 340, no. 6138, pp. 1304–1307, 2013.

37 J. Hao *et al.*, "Optical metamaterial for polarization control," *Phys. Rev. A*, vol. 80, no. 2, 023807, 2009.

38 Y. Zhao and A. Alù, "Tailoring the dispersion of plasmonic nanorods to realize broadband optical meta-waveplates," *Nano Lett.*, vol. 13, no. 3, pp. 1086–1091, 2013.

39 T. Q. Li *et al.*, "Magnetic resonance hybridization and optical activity of microwaves in a chiral metamaterial," *Appl. Phys. Lett.*, vol. 92, no. 13, 131111, 2008.

40 A. Basiri *et al.*, "Nature-inspired chiral metasurfaces for circular polarization detection and full-Stokes polarimetric measurements," *Light Sci. Appl.*, vol. 8, no. 1, p. 78, 2019.

41 S. Maci, "Electromagnetic metamaterials and metasurfaces: A historical journey," *IEEE Antennas Propag. Mag.*, vol. 66, no. 3, pp. 84–101, 2024.

42 A. I. Kuznetsov *et al.*, "Roadmap for optical metasurfaces," *ACS Photon.*, vol. 11, no. 3, pp. 816–865, 2024.

43 Q. Ren, S. Yan, and A. Z. Elsherbeni, *Advances in Time-Domain Computational Electromagnetic Methods*. Wiley, 2022.

44 X. Q. Sheng and W. Song, *Essentials of Computational Electromagnetics*. Singapore: Wiley, 2012.

45 W. Shin, "3D Finite-Difference Frequency-Domain Method for Plasmonics and Nanophotonics," Ph.D., Stanford, CA, 2013. Accessed: Oct. 06, 2022.

46 J.-M. Jin, *The Finite Element Method in Electromagnetics*, 3rd ed. Hoboken, NJ: John Wiley & Sons Inc, 2014.

47 J. H. Lee, T. Xiao, and Q. H. Liu, "A 3-D spectral-element method using mixed-order curl conforming vector basis functions for electromagnetic fields," *IEEE Trans. Microw. Theory Tech.*, vol. 54, no. 1, pp. 437–444, 2006.

48 R. F. Harrington, *Time-Harmonic Electromagnetic Fields.* IEEE, 2001. https://doi.org/10.1109/9780470546710.

49 S. D. Gedney, "Introduction to the finite-difference time-domain (FDTD) Method for electromagnetics," *Synth. Lect. Comput. Electromagn.*, vol. 6, 2011, doi: 10.2200/S00316ED1V01Y201012CEM027. Accessed: Jun. 14, 2022.

50 J. Álvarez González, *A Discontinuous Galerkin Finite Element Method for the Time-Domain Solution of Maxwell Equations.* Universidad de Granada, 2014. Accessed: Sep. 07, 2023.

51 M. Štumpf, *Metasurface Electromagnetics: The Cagniard-DeHoop time-domain approach.* London: SciTech Publishing, 2022.

2

Metasurface Modeling

2.1 Effective Media Modeling Techniques

Analysis of a metasurface with all the details of the microstructure will significantly complicate the issue, which is difficult to solve even if the structures are well defined. Therefore, it is necessary to construct a system without the electrically small structures to simplify the analysis. A feasible method is to determine some equivalent parameters of the metasurface and set the constitutive relations of these equivalent parameters with the surrounding fields. The equivalent parameters obtained in this way can be used to eliminate the electrically small structures to facilitate fast macroscopic analysis.

To achieve this goal, the GSTCs for metasurfaces are introduced using the medium homogenization method in this chapter. Homogenization is a kind of asymptotic analysis technique, which is widely used in mechanics and materials science in addition to electromagnetic theory [1, 2]. Homogenization characterizes the periodic structures with average properties or equivalent parameters. It is a multiscale approach, which deals with the micro- and macro-features of the problem separately.

2.1.1 Asymptotic Analysis of Maxwell's Equations

The asymptotic analysis essentially transforms the PDEs whose variables vary rapidly into those with variables that vary slowly by using the asymptotic expansion with nondimensional parameters [3, 4]. In this section, the asymptotic analysis method is introduced to decompose the total field around the periodic electrically small structure to the p-dependent fields and the p-independent fields (where p denotes the period of electrically small structures). Once these fields are separated, the equivalent transition boundary condition can be obtained. The derivations and conclusions below are mainly based on the work of Holloway and Kuester's team. Readers can refer to [5–7] for comprehensive details.

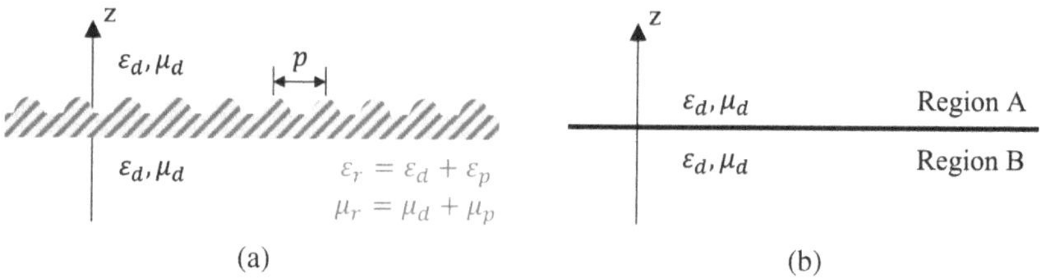

Figure 2.1 Schematic of homogenization. (a) A medium with periodical roughness and (b) zero thickness interface after homogenization.

It is assumed that the electromagnetic field is incident on the two-dimensional metasurface, which is regarded as a medium with periodical roughness, as shown in Figure 2.1(a). The target is to replace the metasurface with a zero-thickness interface, as shown in Figure 2.1(b). The periodic electrically small structure has two spatial length scales, of which L corresponds to the wavelength of interest and p corresponds to the period of the microstructure. To separated macroscopic and microscopic fields, the asymptotic expansion will be performed on a nondimensional ratio $\nu = p/L$. Considering the total fields $\mathbf{E}^T$, $\mathbf{E}^T$ satisfies the Maxwell's equations

$$\nabla \times \mathbf{E}^T = -j\omega\mu_d \mathbf{H}^T \tag{2.1.a}$$

$$\nabla \times \mathbf{H}^T = j\omega\varepsilon_d \mathbf{E}^T \tag{2.1.b}$$

The field intensities $\mathbf{E}$ and $\mathbf{H}$ at any location $\mathbf{r}$ in either regions can be written as functions of slow variable $\hat{r} = \mathbf{r}/L$ and fast variable $\xi = \mathbf{r}/p$ [5]. To distinguish the field variation independent of fast variable, the total fields can be expressed by

$$\mathbf{E}^T(\hat{r}, \xi) = \mathbf{E}(\hat{r}) + \mathbf{e}(\hat{r}, \xi) \tag{2.2.a}$$

$$\mathbf{H}^T(\hat{r}, \xi) = \mathbf{H}(\hat{r}) + \mathbf{h}(\hat{r}, \xi) \tag{2.2.b}$$

where the fields $\mathbf{E}(\hat{r})$ and $\mathbf{H}(\hat{r})$ are the nonboundary layer fields. They are the *macro-response*, which is concerned about in most metasurface research, and they are referred to as the *effective field* [5]. The boundary layer fields $\mathbf{e}(\hat{r}, \xi)$ and $\mathbf{h}(\hat{r}, \xi)$ change periodically with tangential ξ, while decaying exponentially with normal ξ. In asymptotic analysis, differential operators are defined in terms of scaling variables [5], i.e., $\nabla = k_0\nabla_{\hat{r}} + \frac{1}{p}\nabla_\xi$. Substituting the scaled form of the differential operator into the Maxwell's equations leads to

$$\nabla_{\hat{r}} \times \mathbf{E} + \nabla_{\hat{r}} \times \mathbf{e} + \frac{1}{pk_0}\nabla_\xi \times \mathbf{e} = -j\sqrt{\frac{\mu_0}{\varepsilon_0}}(\mu_d\mathbf{H} + \mu_p\mathbf{H} + \mu_d\mathbf{h} + \mu_p\mathbf{h}) \tag{2.3.a}$$

$$\nabla_{\hat{r}} \times \mathbf{H} + \nabla_{\hat{r}} \times \mathbf{h} + \frac{1}{pk_0}\nabla_\xi \times \mathbf{h} = j\sqrt{\frac{\varepsilon_0}{\mu_0}}(\varepsilon_d\mathbf{E} + \varepsilon_p\mathbf{E} + \varepsilon_r\mathbf{e} + \varepsilon_p\mathbf{e}) \tag{2.3.b}$$

Since the boundary layer field decays exponentially in the normal direction of the interface, Eq. (2.3) becomes

$$\nabla_{\hat{r}} \times \mathbf{e} + \frac{1}{pk_0}\nabla_{\xi} \times \mathbf{e} = -j\sqrt{\frac{\mu_0}{\varepsilon_0}}(\mu_p\mathbf{H} + \mu_d\mathbf{h} + \mu_p\mathbf{h}) \tag{2.4.a}$$

$$\nabla_{\hat{r}} \times \mathbf{h} + \frac{1}{pk_0}\nabla_{\xi} \times \mathbf{h} = j\sqrt{\frac{\varepsilon_0}{\mu_0}}(\varepsilon_p\mathbf{E} + \varepsilon_r\mathbf{e} + \varepsilon_p\mathbf{e}) \tag{2.4.b}$$

Since the interested case is in the case that the period is very small compared to the wavelength, which corresponds to that the value of v is close to zero. Thus, the fields in the powers of v can be expanded as

$$\mathbf{E}(\mathbf{r}) = \mathbf{E}^0(\mathbf{r}) + v\mathbf{E}^1(\mathbf{r}) + O(v^2). \tag{2.5.a}$$

$$\mathbf{H}(\mathbf{r}) = \mathbf{H}^0(\mathbf{r}) + v\mathbf{H}^1(\mathbf{r}) + O(v^2). \tag{2.5.b}$$

The numbers in the superscripts denote the order of the expanding coefficients. Each order of the effective fields satisfies Maxwell's equations individually [5, 6]. For v^0 as

$$\nabla_{\hat{r}} \times \mathbf{e}^0 + \nabla_{\xi} \times \mathbf{e}^1 = -j\sqrt{\frac{\mu_0}{\varepsilon_0}}\mu_p\mathbf{H}^0 - j\sqrt{\frac{\mu_0}{\varepsilon_0}}\mu_r\mathbf{h}^0 \tag{2.6.a}$$

$$\nabla_{\hat{r}} \times \mathbf{h}^0 + \nabla_{\xi} \times \mathbf{h}^1 = j\sqrt{\frac{\varepsilon_0}{\mu_0}}\varepsilon_p\mathbf{E}^0 + j\sqrt{\frac{\varepsilon_0}{\mu_0}}\varepsilon_r\mathbf{e}^0 \tag{2.6.b}$$

and for v^1 as

$$\nabla_{\hat{r}} \times \mathbf{e}^1 + \nabla_{\xi} \times \mathbf{e}^2 = -j\sqrt{\frac{\mu_0}{\varepsilon_0}}\mu_p\left(\mathbf{H}^1 + L\xi_z\frac{\partial\mathbf{H}^0}{\partial z}\bigg|_{z=0}\right) - j\sqrt{\frac{\mu_0}{\varepsilon_0}}\mu_r\mathbf{h}^1 \tag{2.7.a}$$

$$\nabla_{\hat{r}} \times \mathbf{h}^1 + \nabla_{\xi} \times \mathbf{h}^2 = j\sqrt{\frac{\varepsilon_0}{\mu_0}}\varepsilon_p\left(\mathbf{E}^1 + L\xi_z\frac{\partial\mathbf{E}^0}{\partial z}\bigg|_{z=0}\right) + j\sqrt{\frac{\varepsilon_0}{\mu_0}}\varepsilon_r\mathbf{e}^1 \tag{2.7.b}$$

For higher-order v, refer to [6]. To solve (2.6) and (2.7), the boundary conditions for zeroth- and first order of fields across the interface of region are needed. On the other hand, the desired equivalent boundary conditions describing the relations between $\mathbf{E}_a^0$, $\mathbf{E}_b^0$, $\mathbf{H}_a^0$, and $\mathbf{H}_b^0$ in Figure 2.1(b) can be obtained once the zeroth-order boundary-layer fields are determined. The boundary conditions will be introduced in Section 2.1.3.

The asymptotic expansion of the Maxwell's equations is the theoretical base of homogenization. It guarantees the equivalent boundary conditions accurately describing the macroscopic response when the periodic electrically small structures are replaced by the equivalent parameters

2.1.2 Electromagnetic Material Properties

In the above section, the total fields are decomposed near the periodic electrical microstructure into the boundary-layer field and the non-boundary-layer field (effective field). The boundary-layer field and effective field are divided into different orders according to a dimensionless ratio coefficient v, in which the zero-order effective fields are the macro-fields mainly concerned. The zero-order and first-order boundary fields are the fields related to the periodic electrically small structures that are very important for the solution of boundary fields. So far, only two simplest material parameters in electromagnetism are used, i.e., ε_r and μ_r. If the conductivity σ is added, these three electromagnetic material parameters are adequate for the traditional full-wave simulations. Note that our goal is to find a way to remove the complex electrically small structures and avoid undersized meshing. To effectively implement these replacements, there should be more material parameters. Computationally, supplementary material parameters are employed to significantly reduce spatial discretization requirements.

To precisely realize this substitution, a thorough understanding of metasurface-wave interactions is essential. These interactions enable the definition of effective electromagnetic parameters that accurately characterize the structure. The parameters presented in this chapter, derived from bulk material considerations, inherently possess three-dimensional properties. However, they are also basically applicable to two-dimensional parameters for two-dimensional (2-D) materials. For example, the volume susceptibility χ_V is used to describe the ratio of the polarization density to the electric field in a very small volume V, which is dimensionless, while the surface susceptibility χ_S is used to describe the ratio of the polarization density to the electric field over a very small area S, whose unit is meter.

2.1.2.1 Constitutive Relations

To make Maxwell's equations determined, six more scalar equations are needed to describe the relationship between **D**, **B**, **E**, and **H**. Such a set of equations is called the constitutive relations, which provides a mathematical description of the electromagnetic media. The most universal constitutive relations are the bianisotropic constitutive relations [8]

$$\mathbf{D} = \overline{\overline{\varepsilon}}\mathbf{E} + \overline{\overline{\xi}}\mathbf{H} \tag{2.8.a}$$

$$\mathbf{B} = \overline{\overline{\zeta}}\mathbf{E} + \overline{\overline{\mu}}\mathbf{H} \tag{2.8.b}$$

where $\overline{\overline{\varepsilon}}$, $\overline{\overline{\mu}}$, $\overline{\overline{\xi}}$, and $\overline{\overline{\zeta}}$ are all 3×3 matrices (tensors) and called constitutive parameters. The units of these four tensors are F/m ($\overline{\overline{\varepsilon}}$), H/m ($\overline{\overline{\mu}}$), S/m ($\overline{\overline{\xi}}$), and S/m ($\overline{\overline{\zeta}}$). Equation (2.8) can be written in terms of susceptibility tensors $\overline{\overline{\chi}}$ as

$$\mathbf{D} = \varepsilon_0(\overline{\overline{I}} + \overline{\overline{\chi}}_{ee})\mathbf{E} + \sqrt{\varepsilon_0\mu_0}\overline{\overline{\chi}}_{em}\mathbf{H} \tag{2.9.a}$$

$$\mathbf{B} = \sqrt{\varepsilon_0\mu_0}\overline{\overline{\chi}}_{me}\mathbf{E} + \mu_0(\overline{\overline{I}} + \overline{\overline{\chi}}_{mm})\mathbf{H} \tag{2.9.b}$$

where ε_0 and μ_0 are the permittivity and permeability in vacuum, and the four tensors, $\overline{\overline{\chi}}_{ee}$, $\overline{\overline{\chi}}_{em}$, $\overline{\overline{\chi}}_{me}$, and $\overline{\overline{\chi}}_{mm}$, correspond to the electric polarizability, magnetoelectric coupling, electromagnetic coupling, and magnetic polarizability tensors, respectively. In addition, the electric flux density and magnetic flux density can also be divided into the parts generated by the free space and the parts caused by the material alone. Thus, the polarization density **P** (C/m^2) and the magnetization density **M** (A/m) are introduced. They satisfy the following relations with the electric flux density and magnetic flux density

$$\mathbf{D} = \varepsilon_0 \mathbf{E} + \mathbf{P} \tag{2.10.a}$$

$$\mathbf{B} = \mu_0 \mathbf{H} + \mu_0 \mathbf{M} \tag{2.10.b}$$

The constitutive relationship between the polarization and magnetization with the volume susceptibilities are

$$\mathbf{P} = \varepsilon_0 \overline{\overline{\chi}}_{ee} \mathbf{E} + \sqrt{\varepsilon_0 \mu_0} \overline{\overline{\chi}}_{em} \mathbf{H} \tag{2.11.a}$$

$$\mathbf{M} = \sqrt{\frac{\varepsilon_0}{\mu_0}} \overline{\overline{\chi}}_{me} \mathbf{E} + \overline{\overline{\chi}}_{mm} \mathbf{H} \tag{2.11.b}$$

2.1.2.2 Dispersion

Dispersion means the retarded response of material to electromagnetic excitation. The metasurfaces whose response to electromagnetic incidence only depends on the current moments are called non-dispersive metasurfaces, while the metasurfaces whose electromagnetic response is affected not only by the current fields but also by the past fields are called temporal dispersive (frequency dispersive) metasurfaces, and the metasurfaces whose electromagnetic response is affected not only by the local fields but also by the fields in the entire region occupied by the constituent are called spatially dispersive metasurfaces [9]. From the perspective of dimensional of meta-atoms and wavelength electromagnetic excitation, when the periodicity of meta-atoms p is much smaller than the wavelength λ, as shown in Figure 2.2(b), the excitation cannot "tell" the physical structure of meta-atoms and hence the coupling between polarization and fields is instantaneous, which is called nondispersive. When the period of the array p increases until some resonant interactions between structure and excitation appear, the coupling between polarization and fields is not instantaneous, as shown in Figure 2.2(c), which is called the temporal dispersive metasurfaces. When p continues to increase until comparable to the wavelength (but still $<\lambda/2$), the response at one point needs both the local fields and the adjacent fields within the metasurfaces, as shown in Figure 2.2(d). These metasurfaces are spatially dispersive metasurfaces, and solving the coupled system requires the field vectors and all their spatially derivatives at this point.

It can be seen that whether a metasurface is dispersive should be judged according to both the frequency range of excitation and the size of the metasurface. From the perspective of equivalent electromagnetic parameters, if susceptibilities are used to describe the coupling between the metasurfaces and excitation, there will

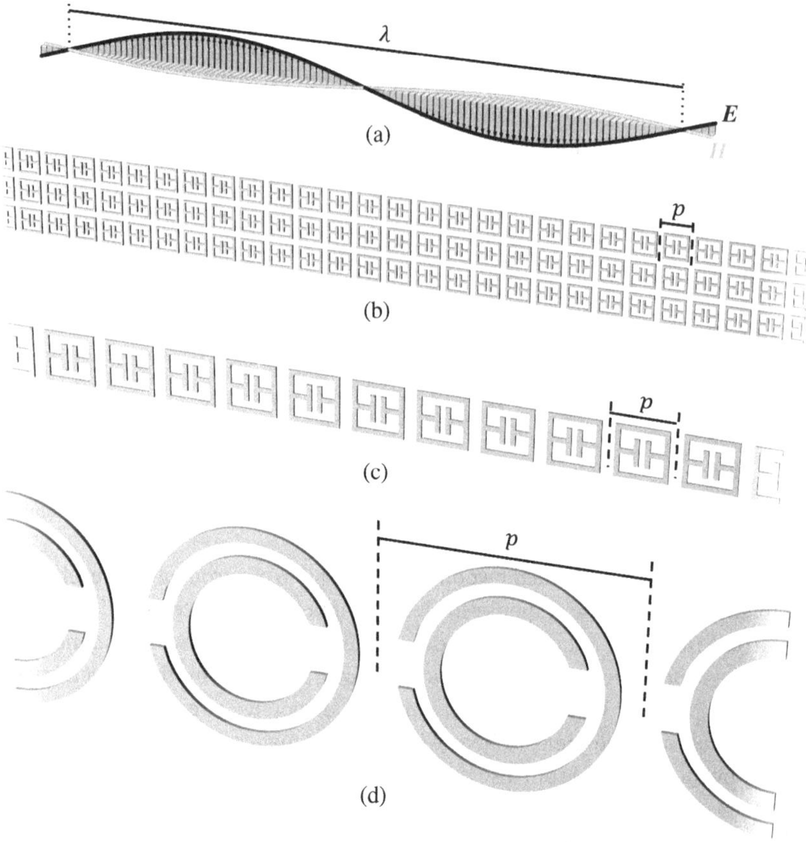

Figure 2.2 The schematic of dispersion in dimensional perspective. (a) The harmonic fields with wavelength λ. (b) The nondispersive metasurface with periodicity p, which is much smaller than λ. (c) The temporal dispersive metasurface whose periodicity p is smaller than λ. (d) The spatially dispersive metasurface whose periodicity p is comparable to λ.

be different dispersion properties of susceptibilities in different frequency ranges for the same metasurface. As shown in Figure 2.3(a), when the susceptibility exhibits negligible variation near the corresponding frequency f_a, it is called the susceptibility nontemporal dispersive (or non-dispersive) within this frequency range and the susceptibility can be regarded as a constant. As the frequency increases, the wavelength gradually becomes smaller (about $\lambda_b \sim 10p$), and the susceptibility begins to show a strong variation with the frequency, which is mostly caused by resonance, as shown in Figure 2.3(b). At this time, the susceptibility can no longer be expressed as a constant around the corresponding frequency f_b, but the susceptibility is temporal dispersive instead.

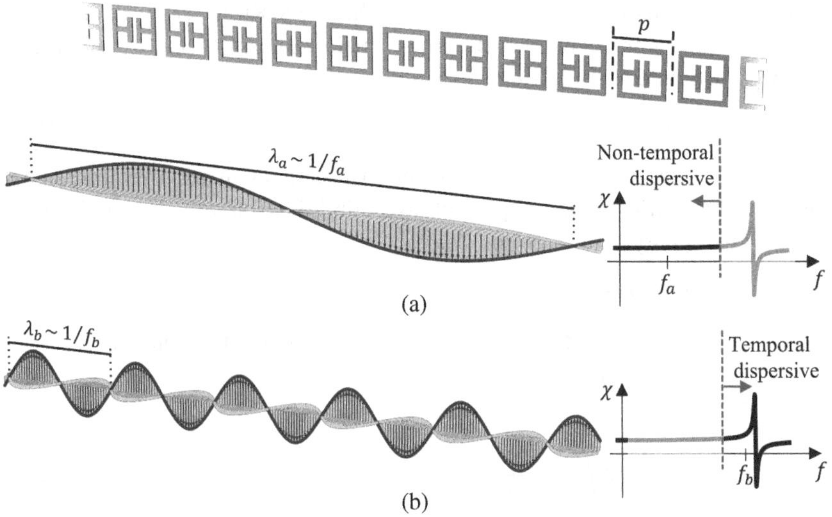

Figure 2.3 Illustrations of (a) non-temporal dispersive and (b) temporal dispersive susceptibility.

The majority of metasurfaces exhibit temporal dispersion owing to their reliance on resonant scattering elements for wave manipulation. This dispersive characteristic manifests as a delayed medium response, where the current electromagnetic reaction depends on prior excitations, mathematically expressed as a convolution in the time domain as

$$\mathbf{P}(t) = \varepsilon_0 \int_{-\infty}^{t} \overline{\overline{\chi}}_{ee}(t-\tau)\mathbf{E}(\tau)d\tau + \sqrt{\varepsilon_0\mu_0}\int_{-\infty}^{t} \overline{\overline{\chi}}_{em}(t-\tau)\mathbf{H}(\tau)d\tau \tag{2.12.a}$$

$$\mathbf{M}(t) = \sqrt{\frac{\varepsilon_0}{\mu_0}} \int_{-\infty}^{t} \overline{\overline{\chi}}_{me}(t-\tau)\mathbf{E}(\tau)d\tau + \int_{-\infty}^{t} \overline{\overline{\chi}}_{mm}(t-\tau)\mathbf{H}(\tau)d\tau \tag{2.12.b}$$

Equation (2.12) can be seen as employing the Fourier transforms on both sides of (2.11). If the susceptibilities are not temporal dispersive, they are frequency-independent, i.e., $\overline{\overline{\chi}}(f) = \overline{\overline{\chi}}_0$ (constant), and its corresponding time-domain function is the impulse function $\overline{\overline{\chi}}(t) = \overline{\overline{\chi}}_0\delta(0)$, and the time-domain responses of **P** and **M** at moment t are

$$\mathbf{P}(t) = \varepsilon_0\overline{\overline{\chi}}_{ee}\mathbf{E}(t) + \sqrt{\varepsilon_0\mu_0}\overline{\overline{\chi}}_{em}\mathbf{H}(t) \tag{2.13.a}$$

$$\mathbf{M}(t) = \sqrt{\frac{\varepsilon_0}{\mu_0}}\overline{\overline{\chi}}_{me}\mathbf{E}(t) + \overline{\overline{\chi}}_{mm}\mathbf{H}(t) \tag{2.13.b}$$

If the susceptibilities are temporal dispersive, the current response of **P** and **M** depends on the fields at previous moments, which implies the concept of causality. The parameters satisfying causality have a certain relationship between the real and imaginary parts, which is called the Kramers–Kronig (K-K) relation [10]. Satisfying causality means that the time-domain expression of susceptibilities is always zero before it is excited as

$$\chi(t) = u(t)A(t) \tag{2.14}$$

where $u(t)$ represents the step function and $A(t)$ denotes an arbitrary time signal. Dividing the Fourier transform pairs of susceptibilities $\widetilde{\chi}(\omega) = \int_{-\infty}^{+\infty} \chi(t)e^{-j\omega t}dt$ into real and imaginary parts leads to

$$\widetilde{\chi}(\omega) = \chi_r(\omega) + j\chi_i(\omega) \tag{2.15}$$

Then, the real and imaginary parts must satisfy the K-K relation as [10]

$$\chi_i(\omega) = \frac{1}{\pi}\mathrm{PV}\int_{-\infty}^{+\infty} \frac{\chi_r(x)}{\omega - x}dx \tag{2.16.a}$$

$$\chi_r(\omega) = -\frac{1}{\pi}\mathrm{PV}\int_{-\infty}^{+\infty} \frac{\chi_i(x)}{\omega - x}dx \tag{2.16.b}$$

where $\mathrm{PV}\int_{-\infty}^{+\infty}\cdot$ represents the Cauchy principal value [11]. The K-K relation shows that changes in the real part of the susceptibilities must be associated with specific changes in its imaginary part. Here, a classical dispersion model is introduced, i.e., the Lorentz dispersion model, initially developed to characterize electron cloud resonance phenomena. This model has become extensively adopted for modeling material temporal dispersion and the corresponding frequency-dependent refractive index [12]. For electron clouds subjected to the Lorentz electromagnetic force, the electric field is associated with the electric polarization density using the Clausius–Mossotti expression [12] as

$$\frac{\partial^2}{\partial t^2}\mathbf{P} + 2\Gamma\frac{\partial}{\partial t}\mathbf{P} + \omega_0^2\mathbf{P} = \frac{Nq_e^2}{m_e}\mathbf{E} \tag{2.17}$$

where Γ, ω_0, N, q_e, and m_e are all constant coefficients of electron clouds, and in the case of time-harmonic wave, (2.17) becomes

$$-\omega^2\mathbf{P} + j2\omega\Gamma\mathbf{P} + \omega_0^2\mathbf{P} = \frac{Nq_e^2}{m_e}\mathbf{E} \tag{2.18}$$

The scalar form of electric susceptibility can be obtained from the above formula as

$$\chi_{ee}(\omega) = \frac{\omega_p^2}{\omega_0^2 + j2\omega\Gamma - \omega^2} \tag{2.19}$$

which naturally satisfies the K-K relation in (2.16).

When the susceptibility is spatially dispersive, it becomes a function of location $\chi(\mathbf{r})$, and sometimes it is also related to the direction of the incident fields $\chi(\mathbf{r}, \mathbf{k}_i)$ [13], where $\mathbf{r}$ denotes the location and $\mathbf{k}_i$ is the wave vector of the incident fields.

From the perspective of the ratio of p to λ, spatially dispersive metasurfaces seem to belong to the category of frequency-selective surfaces (FSS) [9, 14], and the difference of this ratio between FSS and spatially dispersive metasurfaces may be difficult to distinguish quantitatively. The main differences are that spatially dispersive metasurfaces can be aperiodic and the meta-atoms have subwavelength structure. The spatially dispersion is an important property of metasurfaces that the bianisotropy, chirality, and artificial magnetism can be regarded as different orders of spatially dispersion. More theories and applications of spatially dispersion can be referred to [9, 15], and the simulation method with $\mathbf{k}_i$-dependent susceptibilities can be found in [13, 16, 17].

2.1.3 Traditional Boundary Conditions

As described in Section 2.1, the asymptotic analysis is used to decompose the field near the metasurface into the boundary-layer field and effective field. When considering the effective field, an efficient approach is to replace the periodic electrically small structure with effective boundary conditions. This implies two fundamental simplifications: (1) the boundary-layer field consideration is eliminated through effective parameter substitution and (2) the physical thickness of subwavelength periodic structures is abstracted into an infinitesimal surface representation.

Electromagnetic boundary conditions have been studied for over seven decades [18–20], founded on the principle that tangential field relationships at boundaries exhibit local dependence on surrounding medium properties. Our analysis begins with the Leontovich boundary conditions [21], the most prevalent first-order approximation known as standard impedance boundary conditions (IBCs).

2.1.3.1 Standard IBCs

The derivation involves certain approximations and assumptions. For clarity and brevity, this section presents only the essential steps. A more detailed derivation and theoretical background can be found in [19]. The standard IBC is widely used to model the interface of highly conductive bodies (e.g., metals at microwaves and Earth at radio frequencies), which reads

$$\mathbf{E}_{\|} = Z_s \hat{\mathbf{n}} \times \mathbf{H}_{\|} \tag{2.20}$$

where the scalar Z_s denotes the effective surface impedance and the subscript $\|$ denotes the tangential components.

Beginning with the source-less form of Maxwell's equations in frequency domain

$$\nabla \times \mathbf{E} = -j\omega\mu\mathbf{H} \tag{2.21.a}$$

$$\nabla \times \mathbf{H} = j\omega\varepsilon\mathbf{E} \tag{2.21.b}$$

Divide the fields into tangential components ($\mathbf{E}_{\parallel}$, $\mathbf{H}_{\parallel}$) and normal components ($E_n\hat{\mathbf{n}}, H_n\hat{\mathbf{n}}$), and split the differential operator as

$$\nabla = \nabla_{\parallel} + \hat{\mathbf{n}}\frac{\partial}{\partial z} \tag{2.22.a}$$

$$\nabla_{\parallel} = \hat{\mathbf{x}}\frac{\partial}{\partial x} + \hat{\mathbf{y}}\frac{\partial}{\partial y} \tag{2.22.b}$$

Considering normal component of (2.21), the relations about the normal component can be obtained as

$$\hat{\mathbf{n}}H_n = -\frac{1}{j\omega\mu}\nabla_{\parallel} \times \mathbf{E}_{\parallel} \tag{2.23.a}$$

$$\hat{\mathbf{n}}E_n = \frac{1}{j\omega\varepsilon}\nabla_{\parallel} \times \mathbf{H}_{\parallel} \tag{2.23.b}$$

Substituting (2.23) into the tangential components of (2.21) leads to [19]

$$\frac{\partial}{\partial z}\hat{\mathbf{n}} \times \mathbf{E}_{\parallel} = -j\omega\mu\mathbf{H}_{\parallel} - \frac{1}{j\omega\varepsilon}\nabla_{\parallel} \times (\nabla_{\parallel} \times \mathbf{H}_{\parallel}) \tag{2.24}$$

Integrating (2.24) along a half-space medium with $\varepsilon \neq \varepsilon_0$, as shown in Figure 2.4(a). The fields vanish at negative infinity $\mathbf{E}_{\parallel,\ z^{-}\rightarrow -\infty}$ and $\mathbf{H}_{\parallel,\ z^{-}\rightarrow -\infty}$, due to the decaying effect from loss.

For plane wave solutions, the dependence on the transverse coordinates is exponential $e^{-j\mathbf{k}_{\parallel}\cdot\mathbf{r}}$, the differential operator $\nabla_{\parallel}$ can be replaced by $-j\mathbf{k}_{\parallel}$ [19]. Then, (2.24) becomes

$$\hat{\mathbf{n}} \times \mathbf{E}_0 = -j\omega\mu\int_{-\infty}^{0}\mathbf{H}_{\parallel}dz - \frac{1}{j\omega\varepsilon}\mathbf{k}_{\parallel} \times \left(\mathbf{k}_{\parallel} \times \int_{-\infty}^{0}\mathbf{H}_{\parallel}dz\right) \tag{2.25}$$

The magnetic field solution for the negative half space is

$$\mathbf{H}_{\parallel} = \mathbf{H}_0 e^{j\beta z} \tag{2.26}$$

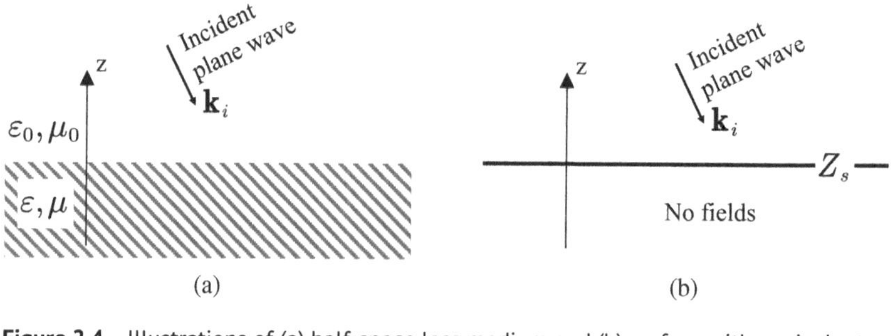

Figure 2.4 Illustrations of (a) half-space loss medium and (b) surface with equivalent impedance Z_s.

where $\mathbf{E}_0$ and $\mathbf{H}_0$ denote the electric and magnetic fields at $z = 0$ and

$$\beta = \sqrt{k^2 - k_{\|}^2} \tag{2.27}$$

Thus, the normal integral of the magnetic field is

$$\int_{-\infty}^{0} \mathbf{H}_{\|} dz = \frac{1}{j\beta}\mathbf{H}_0 \tag{2.28}$$

Substituting (2.28) into (2.25) leads to [19]

$$\mathbf{E}_0 = \sqrt{\frac{\mu}{\varepsilon}} \frac{k}{\sqrt{k^2 - k_{\|}^2}} (\bar{\bar{I}}_{\|} - \bar{\bar{k}}_{\|}) \hat{\mathbf{n}} \times \mathbf{H}_0 \tag{2.29}$$

where the dyadic tensors write

$$\bar{\bar{I}}_{\|} = \begin{bmatrix} 1 & 0 & 0 \\ 0 & 1 & 0 \\ 0 & 0 & 0 \end{bmatrix}, \quad \bar{\bar{k}}_{\|} = \frac{1}{k^2} \begin{bmatrix} k_x^2 & k_x k_y & 0 \\ k_y k_x & k_y^2 & 0 \\ 0 & 0 & 0 \end{bmatrix} \tag{2.30}$$

If the medium in the region $z < 0$ is electrically very dense as compared to that in $z > 0$, the relation $|k| \gg |k_0|$ holds. Besides, when the interface is far from the source, there is $|k| \gg |\mathbf{k}_{\|}|$, which means that the tangential wave vector in (2.30) can be ignored, that is, $k_x, k_y \to 0$. This leads to the standard IBCs

$$\mathbf{E} = \sqrt{\frac{\mu}{\varepsilon}} \hat{\mathbf{n}} \times \mathbf{H} = Z_s \hat{\mathbf{n}} \times \mathbf{H} \tag{2.31}$$

After the boundary with Z_s is used to replace the medium in the negative half space, there is no field value in this region, as shown in the Figure 2.4(b). Despite its conceptual simplicity, the Leontovich boundary condition maintains sufficient accuracy for numerous practical scenarios [22–24]. This standard impedance approximation proves particularly effective when boundary reflection properties remain nearly constant across all incidence angles.

2.1.3.2 Impedance Boundary Conditions

To describe complex electromagnetic phenomena at the interface between materials, many versions of boundary conditions with more parameters have been established by extending (2.31). To streamline the analysis, the IBCs are derived through the equivalence principle, a well-established approach in microwave theory for modeling structural discontinuities using impedance/admittance formulations.

Consider two distinct media separated by a boundary surface S. Medium 1 is located inside S with the fields $\mathbf{E}_1$ and $\mathbf{H}_1$, and medium 2 is located outside of S with the fields of $\mathbf{E}_2$ and $\mathbf{H}_2$. The normal unit vector $\hat{\mathbf{n}}$ points from medium 1 to 2. The surface equivalence theorem establishes that the field discontinuity across the boundary can be characterized by equivalent surface current densities, expressed as

$$\mathbf{J}_s = \hat{\mathbf{n}} \times (\mathbf{H}_2 - \mathbf{H}_1) \tag{2.32.a}$$

$$\mathbf{K}_s = -\hat{\mathbf{n}} \times (\mathbf{E}_2 - \mathbf{E}_1) \tag{2.32.b}$$

where $\mathbf{J}_s$ and $\mathbf{K}_s$ denote the surface electric current and surface magnetic current on S. The average fields $\overline{\mathbf{E}}$ and $\overline{\mathbf{H}}$ on S can then be connected with the surface current $\mathbf{J}_s$ and $\mathbf{K}_s$ through

$$\overline{\mathbf{E}}_{\|} = Z_e \mathbf{J}_s \tag{2.33.a}$$

$$\overline{\mathbf{H}}_{\|} = Y_m \mathbf{K}_s \tag{2.33.b}$$

where Z_e and Y_m are the respective surface impedance and admittance. In addition to the form (2.33.a), there is a bi-isotropic form [25] as

$$\overline{\mathbf{E}}_{\|} = Z_e \mathbf{J}_s - K_{me}(\mathbf{n} \times \mathbf{K}_s) \tag{2.34.a}$$

$$\overline{\mathbf{H}}_{\|} = Y_m \mathbf{K}_s - K_{em}(\mathbf{n} \times \mathbf{J}_s) \tag{2.34.b}$$

where K_{em} and K_{me} are the surface electric-magnetic and magnetoelectric coupling coefficients. Equations (2.33) and (2.34) relate the difference of the electric and magnetic fields across the boundary S to the corresponding effective parameter. The ability to describe bianisotropic structures can be generalized by surface parameters tensors as

$$\overline{\mathbf{E}}_{\|} = \overline{\overline{Z}}_e \mathbf{J}_s - \overline{\overline{K}}_{me}(\mathbf{n} \times \mathbf{K}_s) \tag{2.35.a}$$

$$\overline{\mathbf{H}}_{\|} = \overline{\overline{Y}}_m \mathbf{K}_s - \overline{\overline{K}}_{em}(\mathbf{n} \times \mathbf{J}_s) \tag{2.35.b}$$

These IBCs can effectively describe a wide range of metasurfaces and can also be used as a mathematical model for metasurface design [26].

2.1.3.3 Boundary Conditions Based on Polarizability

The boundary condition based on polarizability directly relates the effects of a single particle to the scattering field generated by a dense array of that particle. Under the condition that scatterer dimensions remain subwavelength, the metasurface can be effectively modeled as a periodic arrangement of coupled electric and magnetic dipole moments **p**, **m**. The total fields scattered by such arrays can be obtained by adding the radiation fields of these dipole moment sources. The electromagnetic properties of a single electrically small scatterer can be modeled with the help of polarizability [19] as

$$\mathbf{p} = \overline{\overline{\hat{\alpha}}}_{ee} \mathbf{E}_i + \overline{\overline{\hat{\alpha}}}_{em} \mathbf{H}_i \tag{2.36.a}$$

$$\mathbf{m} = \overline{\overline{\hat{\alpha}}}_{me} \mathbf{E}_i + \overline{\overline{\hat{\alpha}}}_{mm} \mathbf{H}_i \tag{2.36.b}$$

where $\mathbf{E}_i$ and $\mathbf{H}_i$ are the incident electric and magnetic fields acting on the scattered particles respectively, while $\overline{\overline{\hat{\alpha}}}_{ee}$, $\overline{\overline{\hat{\alpha}}}_{em}$, $\overline{\overline{\hat{\alpha}}}_{me}$, and $\overline{\overline{\hat{\alpha}}}_{mm}$ are effective polarizability tensors. When scattered particles are arranged within a periodic array, the scattered fields $\mathbf{E}_s$ and $\mathbf{H}_s$ corresponding to the sum of dipole moments are

$$\mathbf{E}_s = -\frac{j\omega}{2S}\left(\sqrt{\frac{\mu}{\varepsilon}}\mathbf{p} - \hat{\mathbf{n}} \times \mathbf{m}\right) \tag{2.37.a}$$

$$\mathbf{H}_s = -\frac{j\omega}{2S}\left(\sqrt{\frac{\varepsilon}{\mu}}\mathbf{m} + \hat{\mathbf{n}} \times \mathbf{p}\right) \tag{2.37.b}$$

The boundary condition based on the polarizability is also somewhat related to the IBCs in Section 2.1.3.2. Unlike the well-defined impedance boundary parameters, the effective polarizabilities exhibit greater ambiguity as they originate from individual scatterer responses and their collective impact on metasurface scattering is less straightforward to characterize. However, its merits stem from the simplified polarizability correlations, most evident in elementary geometries and some simple geometries.

2.2 GSTCs for Metasurfaces

In this section, the boundary conditions are introduced mainly used to model metasurfaces in this book, i.e., the GSTCs, and the expression of which are provided as follows [27]:

$$\hat{\mathbf{n}} \times \Delta\mathbf{H} = \frac{\partial \mathbf{P}_{\parallel}}{\partial t} - \hat{\mathbf{n}} \times \nabla M_n \tag{2.38.a}$$

$$\hat{\mathbf{n}} \times \Delta\mathbf{E} = -\mu_0\frac{\partial \mathbf{M}_{\parallel}}{\partial t} - \frac{1}{\varepsilon_0}\hat{\mathbf{n}} \times \nabla P_n \tag{2.38.b}$$

$$\hat{\mathbf{n}} \cdot \Delta\mathbf{D} = -\nabla \cdot \mathbf{P}_{\parallel} \tag{2.38.c}$$

$$\hat{\mathbf{n}} \cdot \Delta\mathbf{B} = -\mu_0\nabla \cdot \mathbf{M}_{\parallel} \tag{2.38.d}$$

where $\mathbf{P}_{\parallel}$ and $\mathbf{M}_{\parallel}$ are tangential *surface* polarization and magnetization, and P_n and M_n are normal *surface* polarization and magnetization. ∇M_n and ∇P_n denote the gradient of normal surface polarization and magnetization. The difference operator Δ characterizes the jump condition of field quantities at the material boundary. Different from the constitutive relations in (2.11) based on volume, the polarization and magnetization in (2.38) are based on surface. The constitutive relationships between **P** and **M** and **E** and **H** have the same mathematical expression as (2.11), except that **P** and **M** are two-dimensional quantities, whose units are (C/m) and (A), respectively, and the units of *surface* susceptibility tensors $\overline{\overline{\chi}}$

are all (m). Note that **P**, **M**, and $\overline{\overline{\chi}}$ are surface-based hereinafter. The constitutive relationship between $\mathbf{P}_{\|}$, $\mathbf{M}_{\|}$, P_n, M_n and the fields on the surface are [27]

$$\begin{bmatrix}\mathbf{P}_{\|}\\ P_n\end{bmatrix} = \varepsilon_0 \begin{bmatrix}\overline{\overline{\chi}}_{ee}^{\|} & \overline{\overline{\chi}}_{ee}^{\|n}\\ \overline{\overline{\chi}}_{ee}^{n\|} & \chi_{ee}^{nn}\end{bmatrix}\begin{bmatrix}\mathbf{E}_{\|}\\ E_n\end{bmatrix} + \sqrt{\varepsilon_0\mu_0}\begin{bmatrix}\overline{\overline{\chi}}_{em}^{\|} & \overline{\overline{\chi}}_{em}^{\|n}\\ \overline{\overline{\chi}}_{em}^{n\|} & \chi_{em}^{nn}\end{bmatrix}\begin{bmatrix}\mathbf{H}_{\|}\\ H_n\end{bmatrix} \tag{2.39.a}$$

$$\begin{bmatrix}\mathbf{M}_{\|}\\ M_n\end{bmatrix} = \sqrt{\frac{\varepsilon_0}{\mu_0}} \begin{bmatrix}\overline{\overline{\chi}}_{me}^{\|} & \overline{\overline{\chi}}_{me}^{\|n}\\ \overline{\overline{\chi}}_{me}^{n\|} & \chi_{me}^{nn}\end{bmatrix}\begin{bmatrix}\mathbf{E}_{\|}\\ E_n\end{bmatrix} + \begin{bmatrix}\overline{\overline{\chi}}_{mm}^{\|} & \overline{\overline{\chi}}_{mm}^{\|n}\\ \overline{\overline{\chi}}_{mm}^{n\|} & \chi_{mm}^{nn}\end{bmatrix}\begin{bmatrix}\mathbf{H}_{\|}\\ H_n\end{bmatrix} \tag{2.39.b}$$

$$\mathbf{E}_{\|} = \hat{\mathbf{n}} \times (\mathbf{E} \times \hat{\mathbf{n}}),\ \ \mathbf{H}_{\|} = \hat{\mathbf{n}} \times (\mathbf{H} \times \hat{\mathbf{n}}),\ \ E_n = \hat{\mathbf{n}} \cdot \mathbf{E},\ \ H_n = \hat{\mathbf{n}} \cdot \mathbf{H} \tag{2.39.c}$$

2.2.1 A Bit of History

At the beginning of the millennium, with the rise of metamaterials, people began to explore the effective representation of metamaterials. Between 2000 and 2010, the joint team of C. L. Holloway and E. F. Kuester demonstrated the inapplicability of bulk-based permittivity ε and permeability μ in describing metasurface and attempted to solve the equivalent parameter problems of metasurfaces from a theoretical perspective [28–31]. Their main point is that, in a bulk-based scheme, the equivalent thickness of a metasurface, d, is not uniquely defined. As shown in Figure 2.5 [14], d can be any value, as long as it is chosen within a reasonable range. Since d is not uniquely defined, the effective material properties obtained in this way have the same ambiguity. Therefore, describing metasurfaces as effective media in this way is inappropriate and leads to misunderstanding and arbitrary selection of d, ε, and μ.

To address this problem, they proposed an alternative model, the surface-based susceptibilities model, to replace the metasurfaces, and combined the GSTC (2.38) to connect the fields and surface polarization and magnetization.

In fact, some early work had very similar forms with GSTC. In 1990, M. M. Idemen analyzed the Maxwell's equations in the sense of distribution and derived general boundary conditions describing discontinuous electromagnetic relations at any interface between two media, which is regarded by many researchers as the prototype of the GSTC [32, 33]. In the Holloway–Kuester (HK) model, metasurfaces are described in terms of six self-coupling axial surface susceptibilities (electrical ones χ_{ee}^{xx}, χ_{ee}^{yy}, and χ_{ee}^{zz} and magnetic ones χ_{mm}^{xx}, χ_{mm}^{yy}, and χ_{mm}^{zz}), which is

Figure 2.5 The non-unique equivalent layer thickness d in bulk-based equivalent parameters. *Source:* Adapted from Holloway et al. [14].

more accurate than the bulk-based model for the metasurfaces [14]. However, it is not universal and only suitable for independent meta-atoms because it does not cover the bianisotropic response of a metasurface. At the same period, another group, directed by S. A. Tretyakov and A. R. Simovski, also worked to develop an alternative method for equivalent parameters of metasurfaces [34–36]. The Tretyakov–Simovski (TS) model, i.e., (2.36), considers the bianisotropy of the meta-atoms. This model is grounded in the polarizability concept of scattering particles, which fundamentally couples to both incident electric and magnetic fields. Compared with the HK model, one disadvantage of the TS model is the lack of a reliable numerically robust retrieval method. A generalized metasurface design methodology based on tensorial surface susceptibilities was introduced by K. Achouri and C. Caloz in 2015 [37]. In this work, they boldly extend the surface susceptibilities model in HK model to a bianisotropic tensor with 36 unknowns ($\overline{\overline{\chi}}_{ee}$, $\overline{\overline{\chi}}_{mm}$, $\overline{\overline{\chi}}_{em}$, and $\overline{\overline{\chi}}_{me}$), according to the bianisotropic property of volume susceptibilities in general electromagnetic theory [8, 38], which is the mostly referred version of the GSTC.

2.2.2 Derivation of GSTC

This section presents two methods for deriving the GSTC, both of which have the same result, i.e., (2.38). The distribution-based derivation, proposed by Idemen [33], is suitable for readers who are familiar with the concept of math distribution theory. While the derivation based on pill-box method is similar to the derivation of the field continuity on the interface of two media in electromagnetic field theory textbooks [39], which is suitable for readers who are familiar with electromagnetism. For readers who are not interested in how the GSTC is derived, you can skip Section 2.2.2 and use (2.38) directly.

2.2.2.1 Derivation Based on Distribution

Early in the eighteenth and nineteenth centuries, when the mature theories of functions and derivatives are not valid to describe the singularities in natural phenomena, some physicists, from time to time, introduced some concepts to derive some formulas that permit the investigation to grasp the certain singularities in physical problems. Among them, the famous English physicist Paul Dirac introduced the famous Dirac function $\delta(x)$, which had the properties as [40]

$$\begin{cases} \delta(x) = 0, & x \neq 0 \\ \delta(x) = \infty, & x = 0 \end{cases} \tag{2.40}$$

and

$$\int_{-\infty}^{\infty} \delta(x)dx = 1 \tag{2.41}$$

With this $\delta(x)$ function, Dirac had presented some conclusions that were adopted by peer physicists. Then many researchers and engineers proposed interesting results with (2.40) and (2.41). These results encouraged mathematicians to establish a rigorous theory for the Dirac function. In 1950, Laurent Schwartz established a well-defined basis for this function scope [42], which guarantees researchers to interpret the results obtained through this function. Since it was early used to represent the distribution of charge, these functions are called distribution functions, and the related theory is called the theory of distribution.

The derivation of GSTCs in the sense of distribution in this section is a brief version of Idemen's work on universal boundary conditions [42]. For more details, refer to Ref. [33]. Begin with some other basic properties of distributions [43], the following is the unit step function of Heaviside

$$U(x) = \begin{cases} 1, & x > 0 \\ 0, & x < 0 \end{cases} \tag{2.42}$$

and the derivative of the unit step function is the Dirac function.

$$U'(x) = \delta(x) \tag{2.43}$$

Then, define a function $f(x)$ as follow, which is continuous everywhere except at $x = 0$ in the sense of distributions as shown in Figure 2.6.

$$f(x) = f_+(x)U(x) + f_-(x)U(-x) \tag{2.44}$$

Then, the derivative of $f(x)$ can be written in the sense of distribution as

$$f'(x) = f'_+(x)U(x) + f'_-(x)U(-x) + (f_+(0) - f_-(0))\delta(x) \tag{2.45}$$

where the sum of first two terms on the right hand side, i.e., $f'_+(x)U(x) + f'_-(x)U(-x)$, denotes an ordinary function that is well defined on $x \neq 0$, and they are called the regular part and will be denoted by $\{f'\}$. While the coefficient of the Dirac function in last term of the right hand side, i.e., $(f_+(0) - f_-(0))$, denotes the jump at $x = 0$, which is called the singular part and will be denoted by $[\![f]\!]$. Thus, the derivative of the function with discontinuity (2.45) can be denoted as

$$f'(x) = \{f'(x)\} + [\![f]\!]\delta(x) \tag{2.46}$$

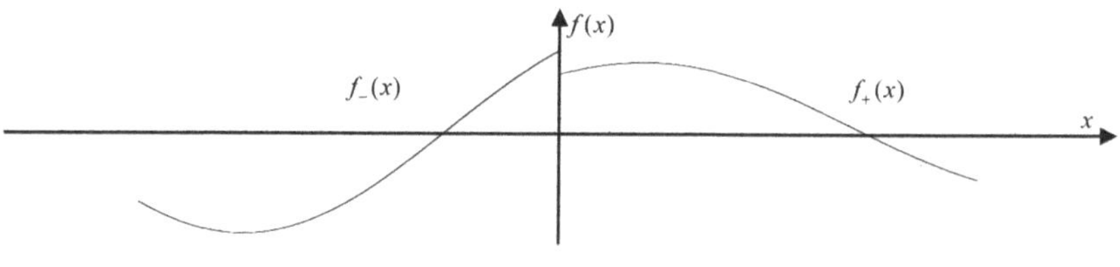

Figure 2.6 A function with discontinuity at $x = 0$.

With the assumption that the Maxwell's equations are valid in the whole computational domain in the sense of distribution, the use of distributions in the expressions of the charge density ρ and the current density $\mathbf{J}$ is natural, and more importantly the fields $\mathbf{E}$, $\mathbf{H}$, $\mathbf{D}$, and $\mathbf{B}$ can be expressed in the sense of distribution, which can be written as [33]

$$\rho = \{\rho\} + \sum_{n=0}^{N} \rho_n \delta^{(n)}(w) \tag{2.47}$$

and

$$\mathbf{E} = \{\mathbf{E}\} + \sum_{n=0}^{N} \mathbf{E}_n \delta^{(n)}(w) \tag{2.48}$$

where N denotes the finite order of the discontinuity, and (u, v, w) denotes a certain orthogonal local coordinate system that is based on the surfaces with jump discontinuities, as shown in Figure 2.7. Equations similar to (2.48) can be derived for $\mathbf{H}$, $\mathbf{D}$, and $\mathbf{B}$. The surface S denotes the region where the jump discontinuity exists. The terms $\delta^{(n)}(w)$ denote the n-th derivative of the Dirac delta function; coefficients ρ_n and $\mathbf{E}_n$ are functions depending on the variables (u, v), but not on the variable w.

Besides, expressing the curl of the field according to (2.48), there is

$$\nabla \times \mathbf{E} = \nabla \times \{\mathbf{E}\} + \sum_{n=0}^{N} [\nabla \times \mathbf{E}_n \delta^{(n)}(w)] \tag{2.49}$$

To separate the normal component from the transverse components, the operator (2.49) can be rewritten using the operator $\nabla_{\parallel} = \hat{\mathbf{u}}\frac{\partial}{\partial u} + \hat{\mathbf{v}}\frac{\partial}{\partial v}$ as [33]

$$\nabla \times \mathbf{E} = \nabla_{\parallel} \times \{\mathbf{E}\} + \hat{\mathbf{w}} \times \frac{\partial\{\mathbf{E}\}}{\partial w} + \sum_{n=0}^{N} \nabla_{\parallel} \times \mathbf{E}_n \delta^{(n)}(w) + \sum_{n=0}^{N} \hat{\mathbf{w}} \times \frac{\partial}{\partial w}[\mathbf{E}_n \delta^{(n)}(w)] \tag{2.50}$$

where $\hat{\mathbf{w}}$ denotes the unit normal vector along w-direction. Furthermore, consider the partial derivative of the regular vector field with respect to w as

$$\frac{\partial\{\mathbf{E}\}}{\partial w} = U(w)\frac{\partial\{\mathbf{E}_+\}}{\partial w} + U(-w)\frac{\partial\{\mathbf{E}_-\}}{\partial w} + (\mathbf{E}_+ - \mathbf{E}_-)\delta(w) = \left\{\frac{\partial \mathbf{E}}{\partial w}\right\} + [\![\mathbf{E}]\!]\delta(w) \tag{2.51}$$

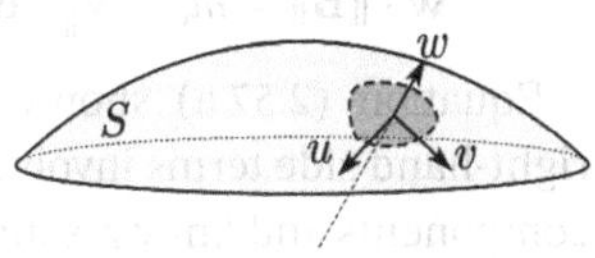

Figure 2.7 The local coordinate system on the surface with jump discontinuities.

Substituting (2.51) into (2.50) leads to

$$\nabla\times\mathbf{E}=\nabla_{\|}\times\{\mathbf{E}\}+\hat{\mathbf{w}}\times\left\{\frac{\partial\mathbf{E}}{\partial w}\right\}+\hat{\mathbf{w}}\times[\![\mathbf{E}]\!]\delta(w)$$
$$+\sum_{n=0}^{N}\nabla_{\|}\times\mathbf{E}_n\delta^{(n)}(w)+\sum_{n=0}^{N}\hat{\mathbf{w}}\times\mathbf{E}_n\delta^{(n+1)}(w) \tag{2.52}$$

Finally, consider the derivative of field with respect to time t

$$\frac{\partial\{\mathbf{E}\}}{\partial t}=U(w)\frac{\partial\{\mathbf{E}_+\}}{\partial t}+U(-w)\frac{\partial\{\mathbf{E}_-\}}{\partial t}+(\mathbf{E}_+-\mathbf{E}_-)\frac{\partial U(w)}{\partial w}\frac{\partial w}{\partial t} \tag{2.53}$$

Because the region S is time-independent, the orthogonal local coordinate system (u, v, w) would be the same and the last term in (2.53) is zero. So, there is

$$\frac{\partial\{\mathbf{E}\}}{\partial t}=\left\{\frac{\partial\mathbf{E}}{\partial t}\right\} \tag{2.54}$$

With (2.48), (2.50), and (2.54), the Ampere's law can be rewritten in the sense of distribution as

$$\nabla_{\|}\times\{\mathbf{H}\}+\hat{\mathbf{w}}\times\left\{\frac{\partial\mathbf{H}}{\partial w}\right\}+\hat{\mathbf{w}}\times[\![\mathbf{H}]\!]\delta(w)+\sum_{n=0}^{N}\nabla_{\|}\times\mathbf{H}_n\delta^{(n)}(w)$$
$$+\sum_{n=0}^{N}\hat{\mathbf{w}}\times\mathbf{H}_n\delta^{(n+1)}(w)$$
$$=\frac{\partial\{\mathbf{D}\}}{\partial t}+\frac{\partial}{\partial t}\sum_{n=0}^{N}\mathbf{D}_n\delta^{(n)}(w)+\{\mathbf{J}\}+\sum_{n=0}^{N}\mathbf{J}_n\delta^{(n)}(w) \tag{2.55}$$

For the coefficients of different regions separated by S, the regular part and the singular parts of different orders must be equal. With this restriction, the equation can be obtained from the $\delta(w)$ part [33]

$$\hat{\mathbf{w}}\times[\![\mathbf{H}]\!]+\nabla_{\|}\times\mathbf{H}_0=\frac{\partial}{\partial t}\mathbf{D}_0+\mathbf{J}_0,\ \delta(w)\ \text{part} \tag{2.56}$$

Dealing with the other three equations in Maxwell's equations, for the zero-order term of $\delta(w)$, there are

$$\hat{\mathbf{w}}\times[\![\mathbf{H}]\!]=-\nabla_{\|}\times\mathbf{H}_0+\frac{\partial}{\partial t}\mathbf{D}_0+\mathbf{J}_0 \tag{2.57.a}$$

$$\hat{\mathbf{w}}\times[\![\mathbf{E}]\!]=-\nabla_{\|}\times\mathbf{E}_0-\frac{\partial}{\partial t}\mathbf{B}_0-\mathbf{K}_0 \tag{2.57.b}$$

$$\hat{\mathbf{w}}\cdot[\![\mathbf{D}]\!]=\rho_0-\nabla_{\|}\cdot\mathbf{D}_0 \tag{2.57.c}$$

$$\hat{\mathbf{w}}\cdot[\![\mathbf{B}]\!]=m_0-\nabla_{\|}\cdot\mathbf{B}_0 \tag{2.57.d}$$

Equation (2.57.a) shows the universal boundary conditions [42], and the right-hand side terms involve only the singular parts of the zero order of the field components and known source.

The universal boundary conditions already provide us a good description of the discontinuity, but **D** and **B** are not very suitable for associating with the meta-atoms of the metasurface. To relate the macroscopic responses with meta-atoms, one may introduce the electric polarization density **P** and the magnetic polarization density **M**. For the first-order surface polarization density, here are the standard constitutive relations

$$\mathbf{D}_0 = \varepsilon \mathbf{E}_0 + \mathbf{P}_0 \tag{2.58.a}$$

$$\mathbf{B}_0 = \mu(\mathbf{H}_0 + \mathbf{M}_0) \tag{2.58.b}$$

For simplicity, ignore the source ($\mathbf{J} = \mathbf{K} = \rho = m = 0$) on surface S and substitute (2.58) into (2.57) to derive the boundary conditions without **D** and **B** as

$$\hat{\mathbf{w}} \times [\![\mathbf{E}]\!] = -\mu \frac{\partial}{\partial t} \mathbf{M}_{0,\parallel} + \frac{1}{\varepsilon} \nabla_{\parallel} \times \mathbf{P}_{0,w} \tag{2.59.a}$$

$$\hat{\mathbf{w}} \times [\![\mathbf{H}]\!] = \frac{\partial}{\partial t} \mathbf{P}_{0,\parallel} + \nabla_{\parallel} \times \mathbf{M}_{0,w} \tag{2.59.b}$$

Omitting the notation 0 for the singular term, replacing $[\![\mathbf{E}]\!]$ by $\Delta\mathbf{E}$ and employing the vector identity $\nabla_{\parallel} \times (\hat{\mathbf{w}} f) = -\hat{\mathbf{w}} \times \nabla f$, one can obtain the final form of GSTCs

$$\hat{\mathbf{w}} \times \Delta\mathbf{E} = -\mu \frac{\partial}{\partial t} \mathbf{M}_{\parallel} - \frac{1}{\varepsilon} \hat{\mathbf{w}} \times \nabla P_w \tag{2.60.a}$$

$$\hat{\mathbf{w}} \times \Delta\mathbf{H} = \frac{\partial}{\partial t} \mathbf{P}_{\parallel} - \hat{\mathbf{w}} \times \nabla M_w \tag{2.60.b}$$

$$\hat{\mathbf{w}} \cdot \Delta\mathbf{D} = -\nabla_{\parallel} \cdot \mathbf{P}_{\parallel} \tag{2.60.c}$$

$$\hat{\mathbf{w}} \cdot \Delta\mathbf{B} = -\mu \nabla_{\parallel} \cdot \mathbf{M}_{\parallel} \tag{2.60.d}$$

Note that equations in (2.60) are the same as those in (2.38) because the vectors $\hat{\mathbf{w}}$ and $\hat{\mathbf{n}}$ denote the same unit vector that is normal to the surface.

2.2.2.2 Derivation Based on the Pill-Box Method

Gauss's divergence theorem describes the relation between the volume integral of the divergence of a vector **F** and the integral of that vector over the surface S enclosing this volume V as

$$\int_V (\nabla \cdot \mathbf{F}) dv = \int_S \mathbf{F} \cdot \mathbf{n} ds \tag{2.61}$$

where the vector **n** is the unit outward vector normal to S. Stokes's theorem describes the relations between the surface integral of the curl of a vector **F** and the integral of that vector over the contour L enclosing this surface S as

$$\int_S (\nabla \times \mathbf{F}) \cdot \mathbf{n} ds = \int_L \mathbf{F} dl \tag{2.62}$$

Gauss's and Stokes's theorems are generally used in deriving the boundary conditions from the Maxwell's equations employing the pill-box method [8, 27, 39].

First of all, the polarization densities should be split into volume polarization densities $\mathbf{P}_v$ and $\mathbf{M}_v$, and surface polarization densities $\mathbf{P}_s$ and $\mathbf{M}_s$ as

$$\mathbf{P} = \mathbf{P}_v + \mathbf{P}_s\delta(s) \tag{2.63.a}$$

$$\mathbf{M} = \mathbf{M}_v + \mathbf{M}_s\delta(s) \tag{2.63.b}$$

where the Dirac function $\delta(s)$ denotes the region where $\mathbf{P}_s$ and $\mathbf{M}_s$ are located.

The GSTCs can be derived from Ampere's law and Faraday's law. Consider an intercepting rectangle A, which is perpendicular to the surface S, as shown in Figure 2.8. According to the mean value theorem of the integral, since the magnetic fields of either side of S are continuous, the integral terms along the tangential direction of S can be expressed as [39]

$$\int_{\overline{p_a p_b}} \mathbf{H}d\mathbf{l} + \int_{\overline{p_c p_d}} \mathbf{H}d\mathbf{l} = \Delta u[H_u(p^+) - H_u(p^-)] \tag{2.64}$$

where Δu is length of the line integral, and points p^+ and p^- represent any random point on their line segments, as illustrated in Figure 2.8. The subscript u denotes the components in the u direction. In a similar way, the integral terms along the direction vertical to S can be expressed as

$$\begin{aligned}\int_{\overline{p_b p_c}} \mathbf{H}d\mathbf{l} + \int_{\overline{p_d p_a}} \mathbf{H}d\mathbf{l} &= \int_{\overline{p_b p_1}} \mathbf{H}d\mathbf{l} + \int_{\overline{p_1 p_c}} \mathbf{H}d\mathbf{l} + \int_{\overline{p_d p_2}} \mathbf{H}d\mathbf{l} \\ &+ \int_{\overline{p_2 p_a}} \mathbf{H}d\mathbf{l} - \int_{\overline{p_b p_c}} \mathbf{M}_s\delta(s)d\mathbf{l} - \int_{\overline{p_d p_a}} \mathbf{M}_s\delta(s)d\mathbf{l} \\ &= \frac{\Delta w}{2}\left[-H_w\left(p_1^+\right) - H_w\left(p_1^-\right) + H_w\left(p_2^-\right) + H_w\left(p_2^+\right)\right] \\ &+ M_{s,w}(p_1) - M_{s,w}(p_2)\end{aligned} \tag{2.65}$$

where Δw is length of the line integral as illustrated in Figure 2.8. The surface integral of $\mathbf{D}$ over A can be divided into three parts

$$\int_S \mathbf{D}d\mathbf{s} = \int_{A^+} \mathbf{D}d\mathbf{s} + \int_{A^-} \mathbf{D}d\mathbf{s} + \int_A \frac{\mathbf{P}_s}{\varepsilon_0}\delta(s)d\mathbf{s}$$

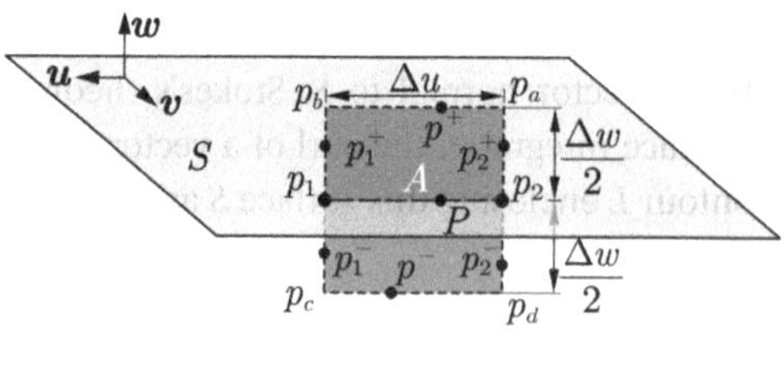

Figure 2.8 The intercepting pill-box for deriving GSTC.

$$= \frac{\Delta u \Delta w}{2}\left(E_v^+ + E_v^-\right) + \frac{\Delta u \Delta w}{2\varepsilon_0}\left(P_{s,v}^+ + P_{s,v}^-\right) + \frac{P_{s,v}}{\varepsilon_0}\Delta u \tag{2.66.a}$$

where A^+ and A^- are the two parts of A separated by S. Substituting (2.64), (2.65), and (2.66.a) into the Ampere's laws leads to

$$\begin{aligned}
&\Delta u(H_u(p^+) - H_u(p^-)) - M_{s,w}(p_1) + M_{s,w}(p_2) \\
&\quad + \frac{\Delta w}{2}\left[-H_w\left(p_1^+\right) - H_w\left(p_1^-\right) + H_w\left(p_2^-\right) + H_w\left(p_2^+\right)\right] \\
&\quad = \frac{\partial}{\partial t}\left[\frac{\Delta u \Delta w}{2}\left(E_v^+ + E_v^-\right) + \frac{\Delta u \Delta w}{2\varepsilon_0}\left(P_{s,v}^+ + P_{s,v}^-\right) + \frac{P_{s,v}}{\varepsilon_0}\Delta u\right]
\end{aligned} \tag{2.66.b}$$

When A is infinitely small, i.e., Δw, $\Delta u \to 0$ and $\{p_a, p_b, \dots, p_1, p_1^{\pm}, \dots\} \to P$. Owing to the continuity of the field on the same side of the surface, there are

$$\lim_{p_1^+, p_2^+ \to P}\left[E_w\left(p_2^+\right) - E_w\left(p_1^+\right)\right] = \lim_{p_1^-, p_2^- \to 0}\left[E_w\left(p_2^-\right) - E_w\left(p_1^-\right)\right] = 0 \tag{2.67.a}$$

$$\lim_{p_1^+, p_2^+ \to P}\left[H_w\left(p_2^+\right) - H_w\left(p_1^+\right)\right] = \lim_{p_1^-, p_2^- \to 0}\left[H_w\left(p_2^-\right) - H_w\left(p_1^-\right)\right] = 0 \tag{2.67.b}$$

Thus, (2.66.b) can be rewritten as

$$\begin{aligned}
H_u(p^+) - H_u(p^-) &= \lim_{\Delta w \to 0}\frac{\partial}{\partial t}\left[\frac{\Delta w}{2}\left(E_v^+ + E_v^-\right) + \frac{\Delta w}{2\varepsilon_0}\left(P_{s,v}^+ + P_{s,v}^-\right)\right] \\
&\quad + \frac{\partial}{\partial t}\frac{P_{s,v}}{\varepsilon_0} + \lim_{\Delta u \to 0}\frac{M_{s,w}(p_2) - M_{s,w}(p_1)}{\Delta u} \\
&= \frac{1}{\varepsilon_0}\frac{\partial}{\partial t}P_{s,v} - \frac{\partial}{\partial u}M_{s,w}
\end{aligned} \tag{2.68}$$

Considering another rectangle A' parallel to the *vow*-plane, another relation can be obtained that

$$\begin{aligned}
-H_v(p^+) + H_v(p^-) &= \lim_{\Delta w \to 0}\frac{\partial}{\partial t}\left[\frac{\Delta w}{2}\left(E_u^+ + E_u^-\right) + \frac{\Delta w}{2\varepsilon_0}\left(P_{s,u}^+ + P_{s,u}^-\right)\right] \\
&\quad + \frac{\partial}{\partial t}\frac{P_{s,u}}{\varepsilon_0} + \lim_{\Delta v \to 0}\frac{M_{s,w}(p_2) - M_{s,w}(p_1)}{\Delta v} \\
&= \frac{1}{\varepsilon_0}\frac{\partial}{\partial t}P_{s,u} + \frac{\partial}{\partial v}M_{s,w}
\end{aligned} \tag{2.69}$$

Equations in (2.68) and (2.69) can be expressed in a more compact form as

$$\hat{\mathbf{w}} \times \Delta \mathbf{H} = \frac{\partial}{\partial t}\mathbf{P}_{\parallel} - \hat{\mathbf{w}} \times \nabla M_w \tag{2.70.a}$$

On the other hand, using the similar method to deal with Faraday's law leads to

$$\hat{\mathbf{w}} \times \Delta \mathbf{E} = -\mu\frac{\partial}{\partial t}\mathbf{M}_{\parallel} - \frac{1}{\varepsilon}\hat{\mathbf{w}} \times \nabla P_w \tag{2.70.b}$$

Equations (2.70.a) and (2.70.b) are the first and second equations of GSTC in (2.38).

2.2.3 Comparison of GSTC and Other Boundary Conditions

The effective parameters in IBCs are the impedance $\overline{\overline{Z}}_e$, admittance $\overline{\overline{Y}}_m$, and coupling coefficients $\overline{\overline{K}}_{em}$ and $\overline{\overline{K}}_{me}$. Since these parameters only describe the relationship between the tangential field components, the values of $\overline{\overline{Z}}_e$, $\overline{\overline{Y}}_m$, $\overline{\overline{K}}_{em}$, and $\overline{\overline{K}}_{me}$ are sometimes related to the propagation direction of the applied field. Therefore, they are not the intrinsic parameters of the two-dimensional surface structure, and their values in different incident scenarios cannot be readily determined. Surface susceptibilities are the more intrinsic parameters for a two-dimensional surface, whose values are independent of the applied electromagnetic fields.

Both IBCs and GSTCs are effective for the case considering only a single incident angle. This is because, according to Huygens' principle, any electromagnetic field on either side of a surface can be created by pure tangential equivalent surface polarization [44]. As shown in Figure 2.9, any surface containing the normal polarized component can be converted into an equivalent (with exactly the same response) surface containing only the tangential polarization.

However, for different wave vectors **k**, this equivalent pure tangential polarization will also be different, that is, the tangential polarization component becomes **k**-dependent.

Therefore, both IBCs and GSTCs can perform the replacement of electrically small structure for a single incident wave case, while GSTC is a better choice for a multi-incident angle situation. On the other hand, the additional normal component provided by the GSTCs have more advantages in metasurface design that the freedom of using the normal component in the 3 × 3 susceptibility tensors is larger than 2 × 2 impedance tensors [45].

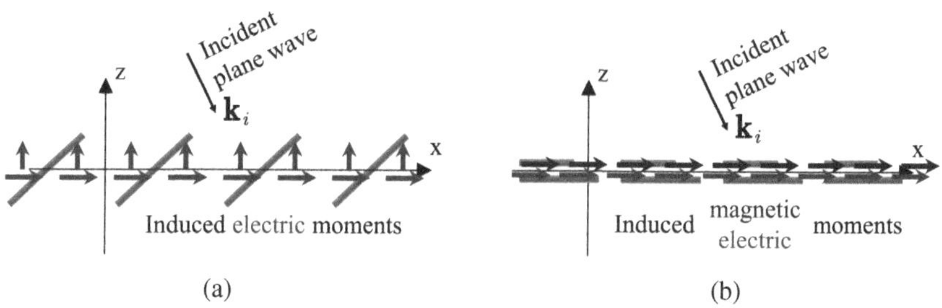

Figure 2.9 Schematic of two different scatterers with equal responses under the same excitation (a) with both tangential and normal components of electrical moments and (b) with only tangential components of electric and magnetic moments.

2.2.4 Analytic Solutions with GSTC

In order to illustrate how the GSTCs (2.38) as a boundary condition affects the solution of Maxwell's equation, in this section, two analytical solutions are discussed in the presence of surface susceptibilities.

2.2.4.1 The Plane Wave Transmission and Reflection Properties of Surface Susceptibilities

Considering a TE polarized plane wave ($E_x = E_z = H_y = 0$) is obliquely incident onto a surface with constant susceptibilities χ_{ee}^{yy}, χ_{mm}^{xx}, and χ_{mm}^{zz}, as shown in Figure 2.10. The incident, reflected, and transmitted electric fields are [31]

$$E_{y,i} = e^{-jk(x\sin\theta + z\cos\theta)} \tag{2.71.a}$$

$$E_{y,r} = Re^{-jk(x\sin\theta - z\cos\theta)} \tag{2.71.b}$$

$$E_{y,t} = Te^{-jk(x\sin\theta + z\cos\theta)} \tag{2.71.c}$$

and the magnetic fields are

$$H_{x,i} = -\frac{k}{\omega\mu_0}\cos\theta e^{-jk(x\sin\theta + z\cos\theta)} \tag{2.72.a}$$

$$H_{z,i} = \frac{k}{\omega\mu_0}\sin\theta e^{-jk(x\sin\theta + z\cos\theta)} \tag{2.72.b}$$

$$H_{x,r} = \frac{k}{\omega\mu_0}R\cos\theta e^{-jk(x\sin\theta - z\cos\theta)} \tag{2.72.c}$$

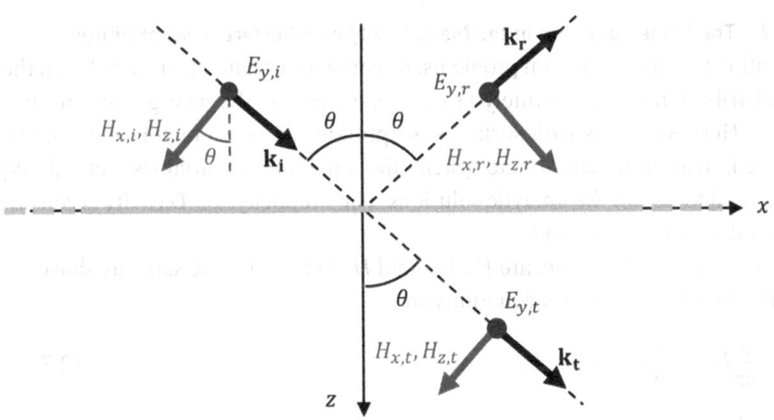

Figure 2.10 TE polarized plane wave incident onto a surface with constant surface susceptibilities. The pink line along x-direction denotes the surface with susceptibilities.

$$H_{z,r} = \frac{k}{\omega\mu_0} R \sin\theta e^{-jk(x\sin\theta - z\cos\theta)} \tag{2.72.d}$$

$$H_{x,t} = -\frac{k}{\omega\mu_0} T \cos\theta e^{-jk(x\sin\theta + z\cos\theta)} \tag{2.72.e}$$

$$H_{z,t} = \frac{k}{\omega\mu_0} T \sin\theta e^{-jk(x\sin\theta + z\cos\theta)} \tag{2.72.f}$$

where R and T are the unknown reflection and transmission coefficients, and k is the wavenumber of free space.

To solve R and T, rewrite (2.38) with the non-zero components

$$H_{x,t} - H_{x,i} - H_{x,r} = \frac{j\omega\varepsilon_0\chi_{ee}^{yy}}{2}(E_{y,t} + E_{y,i} + E_{y,r}) - \frac{\chi_{mm}^{zz}}{2}\frac{\partial}{\partial x}(H_{z,t} + H_{z,i} + H_{z,r}) \tag{2.73.a}$$

$$E_{y,t} - E_{y,i} - E_{y,r} = \frac{j\omega\mu_0\chi_{mm}^{xx}}{2}(H_{x,t} + H_{x,i} + H_{x,r}) \tag{2.73.b}$$

Substituting (2.71) and (2.72) into (2.73) results in [31]

$$T = \frac{4\cos\theta + k^2\cos\theta\chi_{mm}^{xx}\left(\chi_{ee}^{yy} + \sin^2\theta\chi_{mm}^{zz}\right)}{2\left(2 + jk\chi_{mm}^{xx}\cos\theta\right)\left(2\cos\theta + jk\chi_{ee}^{yy} + jk\sin^2\theta\chi_{mm}^{zz}\right)} \tag{2.74.a}$$

$$R = -\frac{2jk\left(\chi_{ee}^{yy} - \cos^2\theta\chi_{mm}^{xx} + 4\sin^2\theta\chi_{mm}^{zz}\right)}{\begin{array}{c}4\cos\theta + 2jk\left(\chi_{ee}^{yy} + \cos^2\theta\chi_{mm}^{xx} + \sin^2\theta\chi_{mm}^{zz}\right)\\ -k^2\cos\theta\chi_{mm}^{xx}\left(\chi_{ee}^{yy} + \sin^2\theta\chi_{mm}^{zz}\right)\end{array}} \tag{2.74.b}$$

2.2.4.2 The Mode Spectrum in the Metal Cavity with Surface Susceptibilities

In addition to the open-region problems, researchers are often concerned with the field distribution inside a finitely large region, such as the waveguides and resonators. Here, solutions with surface susceptibilities inside a metal cavity will be discussed. In a certain coordinate system, the separation of variables method [38] can be used to obtain the analytic solutions. For simplicity, a 2-D cavity is considered, as shown in Figure 2.11.

The non-zero components are E_y, H_x, and H_z, which are the same as above. In this situation, the Maxwell's equations are

$$\frac{\partial}{\partial z}H_x - \frac{\partial}{\partial x}H_z = j\omega\varepsilon_0 E_y \tag{2.75.a}$$

$$\frac{\partial}{\partial x}E_y = -j\omega\mu_0 H_z \tag{2.75.b}$$

$$\frac{\partial}{\partial z}E_y = j\omega\mu_0 H_x \tag{2.75.c}$$

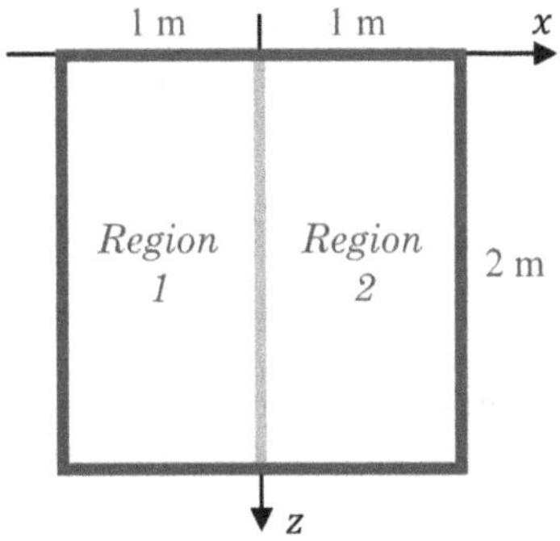

Figure 2.11 A 2-D metal cavity in which an interface ($x = 0, 0 < z < 2$ m) with constant normal surface susceptibilities χ_{mm}^{nn} divides the domain into two regions. Regions 1 and 2 are both free spaces.

which holds for both regions 1 and 2. Substituting (2.75.b) and (2.75.c) into (2.75.a) results in the Helmholtz equation

$$\frac{\partial^2}{\partial x^2}E_y + \frac{\partial^2}{\partial z^2}E_y + k^2 E_y = 0 \tag{2.76}$$

On the metal boundary, i.e., the yellow edge in Figure 2.11, the perfect electric conductor (PEC) imposes the tangential components of E vanishing on it

$$E_y = 0 \tag{2.77}$$

On the interface with χ_{mm}^{xx}, GSTC imposes the discontinuities of field that

$$H_{z,2} - H_{z,1} = \frac{\chi_{mm}^{nn}}{2}\frac{\partial}{\partial z}(H_{x,1} + H_{x,2}) \tag{2.78}$$

where the subscripts 1 and 2 denote the fields in corresponding region. To solve $E_{y,1}$ and $E_{y,2}$, the separation of variables method seeks to find solutions of the forms

$$E_{y,1}(x, z) = X_1(x)Z_1(z) \tag{2.79.a}$$

$$E_{y,2}(x, z) = X_2(x)Z_2(z) \tag{2.79.b}$$

that are solutions which are the products of two functions of one coordinate each. Substitution of (2.79.a) into (2.76) yields

$$\frac{\partial^2}{\partial x^2}X_1 + k_x^2 X_1 = 0 \tag{2.80.a}$$

$$\frac{\partial^2}{\partial z^2}Z_1 + k_z^2 Z_1 = 0 \tag{2.80.b}$$

$$k_x^2 + k_z^2 = k^2 \tag{2.80.c}$$

Equations (2.80.a) and (2.80.b) are called harmonic equations, the solution to them are linear combinations of the harmonic functions [38]

$$h(k_x x) \sim \sin(k_x x), \sinh(k_x x), \cos(k_x x), \cosh(k_x x), e^{-jk_x x}, e^{jk_x x} \tag{2.81}$$

To determine the possible values of k_x and k_z, the boundary conditions must be considered. Considering (2.77) on $\{-1 \leq x \leq 1,\ z = 0\}$ and $\{-1 \leq x \leq 1,\ z = 2\}$ leads to

$$Z_1(z) = Z_2(z) = Z(z) = Z_p \sin(k_z z) \tag{2.82.b}$$

$$k_z = p\frac{\pi}{2},\quad p = \pm 1, \pm 2, \pm 3, \ldots \tag{2.82.c}$$

Then, the possible solutions of k_x are needed.

When k_x is **a pure real number**, the solutions are linear combinations of $\sin(k_x x)$ and $\cos(k_x x)$. Considering (2.77) on $\{x = -1, 0 \leq z \leq 2\}$ leads to

$$X_1(x) = X_q \cos(k_x x) \tag{2.83.a}$$

$$k_x = \left(q + \frac{1}{2}\right)\pi,\quad q = 0, \pm 1, \pm 2, \pm 3, \ldots \tag{2.83.b}$$

or

$$X_1(x) = X_q[\tan(k_x)\cos(k_x x) + \sin(k_x x)] \tag{2.84.a}$$

$$k_x \neq \left(q + \frac{1}{2}\right)\pi,\quad q = 0, \pm 1, \pm 2, \pm 3, \ldots \tag{2.84.b}$$

Equation (2.83) can determine a series of solutions

$$E_y(x, z) = \sum_{p,q} E_{p,q} \sin\left(p\frac{\pi}{2}z\right)\sin(q\pi x) \tag{2.85.a}$$

$$k^2 = (q\pi)^2 + \left(p\frac{\pi}{2}\right)^2 \quad p, q = \pm 1, \pm 2, \pm 3, \ldots \tag{2.85.b}$$

While (2.84) does not actually confirm the value of k_x, GSTC (2.78) are used to confirm more possible values of k_x. First, doing the same derivation for the fields in region 2 leads to

$$X_2(x) = X_q[\tan(k_x)\cos(k_x x) - \sin(k_x x)] \tag{2.86}$$

Substituting (2.84.a) into (2.79.a) and (2.86) into (2.79.b), then substituting the resulting expressions into (2.78) and considering it on $\{x = 0, 0 \leq z \leq 2\}$ yields

$$Z_p \sin(k_z z) X_q \left[2k_x - \chi_{mm}^{nn} k_z^2 \tan(k_x)\right] = 0 \tag{2.87}$$

For a non-zero χ_{mm}^{nn}, (2.87) is a transcendental equation defined in the real field and has quasiperiodic multiple solutions. The first q positive approximate solutions $K_{x,q}$ can be confirmed numerically

$$2K_{x,q} - \chi_{mm}^{nn} k_z^2 \tan(K_{x,q}) \approx 0 \tag{2.88}$$

Then, the solutions can be determined

$$E_{y,1}(x, z) = \sum_{p,q} E_{p,q} \sin\left(p\frac{\pi}{2}z\right)[\tan(K_{x,q})\cos(K_{x,q}x) + \sin(K_{x,q}x)] \tag{2.89.a}$$

$$E_{y,2}(x,z) = \sum_{p,q} E_{p,q} \sin\left(p\frac{\pi}{2}z\right) [\tan(K_{x,q})\cos(K_{x,q}x) - \sin(K_{x,q}x)] \tag{2.89.b}$$

$$k^2 = K_{x,q}^2 + \left(p\frac{\pi}{2}\right)^2, \quad p = \pm 1, \pm 2, \pm 3, \ldots \tag{2.89.c}$$

When k_x is **a pure imaginary number**, i.e., $k_x = j\beta$ with β being a real number, the solutions are linear combinations of $\sinh(k_x x)$ and $\cosh(k_x x)$. Following the method of deriving (2.87) leads to

$$X_q \left[2\beta - \chi_{mm}^{nn} k_z^2 \tanh(\beta)\right] = 0 \tag{2.90}$$

Equation (2.90) is also a transcendental equation defined in the real field. When the equation has a non-zero solution β, which can be confirmed numerically. Then, the solutions can be determined

$$E_{y,1}(x,z) = \sum_{p} E_p \sin\left(p\frac{\pi}{2}z\right) [\tanh(\beta)\cosh(\beta x) + \sinh(\beta x)] \tag{2.91.a}$$

$$E_{y,2}(x,z) = \sum_{p} E_p \sin\left(p\frac{\pi}{2}z\right) [\tanh(\beta)\cosh(\beta x) - \sinh(\beta x)] \tag{2.91.b}$$

$$k^2 = \left(p\frac{\pi}{2}\right)^2 - \beta^2, \quad p = \pm 1, \pm 2, \pm 3, \ldots \tag{2.91.c}$$

When k_x is a complex number, the situation becomes too intricate to analyze, so it will not be analyzed here. So far, three analytical field solutions are derived, i.e., (2.85), (2.89), and (2.91). As a conclusion, the first eight modes are presented with $\chi_{mm}^{nn} = 2$, sorted by the magnitude of k. The electric field distributions are shown in Figure 2.12, and the detailed information is present in Table 2.1.

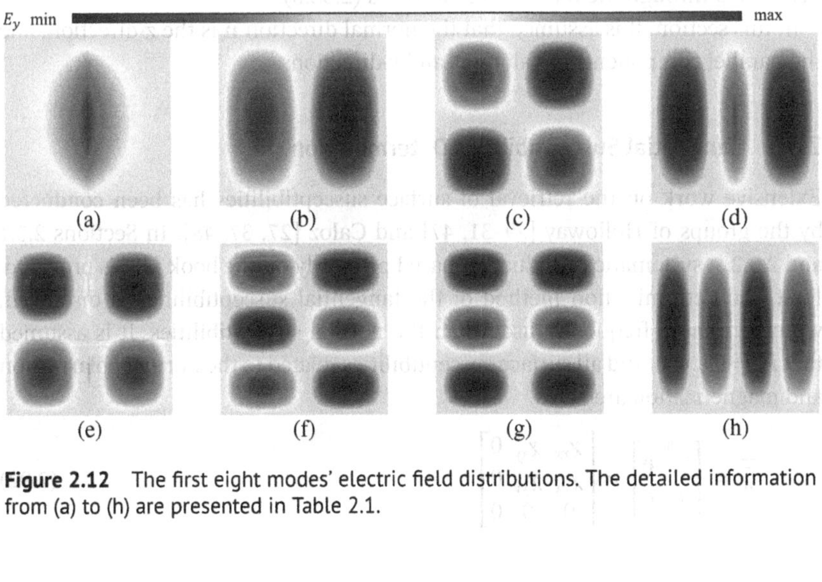

Figure 2.12 The first eight modes' electric field distributions. The detailed information from (a) to (h) are presented in Table 2.1.

Table 2.1 The detailed information of the eight modes corresponding to Figure 2.12.

	(a)	(b)	(c)	(d)	(e)	(f)	(g)	(h)
k	1.85326j	3.51240	4.44288	4.46450	4.6888	5.66359	5.74644	6.4766
q	—	1	1	1	1	1	1	2
p	1	1	2	1	2	3	3	1
Eq.	(2.91)	(2.85)	(2.85)	(2.89)	(2.89)	(2.85)	(2.89)	(2.85)

Note the imaginary k in Table 2.1(a), which implies that mode is a gain mode. It will be a challenge for the time-domain methods that will be introduced in Chapter 4, because it can lead to divergences in the simulation, which looks like instability. This divergence has been reported in [46].

2.3 Susceptibility Determination Method

According to the uniqueness theorem, when the source and tangential electric or magnetic fields are specified on the entire boundary surface of a given region, the solution within that region is unique [8]. Relations (2.38.c) and (2.38.d) are redundant relations in the absence of impressed sources since the transverse components of the fields are sufficient to completely describe the electromagnetic problem. Therefore, the determination of surface susceptibilities is usually performed through the relations (2.38.a) and (2.38.b).

In this section, it is assumed that the normal direction $\hat{\mathbf{n}}$ is the z-direction, and the parallel components are along x- and y-directions.

2.3.1 Tangential Susceptibilities Determination

Extensive work on the retrieval of surface susceptibilities has been conducted by the groups of Holloway [29–31, 47] and Caloz [27, 37, 48]. In Sections 2.3.1 and 2.3.2, a systematic introduction based primarily on the book [48] is provided. First, the determination method of the tangential susceptibilities is discussed, which are more frequently used than the normal susceptibilities. It is assumed that $P_n = M_n = 0$, and all surface susceptibilities related to the normal polarization and magnetization are zeros.

$$\overline{\overline{\chi}} = \begin{bmatrix} \overline{\overline{\chi}}^{\parallel} & \mathbf{0} \\ \mathbf{0} & 0 \end{bmatrix} = \begin{bmatrix} \chi_{xx} & \chi_{xy} & 0 \\ \chi_{yx} & \chi_{yy} & 0 \\ 0 & 0 & 0 \end{bmatrix} \tag{2.92}$$

This assumption is usually accurate enough, and it can easily lead to a closed form of the solution for susceptibilities. In particular, if a uniform metasurface is illuminated by a normally impinged plane wave, the presence of P_n and M_n does not affect the waves it scatters, then this assumption will be exact rather than approximated.

Substituting (2.92) into (2.38.a) and (2.38.b) leads to

$$\hat{\mathbf{n}} \times \Delta\mathbf{H} = j\omega\varepsilon_0 \overline{\overline{\chi}}_{ee}^{\parallel} \mathbf{E}_{\parallel} + jk_0 \overline{\overline{\chi}}_{em}^{\parallel} \mathbf{H}_{\parallel} \tag{2.93.a}$$

$$\hat{\mathbf{n}} \times \Delta\mathbf{E} = -jk_0 \overline{\overline{\chi}}_{me}^{\parallel} \mathbf{E}_{\parallel} - j\omega\mu_0 \overline{\overline{\chi}}_{mm}^{\parallel} \mathbf{H}_{\parallel} \tag{2.93.b}$$

Writing the above equations in a matrix form using the scalar field quantities leads to [48]

$$\begin{pmatrix} \Delta H_y \\ \Delta H_x \\ \Delta E_y \\ \Delta E_x \end{pmatrix} = \begin{pmatrix} -j\omega\varepsilon_0\chi_{ee}^{xx} & -j\omega\varepsilon_0\chi_{ee}^{xy} & -jk_0\chi_{em}^{xx} & -jk_0\chi_{em}^{xy} \\ j\omega\varepsilon_0\chi_{ee}^{yx} & j\omega\varepsilon_0\chi_{ee}^{yy} & jk_0\chi_{em}^{yx} & jk_0\chi_{em}^{yy} \\ jk_0\chi_{me}^{xx} & jk_0\chi_{me}^{xy} & j\omega\mu_0\chi_{mm}^{xx} & j\omega\mu_0\chi_{mm}^{xy} \\ -jk_0\chi_{me}^{yx} & -jk_0\chi_{me}^{yy} & -j\omega\mu_0\chi_{mm}^{yx} & -j\omega\mu_0\chi_{mm}^{yy} \end{pmatrix} \begin{pmatrix} \overline{E}_x \\ \overline{E}_y \\ \overline{H}_x \\ \overline{H}_y \end{pmatrix} \tag{2.94}$$

where the fields with an overline $\overline{E}$ and $\overline{H}$ denote the fields on the surface with susceptibilities. In most situations, the average of fields on both sides of the surface are used to represent them.

$$\overline{E} = \frac{1}{2}(E_1 + E_2) \tag{2.95.a}$$

$$\overline{H} = \frac{1}{2}(H_1 + H_2) \tag{2.95.b}$$

For $\hat{\mathbf{n}}$ points from "1" to "2," the differences of fields are represent as

$$\Delta E = E_2 - E_1 \tag{2.96.a}$$

$$\Delta H = H_2 - H_1 \tag{2.96.b}$$

Equation (2.94) contains 16 unknowns and is underdetermined. To obtain an accurate solution, four sets of 16 equations are constructed to make the equation determined by setting up four independent plane wave incidence tests as [48]

$$\begin{pmatrix} \Delta H_{y,1} & \Delta H_{y,2} & \Delta H_{y,3} & \Delta H_{y,4} \\ \Delta H_{x,1} & \Delta H_{x,2} & \Delta H_{x,3} & \Delta H_{x,4} \\ \Delta E_{y,1} & \Delta E_{y,2} & \Delta E_{y,3} & \Delta E_{y,4} \\ \Delta E_{x,1} & \Delta E_{x,2} & \Delta E_{x,3} & \Delta E_{x,4} \end{pmatrix} = \begin{pmatrix} \hat{\chi}_{ee}^{xx} & \hat{\chi}_{ee}^{xy} & \hat{\chi}_{em}^{xx} & \hat{\chi}_{em}^{xy} \\ \hat{\chi}_{ee}^{yx} & \hat{\chi}_{ee}^{yy} & \hat{\chi}_{em}^{yx} & \hat{\chi}_{em}^{yy} \\ \hat{\chi}_{me}^{xx} & \hat{\chi}_{me}^{xy} & \hat{\chi}_{mm}^{xx} & \hat{\chi}_{mm}^{xy} \\ \hat{\chi}_{me}^{yx} & \hat{\chi}_{me}^{yy} & \hat{\chi}_{mm}^{yx} & \hat{\chi}_{mm}^{yy} \end{pmatrix} \begin{pmatrix} \overline{E}_{x,1} & \overline{E}_{x,2} & \overline{E}_{x,3} & \overline{E}_{x,4} \\ \overline{E}_{y,1} & \overline{E}_{y,2} & \overline{E}_{y,3} & \overline{E}_{y,4} \\ \overline{H}_{x,1} & \overline{H}_{x,2} & \overline{H}_{x,3} & \overline{H}_{x,4} \\ \overline{H}_{y,1} & \overline{H}_{y,2} & \overline{H}_{y,3} & \overline{H}_{y,4} \end{pmatrix} \tag{2.97}$$

where $\hat{\chi}$ represents the normalized susceptibilities and their relations between the susceptibilities are

$$\overline{\overline{\hat{\chi}}}_{ee}^{\parallel} = \begin{bmatrix} -j\omega\varepsilon_0 & 0 \\ 0 & j\omega\varepsilon_0 \end{bmatrix} \cdot \overline{\overline{\chi}}_{ee}^{\parallel} \tag{2.98.a}$$

$$\overline{\overline{\hat{\chi}}}_{em}^{\parallel} = \begin{bmatrix} -jk_0 & 0 \\ 0 & jk_0 \end{bmatrix} \cdot \overline{\overline{\chi}}_{em}^{\parallel} \tag{2.98.b}$$

$$\overline{\overline{\hat{\chi}}}_{me}^{\parallel} = \begin{bmatrix} jk_0 & 0 \\ 0 & -jk_0 \end{bmatrix} \cdot \overline{\overline{\chi}}_{me}^{\parallel} \tag{2.98.c}$$

$$\overline{\overline{\hat{\chi}}}_{mm}^{\parallel} = \begin{bmatrix} j\omega\mu_0 & 0 \\ 0 & -j\omega\mu_0 \end{bmatrix} \cdot \overline{\overline{\chi}}_{mm}^{\parallel} \tag{2.98.d}$$

Equation (2.97) can be written in a compact format as

$$\overline{\overline{D}} = \overline{\overline{\hat{\chi}}}\,\overline{\overline{A}} \tag{2.99}$$

where the tensors $\overline{\overline{D}}$, $\overline{\overline{\hat{\chi}}}$, and $\overline{\overline{A}}$ correspond to the field difference, normalized susceptibilities and the field average, respectively. To solve the 16 tangential susceptibility components in (2.97), the following four plane wave incidence tests are defined

Test 1: y-polarized plane waves propagating in the $+z$-direction,
Test 2: x-polarized plane waves propagating in the $+z$-direction,
Test 3: y-polarized plane waves propagating in the $-z$-direction,
Test 4: x-polarized plane waves propagating in the $-z$-direction.

Then, scattering responses are needed to assemble $\overline{\overline{D}}$ and $\overline{\overline{A}}$. There are two ways to obtain the scattering parameters of a meta-atom, i.e., the microwave scattering experiments [49] or the full-wave simulation. The process of full-wave simulation requires the simulation of the infinite periodic array composed of the meta-atom, which requires the use of periodic boundary conditions (PBC) [50], as shown in Figure 2.13. Most mainstream commercial softwares are equipped with this boundary condition.

The computational domain is truncated by PBC in the x- and y-direction, and the z-direction is truncated by absorbing boundary condition (ABC) [50]. Based on the excitation loaded at the ports and the received response, the scattering parameters of the meta-atom can be obtained. For example, S_{21}^{xy} describes the ratio of the transmitted x-polarized mode at port 2 to the incident y-polarized mode at port 1.

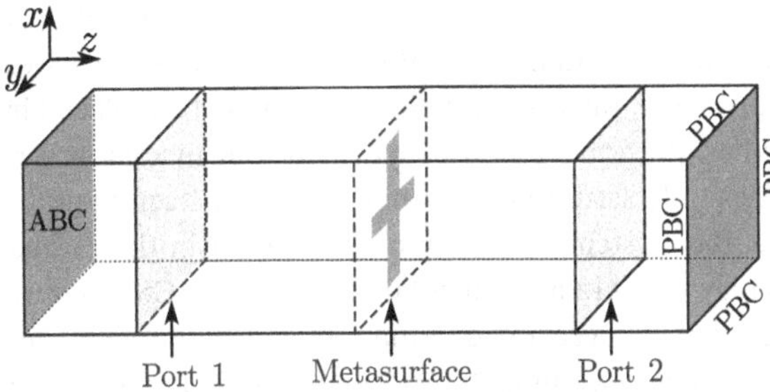

Figure 2.13 Schematic of the full-wave simulation of the meta-atom where PBC denotes the periodic boundary conditions and ABC denotes the absorbing boundary condition.

Tensors $\overline{\overline{D}}$ and $\overline{\overline{A}}$ can be obtained based on the S parameters

$$\mathbb{A} = \frac{1}{2}\begin{bmatrix} 1+S_{11}^{xx}+S_{21}^{xx} & S_{11}^{yx}+S_{21}^{yx} & 1+S_{12}^{xx}+S_{22}^{xx} & S_{12}^{yx}+S_{22}^{yx} \\ -S_{11}^{xy}-S_{21}^{xy} & -1-S_{11}^{yy}-S_{21}^{yy} & -S_{12}^{xy}-S_{22}^{xy} & -1-S_{12}^{yy}-S_{22}^{yy} \\ -S_{11}^{xy}+S_{21}^{xy} & 1-S_{11}^{yy}+S_{21}^{yy} & -S_{12}^{xy}+S_{22}^{xy} & -1-S_{12}^{yy}+S_{22}^{yy} \\ 1-S_{11}^{xx}+S_{21}^{xx} & -S_{11}^{yx}+S_{21}^{yx} & -1-S_{12}^{xx}+S_{22}^{xx} & -S_{12}^{yx}+S_{22}^{yx} \end{bmatrix} \tag{2.100.a}$$

$$\mathbb{D} = \begin{bmatrix} -1+S_{11}^{xx}+S_{21}^{xx} & S_{11}^{yx}+S_{21}^{yx} & -1+S_{12}^{xx}+S_{22}^{xx} & S_{12}^{yx}+S_{22}^{yx} \\ S_{11}^{xy}+S_{21}^{xy} & -1+S_{11}^{yy}+S_{21}^{yy} & S_{12}^{xy}+S_{22}^{xy} & -1+S_{12}^{yy}+S_{22}^{yy} \\ S_{11}^{xy}-S_{21}^{xy} & 1+S_{11}^{yy}-S_{21}^{yy} & S_{12}^{xy}-S_{22}^{xy} & -1+S_{12}^{yy}-S_{22}^{yy} \\ -1-S_{11}^{xx}+S_{21}^{xx} & -S_{11}^{yx}+S_{21}^{yx} & 1-S_{12}^{xx}+S_{22}^{xx} & -S_{12}^{yx}+S_{22}^{yx} \end{bmatrix} \tag{2.100.b}$$

In addition, the process of susceptibility determination can be simplified by artificially reducing the number of unknowns according to external assumptions, such as geometric symmetry or meta-atom properties. The simplified tensors may be $\begin{pmatrix} \widehat{\chi}_{ee}^{xx} & 0 & 0 & 0 \\ 0 & \widehat{\chi}_{ee}^{yy} & 0 & 0 \\ 0 & 0 & \widehat{\chi}_{mm}^{xx} & 0 \\ 0 & 0 & 0 & \widehat{\chi}_{mm}^{yy} \end{pmatrix}$ or $\begin{pmatrix} \widehat{\chi}_{ee}^{xx} & \widehat{\chi}_{ee}^{xy} & 0 & 0 \\ \widehat{\chi}_{ee}^{yx} & \widehat{\chi}_{ee}^{yy} & 0 & 0 \\ 0 & 0 & \widehat{\chi}_{mm}^{xx} & \widehat{\chi}_{mm}^{xy} \\ 0 & 0 & \widehat{\chi}_{mm}^{yx} & \widehat{\chi}_{mm}^{yy} \end{pmatrix}$. For detailed information, refer to [37].

The tangential susceptibilities are effective enough in many situations for the following three reasons: (1) Metasurfaces are usually designed for working within a paraxial approximation, which means that the normal polarization is much less than the tangential one because the field is essentially transversally polarized.

(2) Due to the structure of the meta-atom, the effect of excited normal polarization is negligible compared to the tangential polarization. (3) As mentioned above, according to the Huygens principle, the normal polarization may be replaced by a tangential polarization when the wave vector is specified. But in some cases, this tangential-susceptibilities-only assumption cannot provide an accurate solution anymore. For example, the existence of a conductive circuit in the normal direction of the meta-atom may result in normal polarization, or the GSTC equivalent replacement must be effective over a wide range of angles, in which case the angular response of the metasurface must be considered. In general, as long as the normal component of the polarization or magnetization is not negligible compared to the tangential one, the existence of normal polarization must be reasonably considered; otherwise, the equivalent error will be too large. In the next section, the determination method of normal susceptibilities will be introduced.

2.3.2 Normal Susceptibility Determination

When employing the normal susceptibilities, one should realize that the difficulty of susceptibility determination significantly increases. First, the unknowns increase from 16 to 36. Second, the gradients of the normal polarization are taken into account, which means a class of transformations that involve changing the propagation direction of the wave need to be solved. Equation (2.38.a) becomes a system of PDEs, which means that an analytical solution to the susceptibilities may not be derivable, or the approximate susceptibilities obtained by numerical methods may have to be used. Just like the tangential susceptibility determination, the susceptibilities are simplified to reduce the number of unknowns, such as (1) there is no electromagnetic coupling or it is small enough to be ignored, (2) the meta-atom is symmetric so that some non-diagonal susceptibilities can be ignored, and (3) the thickness of the meta-atom is much smaller than the wavelength and there is no separate charge or current flowing from top to bottom. When these three assumptions are met simultaneously, the susceptibility tensor can be simplified as [47]

$$\overline{\overline{\chi}}_{ee} = \begin{bmatrix} 0 & 0 & 0 \\ 0 & \chi_{ee}^{yy} & 0 \\ 0 & 0 & 0 \end{bmatrix}, \quad \overline{\overline{\chi}}_{mm} = \begin{bmatrix} \chi_{mm}^{xx} & 0 & 0 \\ 0 & 0 & 0 \\ 0 & 0 & \chi_{mm}^{zz} \end{bmatrix}, \quad \overline{\overline{\chi}}_{me} = \overline{\overline{\chi}}_{em} = 0 \tag{2.101}$$

In this situation, susceptibilities can be determined through the following three plane wave incidence tests:

i) x-polarized plane waves propagating in the positive z-direction,
ii) y-polarized plane waves propagating in the positive z-direction,
iii) TE-polarized (s-polarized) plane wave propagation in the direction with an angle of θ_i to the z-axis.

For test (iii), the incident, transmitted, and reflected fields are

$$\mathbf{E}_i = \hat{\mathbf{y}} e^{-jk(x \sin\theta_i + z\cos\theta_i)}, \quad \mathbf{H}_i = (-\cos\theta_i \hat{\mathbf{x}} + \sin\theta_i \hat{\mathbf{z}}) e^{-jk(x\sin\theta_i + z\cos\theta_i)} \tag{2.102.a}$$

$$\mathbf{E}_t = S_{12}^{TE,TE} \hat{\mathbf{y}} e^{-jk(x \sin\theta_i + z\cos\theta_i)}, \quad \mathbf{H}_t = S_{12}^{TE,TE}(-\cos\theta_i \hat{\mathbf{x}} + \sin\theta_i \hat{\mathbf{z}}) e^{-jk(x\sin\theta_i + z\cos\theta_i)} \tag{2.102.b}$$

$$\mathbf{E}_r = S_{11}^{TE,TE} \hat{\mathbf{y}} e^{-jk(x \sin\theta_i - z\cos\theta_i)}, \quad \mathbf{H}_r = S_{11}^{TE,TE}(\cos\theta_i \hat{\mathbf{x}} + \sin\theta_i \hat{\mathbf{z}}) e^{-jk(x\sin\theta_i - z\cos\theta_i)} \tag{2.102.c}$$

With the parameters $S_{11}^{TE,TE}$ and $S_{12}^{TE,TE}$ obtained from (2.102.a) and the parameters S_{11}^{yy} and S_{12}^{yy} from test (ii), one can then calculate the susceptibilities χ_{ee}^{yy} and χ_{mm}^{zz} [51].

When only conditions (1) and (2) are satisfied, the susceptibility tensor can be simplified as

$$\overline{\overline{\chi}}_{ee} = \begin{bmatrix} \chi_{ee}^{xx} & 0 & 0 \\ 0 & \chi_{ee}^{yy} & 0 \\ 0 & 0 & \chi_{ee}^{zz} \end{bmatrix}, \quad \overline{\overline{\chi}}_{mm} = \begin{bmatrix} \chi_{mm}^{xx} & 0 & 0 \\ 0 & \chi_{mm}^{yy} & 0 \\ 0 & 0 & \chi_{mm}^{zz} \end{bmatrix}, \quad \overline{\overline{\chi}}_{me} = \overline{\overline{\chi}}_{em} = 0 \tag{2.103}$$

Then, three additional sets of plane wave tests are needed to determine the values of χ_{ee}^{xx}, χ_{ee}^{zz}, and χ_{mm}^{yy}.

iv) x-polarized plane waves propagating in the negative z-direction,
v) y-polarized plane waves propagating in the negative z-direction, and
vi) TM-polarized (p-polarized) plane wave propagation in the direction with an angle of θ_i to the z-axis.

2.3.3 Transformation from Other Electromagnetic Parameters

The equivalent impedance and admittance tensors can be transformed into tangential susceptibilities, and this kind of transformation between the surface equivalent parameters is direct [48]. On the other hand, surface susceptibilities can also be determined from the transformation of bulk-based electromagnetic parameters. For example, the parameter retrieval methods of the permittivity ε and permeability μ are relatively mature and reliable [52, 53]. Therefore, if the permittivity and permeability of the metamaterials have been obtained, one may determine their corresponding susceptibilities through effective transformation. Considering a metamaterial with thickness d, permittivity ε_2, and permeability μ_2, as shown in Figure 2.14.

Taking the cross-section around the materials and applying the Gauss theorem, the integrals of fields around the cross-section of the bulk-based parameter model

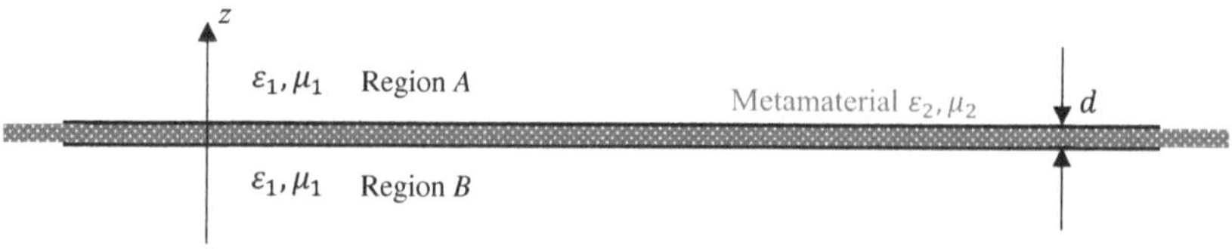

Figure 2.14 A metamaterial with known volume permittivity and permeability.

and the surface susceptibilities model should be equal. The boundary conditions for the effective fields, i.e., the zero-order effective fields, are [6]

$$\begin{aligned}\hat{\mathbf{z}} \times (\mathbf{E}_a - \mathbf{E}_b) = &- j\omega\frac{d}{2}(\mu_2 - \mu_1)[\hat{\mathbf{x}}(H_{a,x} + H_{b,x}) + \hat{\mathbf{y}}(H_{a,y} + H_{b,y})] \\ &- \frac{d}{2}\left(\frac{\varepsilon_2 - \varepsilon_1}{\varepsilon_2}\right)\hat{\mathbf{z}} \times \nabla(E_{a,z} + E_{b,z})\end{aligned} \quad (2.104.a)$$

$$\begin{aligned}\hat{\mathbf{z}} \times (\mathbf{H}_a - \mathbf{H}_b) = &j\omega\frac{d}{2}(\varepsilon_2 - \varepsilon_1)[\hat{\mathbf{x}}(E_{a,x} + E_{b,x}) + \hat{\mathbf{y}}(E_{a,y} + E_{b,y})] \\ &- \frac{d}{2}\left(\frac{\mu_2 - \mu_1}{\mu_2}\right)\hat{\mathbf{z}} \times \nabla(H_{a,z} + H_{b,z})\end{aligned} \quad (2.104.b)$$

The relationship between the coefficients can be found by comparing (2.38.a) and (2.104.a) as

$$\chi_{ee}^{xx} = \chi_{ee}^{yy} = d\left(\frac{\varepsilon_2 - \varepsilon_1}{\varepsilon_0}\right) \quad (2.105.a)$$

$$\chi_{mm}^{xx} = \chi_{mm}^{yy} = d\left(\frac{\mu_2 - \mu_1}{\mu_0}\right) \quad (2.105.b)$$

$$\chi_{ee}^{zz} = d\left(\frac{\varepsilon_1 - \varepsilon_2}{\varepsilon_0\varepsilon_2}\right) \quad (2.105.c)$$

$$\chi_{mm}^{zz} = d\left(\frac{\mu_1 - \mu_2}{\mu_0\mu_2}\right) \quad (2.105.d)$$

In particular, when the background space is vacuum, $\varepsilon_1 = \varepsilon_0$, $\mu_1 = \mu_0$, and the relative permittivity and permeability of the medium are $\varepsilon_2 = \varepsilon_r\varepsilon_0$ and $\mu_2 = \mu_r\mu_0$. Then, the relationship between the coefficient decreases to

$$\chi_{ee}^{xx} = \chi_{ee}^{yy} = d(\varepsilon_r - 1) \quad (2.106.a)$$

$$\chi_{mm}^{xx} = \chi_{mm}^{yy} = d(\mu_r - 1) \quad (2.106.b)$$

$$\chi_{ee}^{zz} = d\left(\frac{1}{\varepsilon_r} - 1\right) \quad (2.106.c)$$

$$\chi_{mm}^{zz} = d\left(\frac{1}{\mu_r} - 1\right) \quad (2.106.d)$$

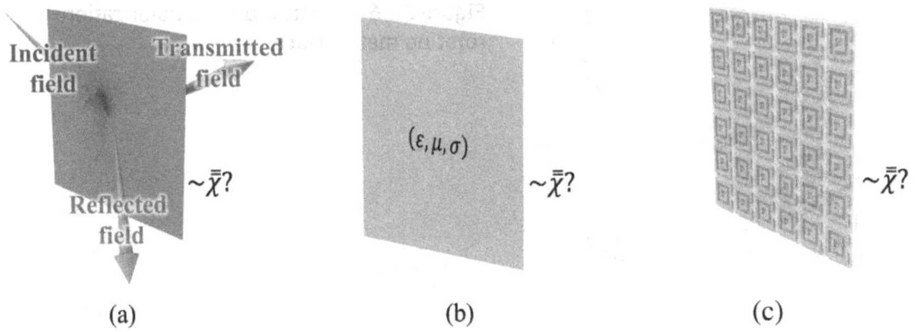

Figure 2.15 Illustration of three kinds of susceptibility determination. (a) Metasurface synthesis, (b) model conversion, and (c) susceptibility retrieval.

2.3.4 Examples of Determination

Figure 2.15 shows the three scenarios of determinations of surface susceptibilities, and detailed examples of each scenario are presented in this section.

2.3.4.1 Susceptibility Synthesis

Figure 2.15(a) shows the synthesis problem, in which the incident wave, the targeted reflected wave and the transmitted wave are all known. Then, the researchers need to synthesize the surface susceptibilities that can produce the desired wave transformation or field discontinuity. The synthesis process usually includes the restriction of the number of independent susceptibilities, because it is an underdetermined problem when the capabilities are richer than needed. For example, when the susceptibilities need to fit one group of wave transformation, like (2.94), there will be a system containing four equations for sixteen unknown susceptibilities. In this situation, the number of independent parameters is manually restricted to four to make the synthesis determined. This reveals a fact that a wave control demand may be obtained by many different combinations of susceptibilities.

A complete synthesis of susceptibilities involves three steps [37]: (1) determining the mathematical transfer function of susceptibilities that produces the demand field, (2) discretizing the transfer function obtained in step one according to a 2-D grid on the surface, and (3) synthesizing the susceptibilities that implement the corresponding transfer function at each point. Some examples are provided below.

A Polarization Rotating Metasurface: Synthesizing a metasurface that converts a normally incident plane wave, which is linearly polarized with an angle of $\pi/8$ with respect to the x-direction into a normally propagating linearly polarized plane wave with an angle of $17\pi/37$ with respect to the x-direction [37], as shown in Figure 2.16. Both reciprocal and non-reciprocal cases are considered.

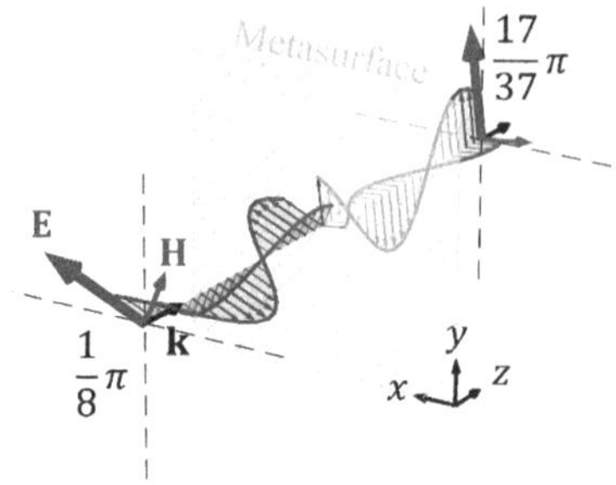

Figure 2.16 Illustration of a polarization rotating metasurface.

The polarization rotation metasurface mentioned above involves only one set of wave transformations, so at least four non-zero plane susceptibilities are required. For the reciprocal metasurface, the magnetoelectric coupling susceptibilities and non-diagonal susceptibilities are required to be zero. Thus the four non-zero susceptibilities are χ_{ee}^{xx}, χ_{ee}^{yy}, χ_{mm}^{xx}, and χ_{mm}^{yy}.

According to the wave rotation requirement, the incident fields are

$$\mathbf{E}_i = E_0\left[\hat{\mathbf{x}}\cos\left(\frac{\pi}{8}\right) + \hat{y}\sin\left(\frac{\pi}{8}\right)\right]e^{-jkz} \tag{2.107.a}$$

$$\mathbf{H}_i = \frac{E_0}{Z_0}\left[-\hat{\mathbf{x}}\sin\left(\frac{\pi}{8}\right) + \hat{y}\cos\left(\frac{\pi}{8}\right)\right]e^{-jkz} \tag{2.107.b}$$

and the transmitted fields are

$$\mathbf{E}_t = E_0\left[\hat{\mathbf{x}}\cos\left(\frac{17\pi}{37}\right) + \hat{y}\sin\left(\frac{17\pi}{37}\right)\right]e^{-jkz} \tag{2.108.a}$$

$$\mathbf{H}_t = \frac{E_0}{Z_0}\left[-\hat{\mathbf{x}}\sin\left(\frac{17\pi}{37}\right) + \hat{y}\cos\left(\frac{17\pi}{37}\right)\right]e^{-jkz} \tag{2.108.b}$$

Substituting the four non-zero surface susceptibilities into (2.94) leads to

$$\chi_{ee}^{xx} = \frac{-\Delta H_y}{j\omega\varepsilon\overline{E}_x} \tag{2.109.a}$$

$$\chi_{ee}^{yy} = \frac{\Delta H_x}{j\omega\varepsilon\overline{E}_y} \tag{2.109.b}$$

$$\chi_{mm}^{xx} = \frac{\Delta E_y}{j\omega\mu\overline{H}_x} \tag{2.109.c}$$

$$\chi_{mm}^{yy} = \frac{-\Delta E_x}{j\omega\mu\overline{H}_y} \tag{2.109.d}$$

Because the designed reflected fields are zero, there are $\Delta\mathbf{E} = \mathbf{E}_t - \mathbf{E}_i$ and $\overline{\mathbf{E}} = \frac{1}{2}(\mathbf{E}_t + \mathbf{E}_i)$. Substituting (2.107) and (2.108) into (2.109) and solving the equations at $z = 0$ with $k = 2\pi$ rad/m leads to

$$\chi_{ee}^{xx} = \chi_{mm}^{yy} = -0.2414j \tag{2.110.a}$$

$$\chi_{ee}^{yy} = \chi_{mm}^{xx} = 0.1411j \tag{2.110.b}$$

For the non-reciprocal metasurface, the non-zero susceptibilities can be the electromagnetic coupling terms or the self-coupling non-diagonal terms. Here, the non-zero terms are assumed as χ_{ee}^{xy}, χ_{ee}^{yx}, χ_{mm}^{xy}, and χ_{mm}^{yx}. Substituting the four non-zero terms into (2.94) leads to

$$\chi_{ee}^{xy} = \frac{-\Delta H_y}{j\omega\varepsilon\overline{E}_y} \tag{2.111.a}$$

$$\chi_{ee}^{yx} = \frac{\Delta H_x}{j\omega\varepsilon\overline{E}_x} \tag{2.111.b}$$

$$\chi_{mm}^{xy} = \frac{\Delta E_y}{j\omega\mu\overline{H}_y} \tag{2.111.c}$$

$$\chi_{mm}^{yx} = \frac{-\Delta E_x}{j\omega\mu\overline{H}_x} \tag{2.111.d}$$

Substituting (2.107) and (2.108) into (2.111) and solving the equations at $z = 0$ with $k = 2\pi$ rad/m leads to

$$\chi_{ee}^{xy} = \chi_{mm}^{xy} = -0.1845j \tag{2.112.a}$$

$$\chi_{ee}^{yx} = \chi_{mm}^{yx} = 0.1845j \tag{2.112.b}$$

2.3.4.2 Susceptibility Transformation

As described in Section 2.3.3, susceptibilities can be transformed from other known electromagnetic parameters, as shown in Figure 2.15(b). Equation (2.104.a) provides a transformation method from permittivity and permeability. Here, an example of transformation from the conductivity model is presented.

Graphene: The graphene, whose carbon atoms are arranged in a hexagonal lattice, has been widely discussed due to its many unusual physical properties [54]. A. Vakil treated the graphene as a very thin layer (with a thickness of around 1 nm) and the effective parameters are the volumetric conductivity [55]. For ease of computation, V. Nayyeri proposed another effective model of graphene using surface conductivity [56]. In the absence of external magnetostatics bias field, the surface conductivity of the graphene is scalar and the fields across the graphene sheet should satisfy

$$\hat{\mathbf{n}} \times \Delta\mathbf{H} = -\sigma_{gra}\hat{\mathbf{n}} \times (\hat{\mathbf{n}} \times \overline{\mathbf{E}}) \tag{2.113}$$

where

$$\sigma_{gra} = \frac{\sigma_{gra0}}{1 + j\omega\tau_{gra}} \tag{2.114}$$

where σ_{gra0} and τ_{gra} are both constants related with graphene [56]. By comparing (2.113) with (2.38.a), the susceptibilities describing graphene in this case can be expressed as

$$\chi_{ee}^{xx} = \chi_{ee}^{yy} = \frac{\sigma_{gra0}}{j\omega\varepsilon_0(1 + j\omega\tau_{gra})} \tag{2.115}$$

2.3.4.3 Susceptibility Retrieval

The simulation of the metasurface plays a crucial role in its development, but it is also a time-consuming process. Simulation primarily involves meta-atom simulations and metasurface array simulations. In terms of time and resource consumption, approximately three quarters of the consumption is dedicated to array simulation owing to its multiscale nature. It should be noted that plenty of computing resources have already been invested in simulating the meta-atoms prior to conducting array simulations. By leveraging the information obtained from previous steps (e.g., S parameters), the effective susceptibilities can be retrieved, as shown in Figure 2.15(c), and replace the electrically small structures with GSTC and equivalent surface susceptibilities. This approach effectively addresses the multiscale challenge associated with array simulation, leading to a substantial reduction in computational time and memory consumption.

The formulas for retrieving the surface susceptibilities from the S parameter have already been provided in Sections 2.3.1 and 2.3.2, and here some practical examples are exhibited.

Simple Two-Dimensional Metal Patch Scatterers: The X band (8–12 GHz) is the most widely utilized microwave frequency range, leading to numerous studies on electromagnetic metasurfaces being applied around 10 GHz. At this frequency, the characteristic wavelength in vacuum is 30 mm, which makes metal patches with dimensions on the order of millimeters commonly employed for metasurface design. Figure 2.17 illustrates several regular two-dimensional metal patch meta-atoms, each with a periodicity (the distance between the centers of adjacent atoms) of 3 mm and a minimal structure length of 0.2 mm. This dimension is considered electrically small for the X band, as it corresponds to 1/150 of the wavelength in vacuum. Accurate full-wave simulations necessitate very fine meshing grids to effectively capture these spatial features, therefore, employing the equivalent surface susceptibilities model for these structures is highly appropriate.

These elements have geometric symmetry, so the susceptibilities can be simplified to contain only three parameters as shown in (2.101) [47]. As shown in Figure 2.13, PBC are employed to mimic the infinite array and obtain the reflection coefficient R and transmission coefficient T of the infinite periodic array composed of these elements, respectively, and then use T and R to obtain the surface

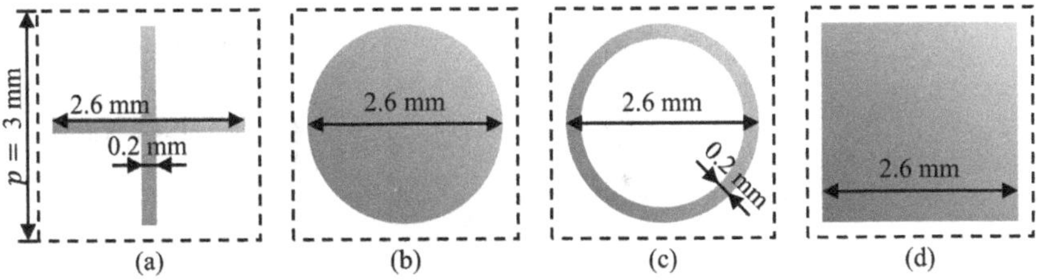

Figure 2.17 Schematics of four regular two-dimensional metal patch meta-atoms, whose periodicity are all 3 mm.

Table 2.2 The corresponding surface susceptibilities of the meta-atoms shown in Figure 2.17.

	$\chi_{ee}^{xx} = \chi_{ee}^{yy}$	χ_{mm}^{zz}
(a)	0.611×10^{-3}	-0.138×10^{-3}
(b)	1.373×10^{-3}	-0.737×10^{-3}
(c)	1.651×10^{-3}	-0.801×10^{-3}
(d)	3.095×10^{-3}	-1.367×10^{-3}

susceptibilities. First, one test with plane wave normally incident on the patch is set up to obtain the tangential susceptibilities as

$$\chi_{ee}^{xx} = \chi_{ee}^{yy} = -\frac{2}{jk}\frac{T_1 + R_1 - 1}{T_1 + R_1 + 1} \tag{2.116}$$

where T_1 and R_1 are the transmission coefficient and reflection coefficient at normal incidence. Then, an oblique incident TE-polarized plane wave, described as (2.102.a), is set up to obtain the normal susceptibilities as

$$\chi_{mm}^{zz} = -2\frac{(T_2 + R_2 - 1)\cos\theta_i + jk(T_2 + R_2 + 1)\chi_{ee}^{yy}}{jk(T_2 + R_2 + 1)\sin^2\theta_i} \tag{2.117}$$

where θ_i, T_2, and R_2 are the incident angle, transmission, and reflection coefficient at oblique incidence. The corresponding parameters of the units are shown in Table 2.2.

Meta-Atom with Strong Normal Response. Some meta-atoms have a strong resonance in the normal direction at oblique incidence, and sometimes, this normal resonance response even exceeds the tangential response. Consider an example of an actual meta-atom [57], as shown in Figure 2.18. This atom is observed to reflect weakly at normal plane wave incidence, while a strong reflection is observed at oblique incidence of the plane wave, as shown in Figure 2.19.

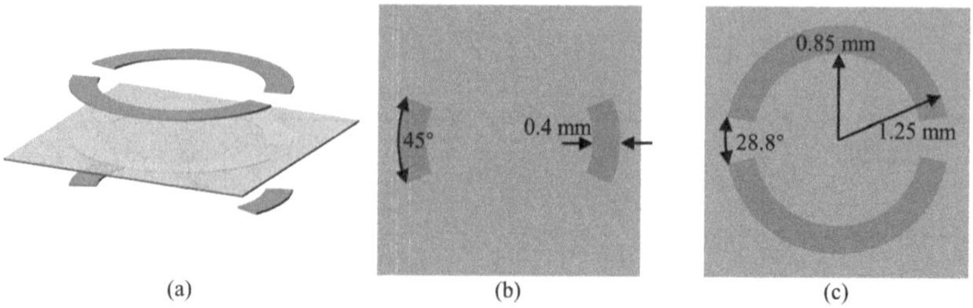

Figure 2.18 Illustration of the meta-atom with strong normal response. (a) Perspective view. (b) Top view. (c) Bottom view.

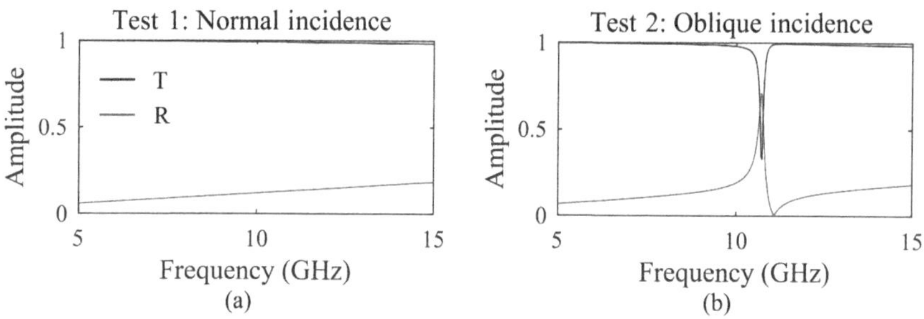

Figure 2.19 Amplitude of S parameters obtained by commercial software high frequency simulation software (HFSS) [58]. (a) Normal incidence. (b) Oblique incidence with incident angle $\theta_i = 30°$.

The surface susceptibilities are now also frequency-dependent, and Equations (2.116) and (2.117) thus become

$$\chi_{ee}^{xx}(f) = \chi_{ee}^{yy}(f) = -\frac{c_0}{j\pi f}\frac{T_1(f) + R_1(f) - 1}{T_1(f) + R_1(f) + 1} \tag{2.118.a}$$

$$\chi_{mm}^{zz}(f) = \frac{jc_0[T_2(f) + R_2(f) - 1]\cos\theta_i - 2\pi f[T_2(f) + R_2(f) + 1]\chi_{ee}^{yy}(f)}{\pi f[T_2(f) + R_2(f) + 1]\sin^2\theta_i} \tag{2.118.b}$$

The susceptibilities retrieved by (2.118) are presented in Figure 2.20, where $\theta_i = 30°$.

To validate the susceptibilities obtained by the aforementioned procedure, reflectivity and transmissivity under the incidence angle of $\theta = 60°$ are predicted by GSTCs and compared with simulation results obtained from commercial software HFSS with actual structures in Figure 2.21.

Ultrawideband Reflection-Type Metasurface for Generating Orbital Angular Momentum. Vortex electromagnetic waves carrying orbital angular

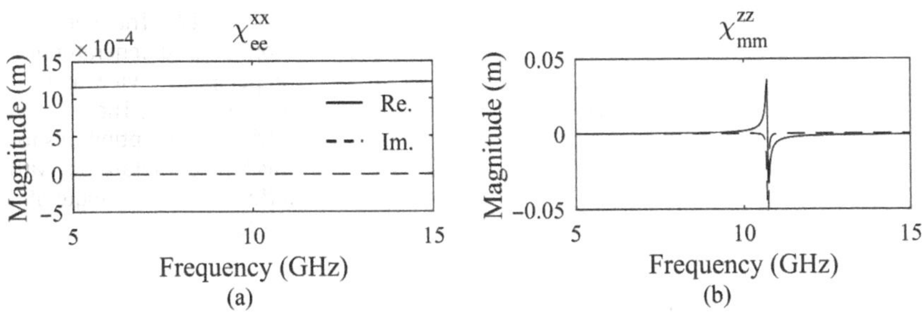

Figure 2.20 Susceptibilities retrieved for GSTCs. (a) Tangential electric susceptibility χ_{ee}^{xx}. (b) Normal magnetic susceptibility χ_{mm}^{zz}.

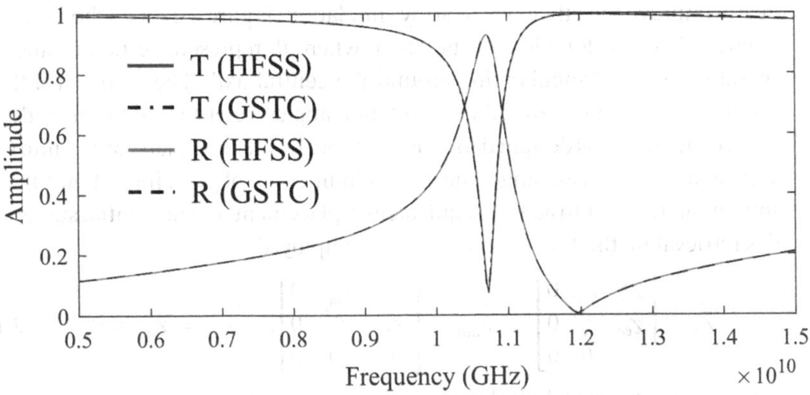

Figure 2.21 Comparison of transmitted and reflected coefficients under 60-degree oblique incidence calculated by the retrieved susceptibility and simulation results by HFSS.

momentum (OAM) can take advantage of the orthogonality of different OAM modes to obtain a larger channel capacity than the common circularly polarized electromagnetic waves carrying spin angular momentum (SAM) [59]. Therefore, it has gradually become a research hotspot in the field of radio, optical and quantum communication. Metasurface is one of the main methods of generating vortex electromagnetic waves carrying OAM.

The reference [60] proposes a reflective metasurface that can efficiently generate OAM waves, the atom of which consists of a three-layer structure, a rectangular ring copper patch, a F4B dielectric substrate and a copper patch ground plate. The minimal geometric size of the upper patch is 0.3 mm, so the simulation of metasurface array composed of 16 × 16 atoms will also face the multiscale problem. Replacing such atoms with zero-thickness sheet with surface susceptibilities can

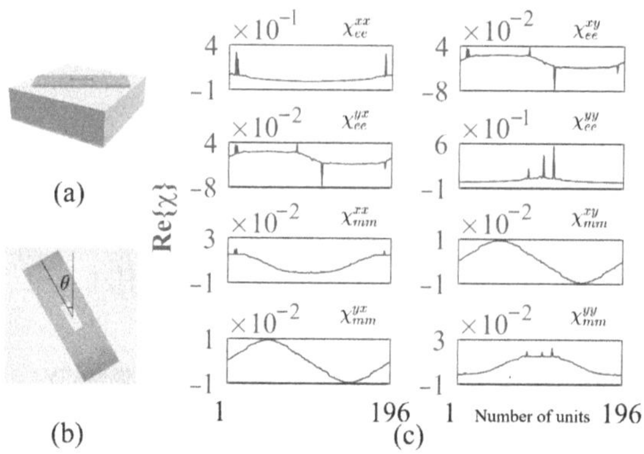

Figure 2.22 The meta-atom of OAM generator at (a) perspective view and (b) top view. (c) The retrieved susceptibilities for this type of meta-atom with different rotation angle of the top loop.

greatly improve the efficiency of array simulation. Figure 2.22(a,b) shows the perspective view and top view of the atom, where θ_i represents different angles of rotation of the rectangular ring around the central axis. The atoms at different positions in the array have different rotation angles; refer to [60] for more details.

Since the surface is designed to convert vertically incident plane waves into OAM electromagnetic waves, tangential susceptibilities are of great importance in array simulation. In order to achieve equivalent replacement, the tangential susceptibilities retrieval methods in Section 2.3.1 are employed.

$$\overline{\overline{\chi}}_{ee} = \begin{bmatrix} \chi_{ee}^{xx} & \chi_{ee}^{xy} & 0 \\ \chi_{ee}^{yx} & \chi_{ee}^{yy} & 0 \\ 0 & 0 & 0 \end{bmatrix}, \quad \overline{\overline{\chi}}_{mm} = \begin{bmatrix} \chi_{mm}^{xx} & \chi_{mm}^{xy} & 0 \\ \chi_{mm}^{yx} & \chi_{mm}^{yy} & 0 \\ 0 & 0 & 0 \end{bmatrix}, \quad \overline{\overline{\chi}}_{em} = \overline{\overline{\chi}}_{me} = 0 \tag{2.119}$$

Because there are eight unknowns, they need two tests:

i) x-polarized normal incident plane wave propagating along the $+z$-direction.
ii) y-polarized normal incident plane wave propagating along the $+z$-direction.

Considering the metal ground of the unit, which means $T = 0$, it is only necessary to record the reflected field to obtain the field average and difference. The expression of susceptibilities are

$$\chi_{ee}^{xx} = \frac{1}{j\varepsilon_0\omega}\frac{\overline{E}_{y,2}\Delta H_{y,1} - \overline{E}_{y,1}\Delta H_{y,2}}{\overline{E}_{x,2}\overline{E}_{y,1} - \overline{E}_{x,1}\overline{E}_{y,2}} \tag{2.120.a}$$

$$\chi_{ee}^{xy} = \frac{1}{j\varepsilon_0\omega}\frac{\overline{E}_{x,1}\Delta H_{y,2} - \overline{E}_{x,2}\Delta H_{y,1}}{\overline{E}_{x,2}\overline{E}_{y,1} - \overline{E}_{x,1}\overline{E}_{y,2}} \tag{2.120.b}$$

$$\chi_{ee}^{yx} = \frac{1}{j\varepsilon_0\omega}\frac{\overline{E}_{y,1}\Delta H_{x,2} - \overline{E}_{y,2}\Delta H_{x,1}}{\overline{E}_{x,2}\overline{E}_{y,1} - \overline{E}_{x,1}\overline{E}_{y,2}} \tag{2.120.c}$$

$$\chi_{ee}^{yy} = \frac{1}{j\varepsilon_0\omega}\frac{\overline{E}_{x,2}\Delta H_{x,1} - \overline{E}_{x,1}\Delta H_{x,2}}{\overline{E}_{x,2}\overline{E}_{y,1} - \overline{E}_{x,1}\overline{E}_{y,2}} \tag{2.120.d}$$

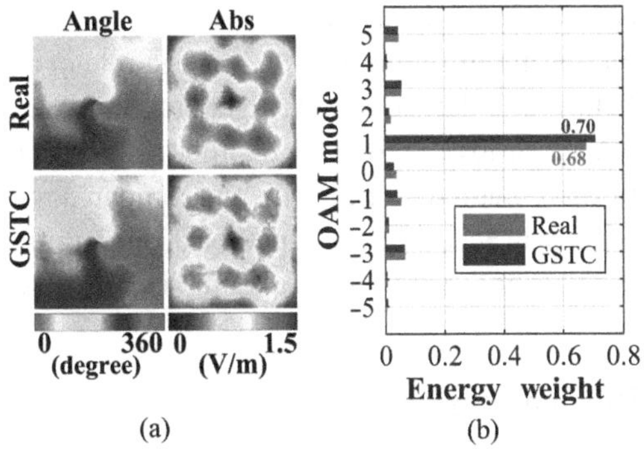

Figure 2.23 Comparison of the result of (a) electric field distribution on an observation plane two wavelengths away from the OAM metasurface generator and (b) the energy weight of different OAM modes, simulated by the retrieved susceptibility shown in Figure 2.22 (c) and by real physical structure.

$$\chi_{mm}^{xx} = \frac{1}{j\mu_0\omega}\frac{\overline{H}_{y,1}\Delta E_{y,2} - \overline{H}_{y,2}\Delta E_{y,1}}{\overline{H}_{x,2}\overline{H}_{y,1} - \overline{H}_{x,1}\overline{H}_{y,2}} \tag{2.120.e}$$

$$\chi_{mm}^{xy} = \frac{1}{j\mu_0\omega}\frac{\overline{H}_{x,2}\Delta E_{y,1} - \overline{H}_{x,1}\Delta E_{y,2}}{\overline{H}_{x,2}\overline{H}_{y,1} - \overline{H}_{x,1}\overline{H}_{y,2}} \tag{2.120.f}$$

$$\chi_{mm}^{yx} = \frac{1}{j\mu_0\omega}\frac{\overline{H}_{y,2}\Delta E_{x,1} - \overline{H}_{y,1}\Delta E_{x,2}}{\overline{H}_{x,2}\overline{H}_{y,1} - \overline{H}_{x,1}\overline{H}_{y,2}} \tag{2.120.g}$$

$$\chi_{mm}^{yy} = \frac{1}{j\mu_0\omega}\frac{\overline{H}_{x,1}\Delta E_{x,2} - \overline{H}_{x,2}\Delta E_{x,1}}{\overline{H}_{x,2}\overline{H}_{y,1} - \overline{H}_{x,1}\overline{H}_{y,2}} \tag{2.120.h}$$

where the subscripts 1 and 2 represent the tests (i) and (ii), respectively. According to (2.120.a), the results of different atoms with different rotation angles θ can be obtained, which are shown in Figure 2.22(c). The angles corresponding to 196 types of units refer to the OAM mode $l = \pm 1$ in Ref. [60]. The susceptibilities are not a continuous function of the rotation angle, which agrees with the expectation, because the units corresponding to each rotation angle are independently simulated in HFSS. Figure 2.23 presents the simulation results obtained by the retrieved susceptibility and the real meta-atom structure. The results of the field distribution and energy weight of different OAM modes obtained by the equivalent surface susceptibility have a good agreement with the results from the real structures.

2.4 Conclusion

In this chapter, the basic theory of surface susceptibility model and GSTC is introduced. Starting from the basic asymptotic analysis of the Maxwell's equations in Section 2.1, the concept of substituting electrically small periodic structure with equivalent boundary conditions is introduced. Then, the properties of susceptibilities are introduced, including the role of constitutive parameters in coupling field and polarization, the temporal and spatially dispersion, and the criterion of gain or loss. In Section 2.2, the GSTC is derived from distribution theory and pill-box method, respectively, and compared with traditional boundary conditions. Finally, in Section 2.3, the method of determining the surface susceptibility tensors is given, and corresponding examples are presented for different applications.

References

1 B. Hassani and E. Hinton, "A review of homogenization and topology optimization III—topology optimization using optimality criteria," *Comput. Struct.*, vol. 69, no. 6, pp. 739–756, 1998.

2 C. C. Mei and B. Vernescu, *Homogenization Methods for Multiscale Mechanics*. World Scientific, 2010.

3 G. Allaire, "Homogenization and two-scale convergence," *SIAM J. Math. Anal.*, vol. 23, no. 6, pp. 1482–1518, Nov. 1992, doi: 10.1137/0523084.

4 C. Miehé and A. Koch, "Computational micro-to-macro transitions of discretized microstructures undergoing small strains," *Arch. Appl. Mech.*, vol. 72, no. 4–5, pp. 300–317, Jul. 2002.

5 C. L. Holloway and E. F. Kuester, "Equivalent boundary conditions for a perfectly conducting periodic surface with a cover layer," *Radio Sci.*, vol. 35, no. 3, pp. 661–681, 2000.

6 A. U. Bhobe, "Derivation of equivalent boundary conditions using the homogenization method and their implementation in time-domain electromagnetics techniques," Ph.D. dissertation, 2003.

7 C. L. Holloway and E. F. Kuester, "A homogenization technique for obtaining generalized sheet-transition conditions for a metafilm embedded in a magnetodielectric interface," *IEEE Trans. Antennas Propag.*, vol. 64, no. 11, pp. 4671–4686, 2016.

8 J. A. Kong, *Electromagnetic Wave Theory*. Cambridge, MA: EMW Publishing, 2008.

9 V. Asadchy, "Spatially dispersive metasurfaces," Ph.D. dissertation, Aalto University, 2017.

10 B. Hu, "Kramers–Kronig in two lines," *Am. J. Phys.*, vol. 57, no. 9, p. 821, 1989.

11 F. W. King, *Hilbert Transforms*. Cambridge, England; New York: Cambridge University Press, 2009.

12 E. J. Rothwell and M. J. Cloud, *Electromagnetics*. CRC Press, 2008.

13 J. G. N. Rahmeier, T. J. Smy, J. Dugan, and S. Gupta, "Zero thickness surface susceptibilities and extended GSTCs—Part I: Spatially dispersive metasurfaces," *IEEE Trans. Antennas Propag.*, vol. 71, no. 7, pp. 5909–5919, Jul. 2023.

14 C. L. Holloway, E. F. Kuester, J. A. Gordon, J. O'Hara, J. Booth, and D. R. Smith, "An overview of the theory and applications of metasurfaces: The two-dimensional equivalents of metamaterials," *IEEE Antennas Propag. Mag.*, vol. 54, no. 2, pp. 10–35, Apr. 2012.

15 F. Capolino, *Theory and Phenomena of Metamaterials.* CRC Press, 2009.

16 T. J. Smy, J. G. N. Rahmeier, J. Dugan, and S. Gupta, "Spatially dispersive metasurfaces – part II: IE-GSTC-SD field solver with extended GSTCs," *IEEE Trans. Antennas Propag.*, vol. 71, no. 7, pp. 5920–5934, Jul. 2023.

17 J. Dugan, J. G. Nizer Rahmeier, T. J. Smy, and S. Gupta, "Part III – spatially dispersive metasurfaces: Zero thickness modeling of periodic and finite non-uniform surfaces," *IEEE Trans. Antennas Propag.*, vol. 71, no. 7, pp. 5935–5945, May 2023.

18 D. J. Hoppe and Y. Rahmat-Samii, *Impedance Boundary Conditions in Electromagnetics.* New York: CRC Press, 1995.

19 S. Tretyakov, *Analytical Modeling in Applied Electromagnetics.* Norwood, MA: Artech House, 2003.

20 I. V. Lindell and A. Sihvola, *Boundary Conditions in Electromagnetics.* Hoboken, NJ: John Wiley & Sons, Inc., 2019.

21 M. A. Leontovich, "Methods of solution for problems of electromagnetic wave propagation along the Earth surface," *Bull. Acad. Sci. USSR, Phys. Ser.*, vol. 8, no. 1, pp. 16–22, 1944. (in Russian)

22 L. N. Medgyesi-Mitschang and J. M. Putnam, "Integral equation formulations for imperfectly conducting scatterers," *IEEE Trans. Antennas Propag.*, vol. 33, no. 2, pp. 206–214, Feb. 1986.

23 P. L. Huddleston, "Scattering by finite open cylinders using approximate boundary conditions," *IEEE Trans. Antennas Propag.*, vol. 37, no. 2, pp. 253–257, Feb. 1989.

24 A. A. Kishk, "Electromagnetic scattering from composite objects using a mixture of exact and impedance boundary conditions," *IEEE Trans. Antennas Propag.*, vol. 39, no. 6, pp. 826–833, June 1991.

25 Y. Radi and S. A. Tretyakov, "Balanced and optimal bianisotropic particles: Maximizing power extracted from electromagnetic fields," *New J. Phys.*, vol. 15, no. 5, 053008, 2013.

26 A. Epstein and G. V. Eleftheriades, "Arbitrary power-conserving field transformations with passive lossless omega-type bianisotropic metasurfaces," *IEEE Trans. Antennas Propag.*, vol. 64, no. 9, pp. 3880–3895, Sep. 2016.

27 K. Achouri and C. Caloz, *Electromagnetic Metasurfaces: Theory and Applications*, 1st ed. Wiley, 2021.

28 E. F. Kuester, M. A. Mohamed, M. Piket-May, and C. L. Holloway, "Averaged transition conditions for electromagnetic fields at a metafilm," *IEEE Trans. Antennas Propag.*, vol. 51, no. 10, pp. 2641–2651, Oct. 2003.

29 C. L. Holloway, A. Dienstfrey, E. F. Kuester, J. F. O'Hara, A. K. Azad, and A. J. Taylor, "A discussion on the interpretation and characterization of metafilms/metasurfaces: The two-dimensional equivalent of metamaterials," *Metamaterials*, vol. 3, no. 2, pp. 100–112, 2009.

30 C. L. Holloway, E. F. Kuester, and D. Novotny, "Waveguides composed of metafilms/metasurfaces: The two-dimensional equivalent of metamaterials," *IEEE Antennas Wirel. Propag. Lett.*, vol. 8, pp. 525–529, 2009.

31 C. L. Holloway, E. F. Kuester, and A. Dienstfrey, "Characterizing metasurfaces/metafilms: The connection between surface susceptibilities and effective material properties," *IEEE Antennas Wirel. Propag. Lett.*, vol. 10, pp. 1507–1511, 2011.

32 M. Idemen, "The Maxwell's equations in the sense of distributions," *IEEE Trans. Antennas Propag.*, vol. 21, no. 5, pp. 736–738, Sep. 1973.

33 M. M. Idemen, *Discontinuities in the Electromagnetic Field.* Hoboken, NJ: John Wiley & Sons, Inc., 2011.

34 C. R. Simovski, M. S. Kondratjev, P. A. Belov, and S. A. Tretyakov, "Interaction effects in two-dimensional bianisotropic arrays," *IEEE Trans. Antennas Propag.*, vol. 47, pp. 1429–1439, 1999.

35 S. I. Maslovski and S. A. Tretyakov, "Full-wave interaction field in two-dimensional arrays of dipole scatterers," *Int. J. Electron. Commun.*, vol. 53, pp. 135–139, 1999.

36 S. A. Tretyakov, A. J. Viitanen, S. I. Maslovski, and I. E. Saarela, "Impedance boundary conditions for regular dense arrays of dipole scatterers," *IEEE Trans. Antennas Propag.*, vol. 51, pp. 2073–2078, 2003.

37 K. Achouri, M. A. Salem, and C. Caloz, "General metasurface synthesis based on susceptibility tensors," *IEEE Trans. Antennas Propag.*, vol. 63, no. 7, pp. 2977–2991, 2015.

38 F. Roger, *Harrington. Time-harmonic Electromagnetic Fields.* Wiley-Interscience, 2001.

39 D. H. Staelin, A. W. Morgenthaler, and J. A. Kong, *Electromagnetic Waves.* Englewood Cliffs, NJ: Prentice Hall, 1994.

40 P. A. M. Dirac, *The Principles of Quantum Mechanics*, 4th ed. Oxford: Clarendon Press, 1967.

41 L. Schwartz, *Théorie des Distributions.* Paris: I Hermann, 1950.

42 M. Idemen, "Universal boundary relations of the electromagnetic field," *J. Phys. Soc. Jpn.*, vol. 59, no. 1, pp. 71–80, Jan. 1990, doi: 10.1143/JPSJ.59.71.

43 I. M. Gel'fand and G. E. Shilov, *Generalized Functions*, vol. 2. New York: Academic Press, 1968.

44 M. Albooyeh, D.-H. Kwon, F. Capolino, and S. A. Tretyakov, "Equivalent realizations of reciprocal metasurfaces: Role of tangential and normal polarization," *Phys. Rev. B*, vol. 95, no. 11, 115435, Mar. 2017.

45 B. Bian *et al.*, "Novel triple-band polarization-insensitive wide-angle ultra-thin microwave metamaterial absorber," *J. Appl. Phys.*, vol. 114, no. 19, 194511, 2013.

46 K. Wu and Q. Ren, "An improved DGTD-GSTC method equipped with normal surface polarizabilities for oblique incidence," *IEEE Trans. Antennas Propag.*, vol. 72, no. 12, pp. 9317–9328, 2024.

47 C. L. Holloway, M. A. Mohamed, E. F. Kuester, and A. Dienstfrey, "Reflection and transmission properties of a metafilm: With an application to a controllable surface composed of resonant particles," *IEEE Trans. Electromagn. Compat.*, vol. 47, no. 4, pp. 853–865, 2005.

48 M. Dehmollaian, G. Lavigne, and C. Caloz, "Comparison of tensor boundary conditions with generalized sheet transition conditions," *IEEE Trans. Antennas Propag.*, vol. 67, no. 12, pp. 7396–7406, 2019.

49 D. R. Smith, R. Dalichaouch, N. Kroll, S. Schultz, S. L. McCall, and P. M. Platzman, "Photonic band structure and defects in one and two dimensions," *JOSA B*, vol. 10, no. 2, pp. 314–321, Feb. 1993.

50 J.-M. Jin, *The Finite Element Method in Electromagnetics*, 3rd ed. Hoboken, NJ: John Wiley & Sons Inc, 2014.

51 X. Jia, F. Yang, X. Liu, M. Li, and S. Xu, "Fast nonuniform metasurface analysis in FDTD using surface susceptibility model," *IEEE Trans. Antennas Propag.*, vol. 68, no. 10, pp. 7121–7130, Oct. 2020.

52 A. M. Nicolson and G. F. Ross, "Measurement of the intrinsic properties of materials by time-domain techniques," *IEEE Trans. Instrum. Meas.*, vol. 19, no. 4, pp. 377–382, Nov. 1970.

53 X. Chen, T. M. Grzegorczyk, B.-I. Wu, J. Pacheco, and J. A. Kong, "Robust method to retrieve the constitutive effective parameters of metamaterials," *Phys. Rev. E*, vol. 70, no. 1, 016608, Jul. 2004.

54 A. K. Geim and K. S. Novoselov, "The rise of graphene," *Nat. Mater.*, vol. 6, pp. 183–191, Mar. 2007.

55 A. Vakil and N. Engheta, "Transformation optics using graphene," *Science*, vol. 332, no. 6035, pp. 1291–1294, Jun. 2011.

56 V. Nayyeri, M. Soleimani, and O. M. Ramahi, "Modeling graphene in the finite-difference time-domain method using a surface boundary condition," *IEEE Trans. Antennas Propag.*, vol. 61, no. 8, pp. 4176–4182, 2013.

57 T. J. Smy, V. Tiukuvaara, and S. Gupta, "IE-GSTC metasurface field solver using surface susceptibility tensors with normal polarizabilities," *IEEE Trans. Antennas Propag.*, vol. 70, no. 6, pp. 4752–4765, Jun. 2022.

58 Ansys, Ansys HFSS | 3D High Frequency Simulation Software, 2025. https://www.ansys.com/products/electronics/ansys-hfss (Sep 23, 2025).

59 S. M. Mohammadi *et al.*, "Orbital angular momentum in radio: A system study," *IEEE Trans. Antennas Propag.*, vol. 58, no. 2, pp. 565–572, 2010.

60 L.-J. Yang, S. Sun, and W. E. I. Sha, "Ultrawideband reflection-type metasurface for generating integer and fractional orbital angular momentum," *IEEE Trans. Antennas Propags*, vol. 68, no. 3, pp. 2166–2175, 2020.

3

Frequency-Domain Method for Metasurface Using GSTCs

Electromagnetic phenomena occur naturally in space and time, as described by Maxwell's equations. But sometimes it's more convenient to analyze electromagnetic problems in other domains, such as the frequency domain, which will be shown in this chapter. A time-harmonic signal $v(t)$ can be mathematically expressed as

$$v(t) = |V| \cos(\omega t + \alpha) \tag{3.1}$$

where $|V|$, ω, and α are the amplitude, angular frequency, and phase of this time-harmonic signal. A time-varying signal can be expressed in terms of time-harmonic components via the Fourier transforms

$$v(t) = \frac{1}{2\pi} \int_{-\infty}^{+\infty} V(\omega) e^{j\omega t}\, d\omega \tag{3.2.a}$$

$$V(\omega) = \int_{-\infty}^{+\infty} v(t) e^{-j\omega t}\, dt \tag{3.2.b}$$

With the Fourier transform, Maxwell's equations can be solved in the frequency domain by replacing the time derivative $\frac{\partial}{\partial t}$ with $j\omega$ as

$$\nabla \times \mathbf{E}(\omega) = -j\omega\mu \mathbf{H}(\omega) - \mathbf{M}(\omega) \tag{3.3.a}$$

$$\nabla \times \mathbf{H}(\omega) = j\omega\varepsilon \mathbf{E}(\omega) + \mathbf{J}(\omega) \tag{3.3.b}$$

where

$$\mathbf{E}(\omega) = E_x(\omega)\mathbf{i_x} + E_y(\omega)\mathbf{i_y} + E_z(\omega)\mathbf{i_z} \tag{3.4}$$

$\mathbf{H}(\omega)$, $\mathbf{M}(\omega)$, and $\mathbf{J}(\omega)$. Each component of $\mathbf{E}(\omega)$ can be related to a time-harmonic field. For example, the x-components of $\mathbf{E}(\omega)$ and $\mathbf{E}(t)$ are related by

$$E_x(t) = |E_x| \cos(\omega t + \alpha_x) = \mathrm{Re}\left\{|E_x| e^{j\alpha_x} e^{j\omega t}\right\} = \mathrm{Re}\{E_x(\omega) e^{j\omega t}\} \tag{3.5}$$

where $E_x(\omega) = |E_x| e^{j\alpha_x}$ is the phasor quantity. Similar equations hold for the y- and z-components and other field quantities, such as $\mathbf{H}(\omega)$, $\mathbf{M}(\omega)$, and $\mathbf{J}(\omega)$.

The advantages of the frequency-domain method are obvious. First, the complexity of the problem is reduced by one dimension (the time dimension) by replacing the time derivative with the multiplication by $j\omega$. Second, the solutions of (3.3.a) are more convenient to investigating steady states of the electromagnetic system. Finally, the dispersive materials can be treated more easily in the frequency-domain method. Since the constitutive parameters of materials (such as ε_r, μ_r, and σ) are mostly given explicitly as a function of frequency or a table versus frequency, the electromagnetic analysis of dispersive materials using the frequency-domain method is natural and simple. According to Section 2.3, the surface susceptibilities of metasurfaces, i.e., $\overline{\overline{\chi}}_{ee}$, $\overline{\overline{\chi}}_{mm}$, $\overline{\overline{\chi}}_{me}$, and $\overline{\overline{\chi}}_{em}$, are all synthesized and extracted in the frequency domain. Therefore, developing frequency-domain numerical methods for metasurfaces is significant.

In the following sections, three frequency-domain methods will be introduced, and it will be shown how to incorporate the GSTCs. Section 3.1 is about the FDFD method, which is a simple-to-implement CEM algorithm. In this section, the strategy to incorporate GSTC in the framework of the finite difference technique is introduced, and the ability of the surface susceptibility model in describing complex wave control is illustrated by simulation examples. Section 3.2 is about the FEM, which is particularly useful in problems involving complex geometries. This section will introduce how to incorporate GSTC in the framework of the finite element technique. The advantages of using homogenization instead of metasurface structures can be exhibited directly in this section. Section 3.3 introduces the strategies for integrating GSTCs in the SEM, which is a high-order version of FEM. Section 3.4 is about the IE method. The IE method is suitable for open-region problems, which is the main working scenario of metasurface arrays. The commercial software FEKO is famous for the IE-based solver, but it has not developed the GSTC-related solvers yet. This section introduces how to incorporate GSTC in the framework of the IE method. Conclusions are provided in Section 3.5.

3.1 FDFD Method with GSTCs

3.1.1 Introduction of FDFD Method

The finite-difference method approximates the PDEs by differences, transforming the original problem into a discrete form. By solving the resulting discrete system, the approximate numerical solution of the original problem is obtained. In 1966, K. S. Yee proposed the basic form of the finite difference method in the time domain [1]. He also introduced the famous Yee's grid with staggered sampling points for the electric and magnetic field intensities.

The FDFD method is the frequency domain form of the finite difference method. Compared to Maxwell's equation in the time domain, the frequency-domain version does not involve partial derivatives with time. The solving process of the FDFD method generally involves

1) Discretizing the space with Yee's grids and establishing difference equations at each sampling node within the grids.
2) Assembling all the differential equations into a matrix equation $\mathbf{Ax} = \mathbf{b}$.
3) Solving the matrix equation and giving the corresponding field values at the discrete nodes.

Although the FDFD method calculates directly in the frequency domain and there is no stability issue, it has the drawback of generating a large and sparse matrix equation that is difficult to solve as the problem size increases. In the finite difference method, the size of the generated matrix ultimately depends on the number of grid points after spatial discretization. Therefore, one key research direction is how to reduce the size of the matrix by accurately describing the problem with fewer spatial sampling points.

The FDFD-GSTC method to be introduced in this section can solve the multi-scale problem in metasurface simulations. Considering the ultra-thin property of metasurfaces in one direction, it reasonably models the metasurface as a generalized sheet with zero thickness, and the macro response produced by the structure will be reproduced by the equivalent surface susceptibilities. This replacement eliminates the need for the dense grid to capture the geometric structures of each meta-atom of the metasurface.

3.1.1.1 Finite-Difference Method and Yee Algorithm

The frequency-domain Maxwell's equations in the Cartesian coordinate system are written as

$$\frac{\partial E_z}{\partial y} - \frac{\partial E_y}{\partial z} = -j\omega\mu H_x - M_x \tag{3.6.a}$$

$$\frac{\partial E_x}{\partial z} - \frac{\partial E_z}{\partial x} = -j\omega\mu H_y - M_y \tag{3.6.b}$$

$$\frac{\partial E_y}{\partial x} - \frac{\partial E_x}{\partial y} = -j\omega\mu H_z - M_z \tag{3.6.c}$$

$$\frac{\partial H_z}{\partial y} - \frac{\partial H_y}{\partial z} = j\omega\varepsilon E_x + J_x \tag{3.6.d}$$

$$\frac{\partial H_x}{\partial z} - \frac{\partial H_z}{\partial x} = j\omega\varepsilon E_y + J_y \tag{3.6.e}$$

$$\frac{\partial H_y}{\partial x} - \frac{\partial H_x}{\partial y} = j\omega\varepsilon E_z + J_z \tag{3.6.f}$$

Each derivative can be approximated by finite difference as

$$\frac{E_z^{i,j+1,k} - E_z^{i,j,k}}{\Delta y} - \frac{E_y^{i,j,k+1} - E_y^{i,j,k}}{\Delta z} = -j\omega\mu_x^{i,j,k} H_x^{i,j,k} - M_x^{i,j,k} \tag{3.7.a}$$

$$\frac{E_x^{i,j,k+1} - E_x^{i,j,k}}{\Delta z} - \frac{E_z^{i+1,j,k} - E_z^{i,j,k}}{\Delta x} = -j\omega\mu_y^{i,j,k} H_y^{i,j,k} - M_y^{i,j,k} \tag{3.7.b}$$

$$\frac{E_y^{i+1,j,k} - E_y^{i,j,k}}{\Delta x} - \frac{E_x^{i,j+1,k} - E_x^{i,j,k}}{\Delta y} = -j\omega\mu_z^{i,j,k} H_z^{i,j,k} - M_z^{i,j,k} \tag{3.7.c}$$

$$\frac{H_z^{i,j,k} - H_z^{i,j-1,k}}{\Delta y} - \frac{H_y^{i,j,k} - H_y^{i,j,k-1}}{\Delta z} = j\omega\varepsilon_x^{i,j,k} E_x^{i,j,k} + J_x^{i,j,k}, \tag{3.7.d}$$

$$\frac{H_x^{i,j,k} - H_x^{i,j,k-1}}{\Delta z} - \frac{H_z^{i,j,k} - H_z^{i-1,j,k}}{\Delta x} = j\omega\varepsilon_y^{i,j,k} E_y^{i,j,k} + J_y^{i,j,k} \tag{3.7.e}$$

$$\frac{H_y^{i,j,k} - H_y^{i-1,j,k}}{\Delta x} - \frac{H_x^{i,j,k} - H_x^{i,j-1,k}}{\Delta y} = j\omega\varepsilon_z^{i,j,k} E_z^{i,j,k} + J_z^{i,j,k} \tag{3.7.f}$$

where Δv $(v = x, y, z)$ is the step size in space, $\varepsilon_v^{i,j,k}$ and $\mu_v^{i,j,k}$ are the electric permittivity and magnetic permeability at the locations of $E_v^{i,j,k}$ and $H_v^{i,j,k}$, respectively. Superscripts i, j, and k represent the positions of the electric field (magnetic field) in the x-, y-, and z-directions as $i\Delta x$, $j\Delta y$, and $k\Delta z$, respectively. Yee algorithm assigns $E_v^{i,j,k}$ and $H_v^{i,j,k}$ in Yee's finite-difference grid cell, as shown in Figure 3.1.

The computational domain is first discretized with an adequate number of Yee grids, as shown in Figure 3.1. Establishing the differential equations at each node and assembling all the difference equations into a matrix equation leads to [2]

$$\mathbf{D}_y^e \mathbf{e}_z - \mathbf{D}_z^e \mathbf{e}_y = \boldsymbol{\mu}_{xx} \mathbf{h}_x - \mathbf{m}_x \tag{3.8.a}$$

$$\mathbf{D}_z^e \mathbf{e}_x - \mathbf{D}_x^e \mathbf{e}_z = \boldsymbol{\mu}_{yy} \mathbf{h}_y - \mathbf{m}_y \tag{3.8.b}$$

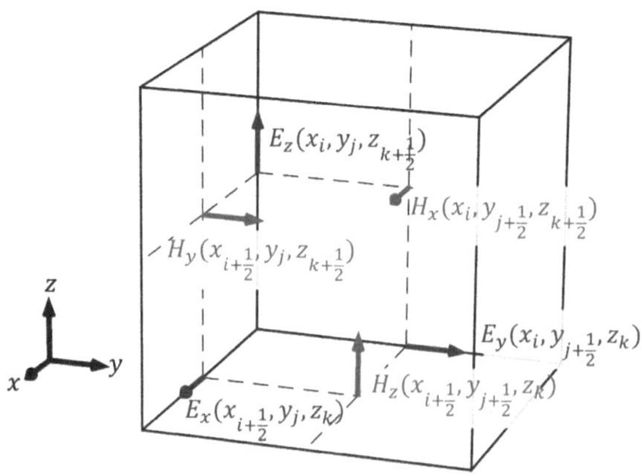

Figure 3.1 Yee's finite-difference grid cell and its electric and magnetic field nodes.

$$\mathbf{D}_x^e\mathbf{e}_y - \mathbf{D}_y^e\mathbf{e}_x = \boldsymbol{\mu}_{zz}\mathbf{h}_z - \mathbf{m}_z \tag{3.8.c}$$

$$\mathbf{D}_y^h\mathbf{h}_z - \mathbf{D}_z^h\mathbf{h}_y = \boldsymbol{\varepsilon}_{xx}\mathbf{e}_x + \mathbf{j}_x \tag{3.8.d}$$

$$\mathbf{D}_z^h\mathbf{h}_x - \mathbf{D}_x^h\mathbf{h}_z = \boldsymbol{\varepsilon}_{yy}\mathbf{e}_y + \mathbf{j}_y \tag{3.8.e}$$

$$\mathbf{D}_x^h\mathbf{h}_y - \mathbf{D}_y^h\mathbf{h}_x = \boldsymbol{\varepsilon}_{zz}\mathbf{e}_z + \mathbf{j}_z \tag{3.8.f}$$

where

$$\mathbf{e}_v = \begin{bmatrix} E_v^{i_1,j_1,k_1} \\ \vdots \\ E_v^{i_x,j_y,k_z} \end{bmatrix}, \quad \mathbf{h}_v = \begin{bmatrix} H_v^{i_1,j_1,k_1} \\ \vdots \\ H_v^{i_x,j_y,k_z} \end{bmatrix}, \quad \mathbf{j}_v = \begin{bmatrix} J_v^{i_1,j_1,k_1} \\ \vdots \\ J_v^{i_x,j_y,k_z} \end{bmatrix}, \quad \mathbf{m}_v = \begin{bmatrix} M_v^{i_1,j_1,k_1} \\ \vdots \\ M_v^{i_x,j_y,k_z} \end{bmatrix} \tag{3.9}$$

are the column vectors representing the unknown field components. $\mathbf{D}_v^e$ and $\mathbf{D}_v^h$ are the curl operator matrices, while $\boldsymbol{\mu}$ and $\boldsymbol{\varepsilon}$ are diagonal matrices from the permeability and permittivity. To solve these six matrix equations, the first three and the last three can be grouped.

$$\begin{bmatrix} 0 & -\mathbf{D}_z^e & \mathbf{D}_y^e \\ \mathbf{D}_z^e & 0 & -\mathbf{D}_x^e \\ -\mathbf{D}_y^e & \mathbf{D}_x^e & 0 \end{bmatrix} \begin{bmatrix} \mathbf{e}_x \\ \mathbf{e}_y \\ \mathbf{e}_z \end{bmatrix} = \begin{bmatrix} \boldsymbol{\mu}_{xx} & 0 & 0 \\ 0 & \boldsymbol{\mu}_{yy} & 0 \\ 0 & 0 & \boldsymbol{\mu}_{zz} \end{bmatrix} \begin{bmatrix} \mathbf{h}_x \\ \mathbf{h}_y \\ \mathbf{h}_z \end{bmatrix} + \begin{bmatrix} \mathbf{m}_x \\ \mathbf{m}_y \\ \mathbf{m}_z \end{bmatrix} \tag{3.10}$$

$$\begin{bmatrix} 0 & -\mathbf{D}_z^h & \mathbf{D}_y^h \\ \mathbf{D}_z^h & 0 & -\mathbf{D}_x^h \\ -\mathbf{D}_y^h & \mathbf{D}_x^h & 0 \end{bmatrix} \begin{bmatrix} \mathbf{h}_x \\ \mathbf{h}_y \\ \mathbf{h}_z \end{bmatrix} = \begin{bmatrix} \boldsymbol{\varepsilon}_{xx} & 0 & 0 \\ 0 & \boldsymbol{\varepsilon}_{yy} & 0 \\ 0 & 0 & \boldsymbol{\varepsilon}_{zz} \end{bmatrix} \begin{bmatrix} \mathbf{e}_x \\ \mathbf{e}_y \\ \mathbf{e}_z \end{bmatrix} + \begin{bmatrix} \mathbf{j}_x \\ \mathbf{j}_y \\ \mathbf{j}_z \end{bmatrix} \tag{3.11}$$

Ignoring $\mathbf{m}$ in (3.10) and substituting it into (3.11) leads to

$$\mathbf{Ax} = \mathbf{b} \tag{3.12}$$

where

$$\mathbf{A} = \begin{bmatrix} 0 & -\mathbf{D}_z^h & \mathbf{D}_y^h \\ \mathbf{D}_z^h & 0 & -\mathbf{D}_x^h \\ -\mathbf{D}_y^h & \mathbf{D}_x^h & 0 \end{bmatrix} \begin{bmatrix} \boldsymbol{\mu}_{xx} & 0 & 0 \\ 0 & \boldsymbol{\mu}_{yy} & 0 \\ 0 & 0 & \boldsymbol{\mu}_{zz} \end{bmatrix}^{-1} \begin{bmatrix} 0 & -\mathbf{D}_z^e & \mathbf{D}_y^e \\ \mathbf{D}_z^e & 0 & -\mathbf{D}_x^e \\ -\mathbf{D}_y^e & \mathbf{D}_x^e & 0 \end{bmatrix} - \begin{bmatrix} \boldsymbol{\varepsilon}_{xx} & 0 & 0 \\ 0 & \boldsymbol{\varepsilon}_{yy} & 0 \\ 0 & 0 & \boldsymbol{\varepsilon}_{zz} \end{bmatrix} \tag{3.13.a}$$

$$\mathbf{x} = \begin{bmatrix} \mathbf{e}_x \\ \mathbf{e}_y \\ \mathbf{e}_z \end{bmatrix}, \quad \mathbf{b} = \begin{bmatrix} \mathbf{j}_x \\ \mathbf{j}_y \\ \mathbf{j}_z \end{bmatrix} \tag{3.13.b}$$

3.1.1.2 Iterative Methods to Solve Ax = b

Direct methods and iterative methods are two categories of approaches to solve the system of linear equations in (3.12). Direct methods produced a solution in a fixed number of steps by calculating $\mathbf{A}^{-1}\mathbf{b}$. Many factorization methods such as the Cholesky, lower-upper (LU), lower-diagonal-upper (LDU), QR, and SVD factorizations are used to get $\mathbf{A}^{-1}$.

The matrix **A** of Maxwell equations is extremely large. For example, the matrix of a problem with 40 × 40 × 40 Yee cells has 375,000 rows and columns and 4,785,300 non-zero elements. For such a large and sparse matrix, iterative methods are more suitable. Iterative methods start with an initial guess of the solution $\mathbf{x_0}$ and then iteratively refine the solution until it converges to a desired level of accuracy. Usually the residual vector $\mathbf{r_m} = \mathbf{b} - \mathbf{A}\mathbf{x_m}$ satisfies $\frac{\|\mathbf{r_m}\|}{\|\mathbf{b}\|} < \tau$, where $\mathbf{x_m}$ is the approximate solution at the m-th iteration, $\|\cdot\|$ is the norm of a vector, and τ is the tolerance defined by the user.

Since the matrix **A** from FDFD-GSTC is **NOT** complex-Hermitian or real-symmetric, some generic iterative methods are needed to handle the matrices. The biconjugate gradient (BiCG) and quasi-minimal residual (QMR) are both suitable methods [3]. The convergence speed is the key to determine the total solution time when using iterative methods.

3.1.2 Strategies for Integrating GSTCs in FDFD

3.1.2.1 GSTC Implementation in Two-Dimensional (2-D) FDFD

It is always a good choice when a problem can be modeled in two dimensions. This happens when research targets are uniform in one direction and no wave propagates along this uniform direction. Without loss of generality, the TE problem is considered here: $(H_y = E_x = E_z = 0)$. Equation (3.8.a) is simplified to

$$-\mathbf{D}_z^e \mathbf{e}_y = \boldsymbol{\mu}_{xx} \mathbf{h}_x \tag{3.14.a}$$

$$\mathbf{D}_x^e \mathbf{e}_y = \boldsymbol{\mu}_{zz} \mathbf{h}_z \tag{3.14.b}$$

$$\mathbf{D}_z^h \mathbf{h}_x - \mathbf{D}_x^h \mathbf{h}_z = \boldsymbol{\varepsilon}_{yy} \mathbf{e}_y + \mathbf{j}_y \tag{3.14.c}$$

and the operator matrix **A** of the final linear equation is simplified to

$$\mathbf{A} = \mathbf{D}_z^h \boldsymbol{\mu}_{xx}^{-1} \mathbf{D}_z^e + \mathbf{D}_x^h \boldsymbol{\mu}_{zz}^{-1} \mathbf{D}_x^e + \boldsymbol{\varepsilon}_{yy} \tag{3.15}$$

Since surface polarization and magnetization create discontinuities in both the electric and magnetic fields, the zero-thickness equivalent sheet used to replace the metasurface cannot be placed at a grid node of either an electric or magnetic node. The space between two adjacent electric nodes and the magnetic nodes of Yee's grid is the best placement for the generalized sheet. Special GSTC equations will be applied at the grid nodes surrounding the generalized sheet [4]. Here, the TE wave propagating along the z-direction is considered, and a metasurface is placed normally to it. The generalized sheet is assumed to be placed between $z = k_m dz$ and $z = (k_m + 1/2)dz$, as shown in Figure 3.2. Therefore, there will be field discontinuities between $H_x^{i_0,k_m}$ and $H_x^{i_0,k_m+1}$ and between $E_y^{i_0,k_m}$ and $E_y^{i_0,k_m+1}$, which means (3.14.a) and (3.14.c) cannot describe $H_x^{i_0,k_m}$ and $E_y^{i_0,k_m+1}$ any more, while (3.14.b) can still describe H_z^{i,k_m}.

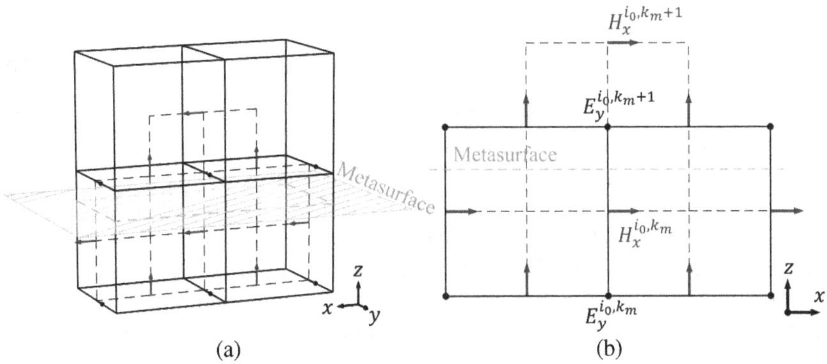

Figure 3.2 GSTCs modeling metasurface (a) in three-dimensional (3-D) Yee's grid and (b) in 2-D Yee's grid.

The GSTC equations applied at $H_x^{i_0,k_m}$ and $E_y^{i_0,k_m+1}$ are

$$\Delta H_x = j\omega\varepsilon_0\chi_{ee}^{yy}E_y \tag{3.16.a}$$

$$\Delta E_y = j\omega\mu_0\chi_{mm}^{xx}H_x \tag{3.16.b}$$

Discretizing (3.16.a) and (3.16.b) at $z = \left(k_m + \frac{3}{4}\right)dz$ leads to

$$H_x^{i,k_m+1} - H_x^{i,k_m} = \frac{j\omega\varepsilon_0\chi_{ee,yy}^{i,k_m}}{2}\left(E_y^{i,k_m+1} + E_y^{i,k_m}\right) \tag{3.17.a}$$

$$E_y^{i,k_m+1} - E_y^{i,k_m} = \frac{j\omega\mu_0\chi_{mm,xx}^{i,k_m}}{2}\left(H_x^{i,k_m+1} + H_x^{i,k_m}\right) \tag{3.17.b}$$

The implementation of (3.17) requires the replacement of the corresponding equation of $H_x^{i_0,k_m}$ and $E_y^{i_0,k_m+1}$. Matrices $\mathbf{D}_z^e$, $\mathbf{D}_z^h$, $\mathbf{D}_x^h$, $\boldsymbol{\mu}_{xx}$, and $\boldsymbol{\varepsilon}_{yy}$ should be adjusted according to (3.17.a)

$$\mathbf{G}_z^h\mathbf{h}_x = \boldsymbol{\chi}_{ee}^{yy}\mathbf{e}_y \tag{3.18.a}$$

$$\mathbf{G}_z^e\mathbf{e}_y = \boldsymbol{\chi}_{mm}^{xx}\mathbf{h}_x \tag{3.18.b}$$

where the matrices $\mathbf{G}_z^e$ and $\mathbf{G}_z^h$ are the difference operators and $\boldsymbol{\chi}_{ee}^{yy}$ and $\boldsymbol{\chi}_{mm}^{xx}$ are susceptibility matrices for $E_y^{i_0,k_m}$, $E_y^{i_0,k_m+1}$ and $H_x^{i_0,k_m}$, $H_x^{i_0,k_m}$, respectively. Equation (3.15) needs to be modified as

$$\mathbf{D}_z'^e = \mathbf{D}_z^e - \mathbf{D}_z^e(p) + \mathbf{G}_z^e \tag{3.19.a}$$

$$\mathbf{D}_z'^h = \mathbf{D}_z^h - \mathbf{D}_z^h(p) + \mathbf{G}_z^h \tag{3.19.b}$$

$$\mathbf{D}_x'^h = \mathbf{D}_x^h - \mathbf{D}_x^h(p) \tag{3.19.c}$$

$$\boldsymbol{\mu}'_{xx} = \boldsymbol{\mu}_{xx} - \boldsymbol{\mu}_{xx}(p) + \boldsymbol{\chi}_{mm}^{xx} \tag{3.20.a}$$

$$\boldsymbol{\varepsilon}'_{yy} = \boldsymbol{\varepsilon}_{yy} - \boldsymbol{\varepsilon}_{yy}(p) + \boldsymbol{\chi}_{ee}^{yy} \tag{3.20.b}$$

where p is the index corresponding to $H_x^{i_0,k_m}$ and $E_y^{i_0,k_m+1}$. Note that $\mathbf{D}_x^e$ and $\boldsymbol{\mu}_{zz}$ remain unchanged. The new matrix $\mathbf{A}'$ is

$$\mathbf{A}' = \mathbf{D}_z'^h \boldsymbol{\mu}_{xx}'^{-1} \mathbf{D}_z'^e + \mathbf{D}_x'^h \boldsymbol{\mu}_{zz}^{-1} \mathbf{D}_x^e + \boldsymbol{\varepsilon}'_{yy} \tag{3.21}$$

As a general conclusion, using GSTCs to simulate a metasurface as a zero-thickness sheet requires changing the standard FDFD equations with GSTC equations at the surrounding electric field and magnetic field nodes of Yee's grid of the metasurface. This change results in a replacement in traditional differential operator matrices and material matrices.

3.1.2.2 GSTC Implementation in 3-D FDFD

In the 3-D FDFD-GSTC method, metasurfaces are modeled as 2-D generalized sheets. Without loss of generality, the zero-thickness sheet is assumed to be parallel to the XoY-plane between $z = (k_m + 1/2)dz$ and $z = (k_m + 1)dz$, just like in the 2-D FDFD-GSTC method. The generalized sheet with surface susceptibility tensor will cause discontinuities between E_x^{i,j,k_m+1} and E_x^{i,j,k_m}, E_y^{i,j,k_m+1} and E_y^{i,j,k_m}, H_x^{i,j,k_m+1} and H_x^{i,j,k_m}, and H_y^{i,j,k_m+1} and H_y^{i,j,k_m}, which means that these field points cannot be calculated using (3.8.a). According to (2.97), the discontinuities satisfy

$$-\Delta E_x = j\omega\mu\left(\chi_{mm}^{yx} H_x + \chi_{mm}^{yy} H_y\right) \tag{3.22.a}$$

$$\Delta E_y = j\omega\mu\left(\chi_{mm}^{xx} H_x + \chi_{mm}^{xy} H_y\right) \tag{3.22.b}$$

$$\Delta H_x = j\omega\varepsilon\left(\chi_{ee}^{yx} E_x + \chi_{ee}^{yy} E_y\right) \tag{3.22.c}$$

$$-\Delta H_y = j\omega\varepsilon\left(\chi_{ee}^{xx} E_x + \chi_{ee}^{xy} E_y\right) \tag{3.22.d}$$

where the electric–magnetic coupling susceptibilities are ignored for simplicity. The corresponding discrete forms of (3.22) are

$$\begin{aligned} E_x^{i,j,k_m} - E_x^{i,j,k_m+1} &= \frac{j\omega\mu\chi_{mm,yy}^{i,j,k_m}}{2}\left(H_y^{i,j,k_m} + H_y^{i,j,k_m+1}\right) \\ &+ \frac{j\omega\mu\chi_{mm,yx}^{i,j,k_m}}{8}(H_x^{i,j,k_m} + H_x^{i,j-1,k_m} + H_x^{i+1,j,k_m} + H_x^{i+1,j-1,k_m} \\ &+ H_x^{i,j,k_m+1} + H_x^{i,j-1,k_m+1} + H_x^{i+1,j,k_m+1} + H_x^{i+1,j-1,k_m+1}), \end{aligned} \tag{3.23.a}$$

$$\begin{aligned} E_y^{i,j,k_m+1} - E_y^{i,j,k_m} &= \frac{j\omega\mu\chi_{mm,xx}^{i,j,k_m}}{2}\left(H_x^{i,j,k_m} + H_x^{i,j,k_m+1}\right) \\ &+ \frac{j\omega\mu\chi_{mm,xy}^{i,j,k_m}}{8}(H_y^{i,j,k_m} + H_y^{i-1,j,k_m} + H_y^{i,j+1,k_m} + H_y^{i-1,j+1,k_m} \\ &+ H_y^{i,j,k_m+1} + H_y^{i-1,j,k_m+1} + H_y^{i,j+1,k_m+1} + H_y^{i-1,j+1,k_m+1}), \end{aligned} \tag{3.23.b}$$

$$
\begin{aligned}
H_x^{i,j,k_m+1} - H_x^{i,j,k_m} &= \frac{j\omega\varepsilon\chi_{ee,yy}^{i,j,k_m}}{2}\left(E_y^{i,j,k_m} + E_y^{i,j,k_m+1}\right) \\
&+ \frac{j\omega\varepsilon\chi_{ee,yx}^{i,j,k_m}}{8}(E_x^{i,j,k_m} + E_x^{i,j+1,k_m} + E_x^{i-1,j,k_m} + E_x^{i-1,j+1,k_m} \\
&+ E_x^{i,j,k_m+1} + E_x^{i,j+1,k_m+1} + E_x^{i-1,j,k_m+1} + E_x^{i-1,j+1,k_m+1}),
\end{aligned} \tag{3.23.c}
$$

$$
\begin{aligned}
H_y^{i,j,k_m} - H_y^{i,j,k_m+1} &= \frac{j\omega\varepsilon\chi_{ee,xx}^{i,j,k_m}}{2}\left(E_x^{i,j,k_m} + E_x^{i,j,k_m+1}\right) \\
&+ \frac{j\omega\varepsilon\chi_{ee,xy}^{i,j,k_m}}{8}(E_y^{i,j,k_m} + E_y^{i,j-1,k_m} + E_y^{i+1,j,k_m} + E_y^{i+1,j-1,k_m} \\
&+ E_y^{i,j,k_m+1} + E_y^{i,j-1,k_m+1} + E_y^{i+1,j,k_m+1} + E_y^{i+1,j-1,k_m+1}).
\end{aligned} \tag{3.23.d}
$$

Figure 3.3(a–d) shows the relative positions of electric and magnetic fields corresponding to (3.23) in Yee grid, respectively. Assemble (3.23.a) into matrix equations

$$-\mathbf{G}_z^e\mathbf{e}_x = \chi_{mm}^{yx}\mathbf{h}_x + \chi_{mm}^{yy}\mathbf{h}_y \tag{3.24.a}$$

$$\mathbf{G}_z^e\mathbf{e}_y = \chi_{mm}^{xx}\mathbf{h}_x + \chi_{mm}^{xy}\mathbf{h}_y \tag{3.24.b}$$

$$\mathbf{G}_z^h\mathbf{h}_x = \chi_{ee}^{yx}\mathbf{e}_x + \chi_{ee}^{yy}\mathbf{e}_y \tag{3.24.c}$$

$$-\mathbf{G}_z^h\mathbf{h}_y = \chi_{ee}^{xx}\mathbf{e}_x + \chi_{ee}^{xy}\mathbf{e}_y \tag{3.24.d}$$

Eliminate the GSTC-related rows (denoted by index p) of the operator matrix $\mathbf{D}$, the permittivity matrix $\boldsymbol{\varepsilon}$, and the permeability matrix $\boldsymbol{\mu}$.

$$\mathbf{D}_u^{\prime v} = \mathbf{D}_u^v - \mathbf{D}_u^v(p) \tag{3.25.a}$$

$$\boldsymbol{\mu}_u^{\prime v} = \boldsymbol{\mu}_u^v - \boldsymbol{\mu}_u^v(p) \tag{3.25.b}$$

$$\boldsymbol{\varepsilon}_u^{\prime v} = \boldsymbol{\varepsilon}_u^v - \boldsymbol{\varepsilon}_u^v(p) \tag{3.25.c}$$

So that the final operator matrix $\mathbf{A}'$ for 3-D FDFD-GSTC can be obtained

$$\mathbf{A}' = [\mathbf{DE}'][\boldsymbol{\mu}']^{-1}[\mathbf{DH}'] - [\boldsymbol{\varepsilon}'] \tag{3.26}$$

where

$$[\mathbf{DE}'] = \begin{bmatrix} \chi_{ee}^{yx} & -\mathbf{D}_z^{\prime e} + \chi_{ee}^{yy} & \mathbf{D}_y^{\prime e} \\ \mathbf{D}_z^{\prime e} + \chi_{ee}^{xx} & \chi_{ee}^{xy} & -\mathbf{D}_x^{\prime e} \\ -\mathbf{D}_y^e & \mathbf{D}_x^e & 0 \end{bmatrix} \tag{3.27.a}$$

$$[\mathbf{DH}'] = \begin{bmatrix} \chi_{mm}^{yx} & -\mathbf{D}_z^{\prime h} + \chi_{mm}^{yy} & \mathbf{D}_y^{\prime h} \\ \mathbf{D}_z^{\prime h} + \chi_{mm}^{xx} & \chi_{mm}^{xy} & -\mathbf{D}_x^{\prime h} \\ -\mathbf{D}_y^h & \mathbf{D}_x^h & 0 \end{bmatrix} \tag{3.27.b}$$

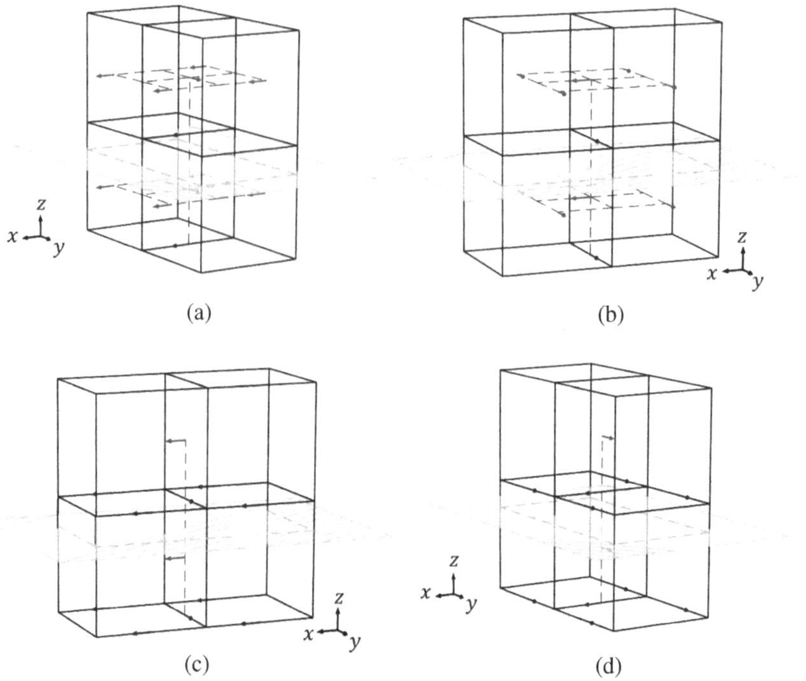

Figure 3.3 GSTCs modeling metasurface in 3-D Yee's grid. Coupling nodes between (a) ΔE_x, H_x, and H_y; (b) ΔE_y, H_x, and H_y; (c) ΔH_x, E_x, and E_y; and (d) ΔH_y, E_x, and E_y.

$$[\boldsymbol{\mu}'] = \begin{bmatrix} \boldsymbol{\mu}'_{xx} + \mathbf{G}_z^h & 0 & 0 \\ 0 & \boldsymbol{\mu}'_{yy} - \mathbf{G}_z^h & 0 \\ 0 & 0 & \boldsymbol{\mu}_{zz} \end{bmatrix} \tag{3.27.c}$$

$$[\boldsymbol{\varepsilon}'] = \begin{bmatrix} \boldsymbol{\varepsilon}'_{xx} - \mathbf{G}_z^e & 0 & 0 \\ 0 & \boldsymbol{\varepsilon}'_{yy} + \mathbf{G}_z^e & 0 \\ 0 & 0 & \boldsymbol{\varepsilon}_{zz} \end{bmatrix} \tag{3.27.d}$$

There is a difference between the 3-D and 2-D methods; the matrices **G** are no longer combined with the operator matrices **D**, but with the matrices $\boldsymbol{\mu}$ and $\boldsymbol{\varepsilon}$. This is because when constructing the matrix **A**, it is necessary to invert the matrix $\boldsymbol{\mu}$, and the combination of the matrices $\boldsymbol{\chi}$ and $\boldsymbol{\mu}$ greatly increases the difficulty of inversion. In contrast, the combination of **G** and $\boldsymbol{\mu}$ has almost no effect.

3.1.3 Simulation Examples

In the previous section, the FDFD method and the strategies for integrating GSTCs in FDFD are introduced. Now this FDFD-GSTC scheme is employed to simulate two examples to verify the accuracy of the proposed 2-D and 3-D FDFD-GSTC.

3.1.3.1 Deflecting Metasurfaces

In the first example, a deflecting metasurface is presented. Here, the metasurface is designed to have the ability to deflect electromagnetic waves. Refraction of an electromagnetic wave typically occurs at the interface between media with different permittivity, whereas the metasurface achieves this within the same medium through its structure. Consider a TE wave incident with an angle of ϑ_i on a metasurface placed normal to the x-axis. The metasurface then deflects it into a reflected wave with an amplitude of R and angle of ϑ_r, as well as a transmitted wave with an amplitude of T and angle of ϑ_t. All examples in this section are with $f = 10\,\text{GHz}$, and the spatial difference of the Yee grid in the FDFD-GSTC cases is 0.001 m.

First, a fully absorbing metasurface can absorb completely the y-polarized TE waves that are incident normally, without generating any reflected or transmitted waves ($\vartheta_i = 0$, $T = 0$, $R = 0$). According to (2.97), the non-zero terms in the susceptibility tensor can be obtained as follows

$$\chi_{ee}^{zz} = \chi_{mm}^{yy} = -2j/k_0 \tag{3.28}$$

Figure 3.4 shows the simulation result of a fully absorbing metasurface under matched normal illumination, using the FDFD-GSTC method (left) and a comparison from COMSOL (right). In COMSOL, a thin slab is preset according to (2.105) to produce the same effect, whose width is $d = \lambda/100 = 3\times10^{-5}$ m, with

Figure 3.4 FDFD-GSTC results for a fully absorbing ($R = 0$, $T = 0$) metasurface illuminated by a matched normally incident Gaussian beam, and comparison with COMSOL. The shown field component is the real part value of E_z. The left figure is the simulation result of the proposed FDFD-GSTC, the metasurface is modeled by the proposed sheet with susceptibility (χ), and the area in black is PML used to truncate the computation domain. The right figure is the simulation result of COMSOL. The metasurface is modeled by an extremely thin slab with 3-D susceptibility (χ/d), as the white line shown in the figure, and the same PML is used.

d-dependent relative permittivity $\left(\varepsilon_r^{zz} = 1 + \chi_{ee}^{zz}/d\right)$ and relative permeability $\left(\mu_r^{zz} = 1 + \chi_{mm}^{yy}/d\right)$. The left image shows the result from FDFD-GSTC, while the right image shows the reference result from COMSOL. Both simulations were truncated by a perfectly matched layer (PML). It can be seen that the metasurface in the FDFD-GSTC result absorbs the incident wave perfectly as specified, without generating any reflected or transmitted fields. The impedance between the thin sheet and its surrounding free space is mismatched in the COMSOL simulation, which causes multiple reflections between the thin sheet and the surrounding free space, resulting in some weak transmitted field.

Next, an oblique incident wave is used, which is mismatched with the metasurface ($\theta_i = \pi/3 \neq \vartheta_i$) to illuminate it. According to the specified function of the metasurface, waves propagating normally to the metasurface in oblique incidence will be perfectly absorbed, while waves propagating parallel to the metasurface will not be affected by the metasurface. Therefore, there will be no transmitted field, and the field that is not absorbed will propagate along the $+y$-direction. This appears to be reflection; although it is no longer a Gaussian beam like the incident field, it still satisfies the homogeneous wave equation. It should be noted that this "reflection" is different from the reflection produced when waves are incident on a perfect electrical conductor (PEC).

Figure 3.5 shows the simulation results of the above situation. It can be seen that the results of FDFD-GSTC (left) are consistent with COMSOL (right), and the reason for the weak transmitted field in COMSOL is the same as in the case of normal incidence. In addition, a magnified view of a localized region near the metasurface is provided with mesh discretization from FDFD-GSTC and COMSOL. Although the mesh size of the finite difference method and the FEM (COMSOL) is not comparable, the GSTC method allowing the near-metasurface region to use the same mesh size as other regions without the need for special fine meshing is demonstrated. The advantages of the GSTCs method in finite-element-based meshing can be more pronounced in the subsequent FEM-GSTC method.

Second, a metasurface can deflect an oblique incident wave into a specified reflected and transmitted wave ($\vartheta_i = \pi/6$, $\vartheta_t = -\pi/4$, $\vartheta_r = \pi/3$, $T = 0.5$, $R = 0.5$). According to (2.97), the non-zero terms in the susceptibility tensor can be obtained as follows

$$\chi_{ee}^{zz}(y) = \frac{2}{jk_0} \frac{\cos\vartheta_i e^{-jk_0 y \sin\vartheta_i} - T\cos\vartheta_t e^{-jk_0 y \sin\vartheta_t} - R\cos\vartheta_r e^{-jk_0 y \sin\vartheta_r}}{e^{-jk_0 y \sin\vartheta_i} + Te^{-jk_0 y \sin\vartheta_t} + \mathrm{R}e^{-jk_0 y \sin\vartheta_r}} \tag{3.29.a}$$

$$\chi_{mm}^{yy}(y) = \frac{2}{jk_0} \frac{e^{-jk_0 y \sin\vartheta_i} - Te^{-jk_0 y \sin\vartheta_t} + \mathrm{R}e^{-jk_0 y \sin\vartheta_r}}{\cos\vartheta_i e^{-jk_0 y \sin\vartheta_i} + T\cos\vartheta_t e^{-jk_0 y \sin\vartheta_t} - R\cos\vartheta_r e^{-jk_0 y \sin\vartheta_r}} \tag{3.29.b}$$

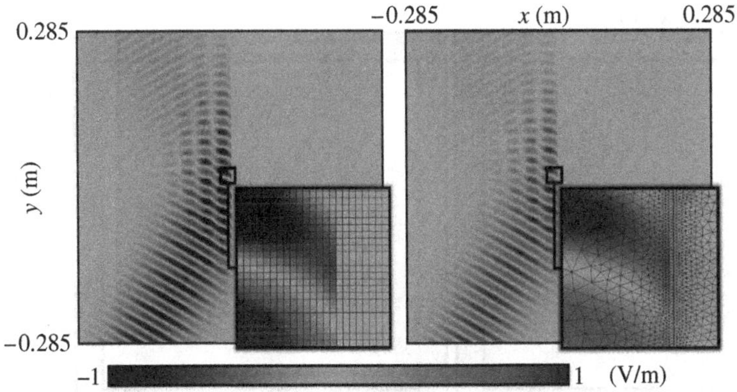

Figure 3.5 FDFD-GSTC results for a fully absorbing ($R = 0$, $T = 0$) metasurface illuminated by a mismatched obliquely incident Gaussian beam and a comparison with COMSOL. The shown field component is the real part value of E_z. The left figure is the simulation result of the proposed FDFD-GSTC; the metasurface is modeled by the proposed sheet with susceptibility (χ), and the area in black is the PML used to truncate the computation domain. The right figure is the simulation result of COMSOL. The metasurface is modeled by an extremely thin slab with 3-D susceptibility (χ/d), as the white line shown in the figure, and the same PML is used.

Figure 3.6 shows the simulation results, where the setup method for the metasurface in COMSOL is the same as in the previous example. Both (a) and (b) are matched incident waves, while (c) and (d) are mismatched incident waves. (a) and (c) are the results of FDFD-GSTC, while (b) and (d) are the results of COMSOL. The simulation results of FDFD-GSTC are the same as those of COMSOL, although simulation results from COMSOL show a decrease in amplitude, which is caused by the impedance mismatch between the medium and free space. For matched incident waves, both the reflected waves and the transmitted waves have identical amplitude and direction as specified, and they are both Gaussian beams that are the same as the incident waves. For mismatched incident waves, there is no focused reflected or transmitted wave like the incident Gaussian beam.

3.1.3.2 Polarization Selective Metasurface

The second example presents the simulation of a polarization-selective metasurface, which is representative due to its different sensitivity coefficients on the diagonal. The metasurface is placed on the XoY-plane and can transmit x-polarized waves while fully absorbing y-polarized waves ($T_x = 0.5$, $T_y = 0$, $R_x = 0$, $R_y = 0$). In the FDFD-GSTC method, this metasurface is modeled as a zero-thickness sheet with dimensions of 0.04 m × 0.04 m. According to (2.97), the susceptibility tensor

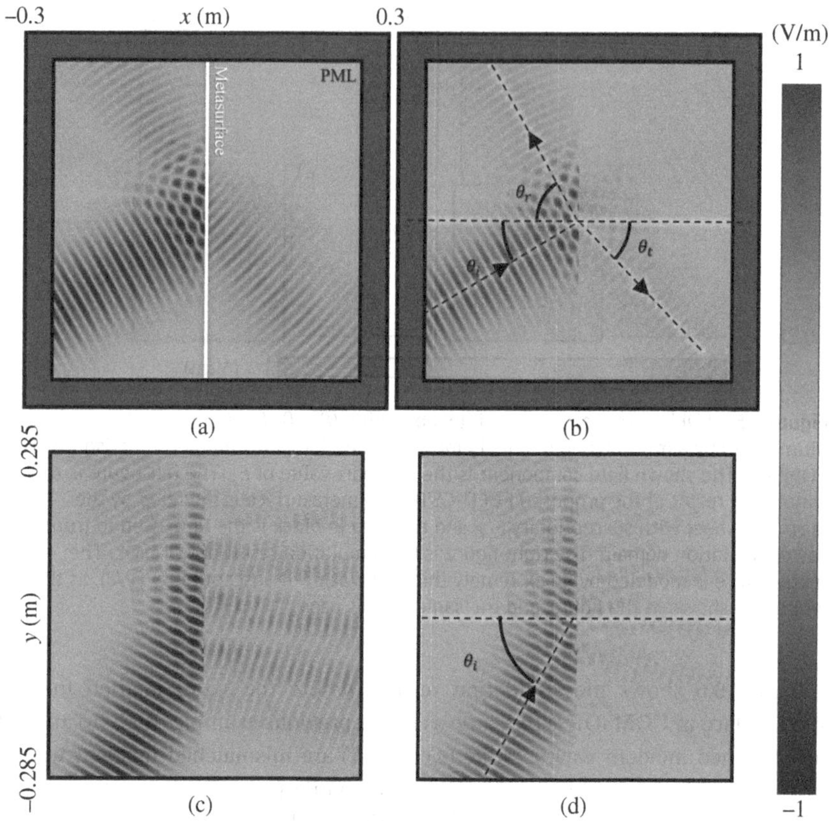

Figure 3.6 FDFD-GSTC results for a deflection ($\vartheta_i = \pi/6$, $\vartheta_r = \pi/3$, $\vartheta_t = -\pi/4$, $T = 0.5$, $R = 0.5$) metasurface illuminated by a matched obliquely incident Gaussian beam ($\theta_i = \pi/6$) and a mismatched obliquely incident Gaussian beam ($\theta_i = \pi/3$), and comparisons with COMSOL. The shown field component is the real part value of E_z. (a) The FDFD-GSTC result for the matched incident wave: the white line is the zero-thickness sheet modeling the metasurface, and the areas in black are PML used to truncate the computation domain. (b) The COMSOL result for the matched incident wave. Terms θ_i, θ_r, and θ_t are angles of incident, reflected, and transmitted waves. (c) The FDFD-GSTC result for mismatched incident wave, without PML region. (d) The COMSOL result for mismatched incident wave.

of this sheet can be obtained as follows

$$\begin{bmatrix} \chi_{ee}^{xx} & \chi_{ee}^{yx} \\ \chi_{ee}^{xy} & \chi_{ee}^{yy} \end{bmatrix} = \begin{bmatrix} 2/3 & 0 \\ 0 & 2 \end{bmatrix} \frac{1}{jk_0} \tag{3.30.a}$$

$$\begin{bmatrix} \chi_{mm}^{xx} & \chi_{mm}^{yx} \\ \chi_{mm}^{xy} & \chi_{mm}^{yy} \end{bmatrix} = \begin{bmatrix} 2 & 0 \\ 0 & 2 \end{bmatrix} \frac{1}{jk_0} \tag{3.30.b}$$

In COMSOL, this metasurface is modeled as a thin slab with a certain thickness $d = \lambda/100$, just the same as in the 2-D simulation. The sheet has the dimensions of

0.04 m × 0.04 m × d m. Its relative permittivity and relative permeability are diagonal and anisotropic

$$\overline{\overline{\varepsilon}}_r = \begin{bmatrix} 1+\frac{2}{3}\frac{1}{jk_0 d} & 0 & 0 \\ 0 & 1+\frac{2}{jk_0 d} & 0 \\ 0 & 0 & 1 \end{bmatrix} \tag{3.31.a}$$

$$\overline{\overline{\mu}}_r = \begin{bmatrix} 1+\frac{2}{jk_0 d} & 0 & 0 \\ 0 & 1+\frac{2}{jk_0 d} & 0 \\ 0 & 0 & 1 \end{bmatrix} \tag{3.31.b}$$

Both simulation computational domains are truncated by PML. Figure 3.7 shows the simulation results of a right-handed circularly polarized Gaussian beam

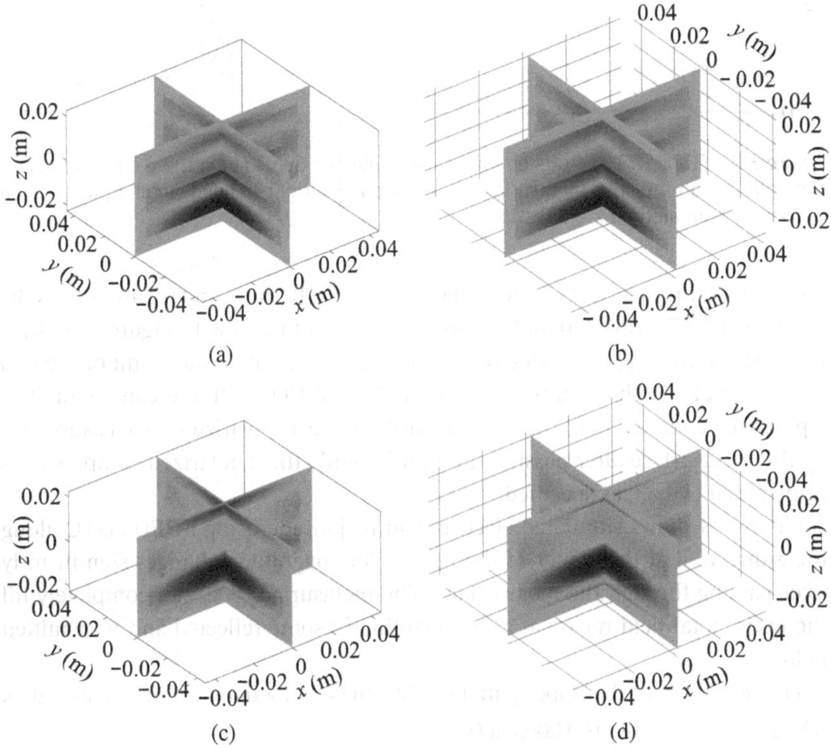

Figure 3.7 FDFD-GSTC results for a polarization-selective metasurface illuminated by a right-handed circular polarization Gaussian beam and comparisons with COMSOL. (a) The FDFD-GSTC and (b) COMSOL results of the real part of the x-component of the electric field. (c) The FDFD-GSTC and (d) COMSOL results of the real part of the y-component of the electric field.

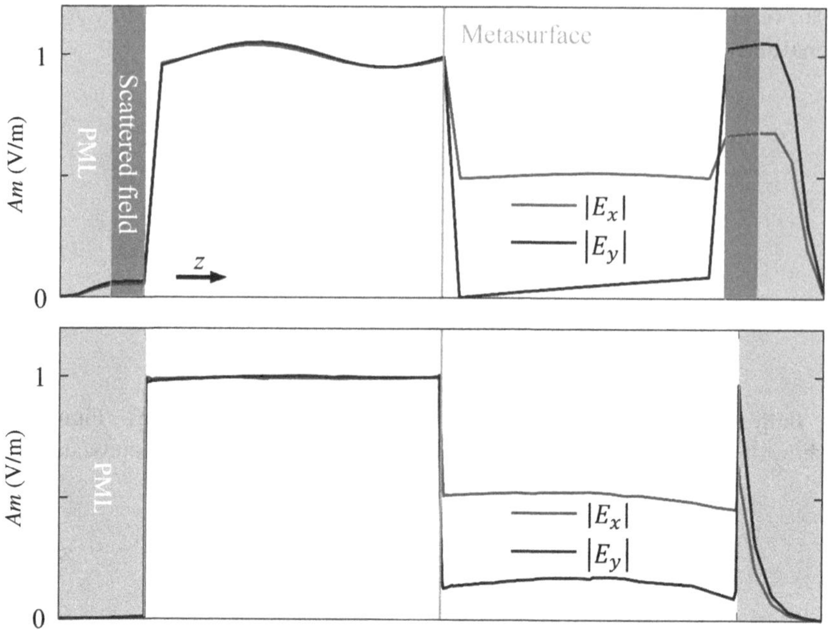

Figure 3.8 FDFD-GSTC results of amplitude of electric fields along the z-direction for the polarization selective metasurface illuminated by a right-handed circular polarization Gaussian beam along the z-direction.

incident on this metasurface along the z-direction. Both (a) and (c) are the results of FDFD-GSTC, while (b) and (d) are the results of COMSOL. Figure 3.7(a,b) is the result of the real part value of E_x, and Figure 3.7(c,d) is the result of the real part value of E_y. The results of FDFD-GSTC and COMSOL are consistent. The y-polarized component of the right-handed circularly polarized wave is absorbed by the metasurface after passing through it, while the x-polarized component is transmitted by half, as specified.

Figure 3.8 shows simulated electric field magnitude using FDFD-GSTC along the z-direction at the center of the beam. The magnitude changes significantly after passing through the metasurface. The metasurface does not completely fill the entire total field region at $z = 0$, resulting in some reflected and transmitted fields.

The mesh size of the upper panel is 0.001 m ($= \frac{\lambda}{30}$, λ denotes the wavelength), while the lower panel is 0.0003 m ($= \frac{\lambda}{100}$).

3.2 FEM with GSTCs

3.2.1 Introduction of FEM

The FEM is a numerical technique for obtaining approximate solutions to boundary value problems governed by PDEs. Originating in the 1950s, FEM was initially developed by engineers and mathematicians in aerospace and civil engineering to address stress analysis in complex structures. Its earliest roots, however, can be traced back to the 1940s, when Richard Courant employed variational methods to solve certain physics problems [5]. Over the following decades, FEM rapidly expanded and found widespread application across diverse engineering domains, including electromagnetics, aerospace, automotive, and mechanics. It is particularly effective for addressing problems involving complex geometries, nonlinear material behavior, and multiphysics interactions. Its popularity also stems from the availability of efficient, user-friendly software packages, such as ANSYS and COMSOL, which greatly facilitate its application to real-world engineering and scientific problems.

The core idea of the FEM is to discretize a complex continuous problem into a finite number of simpler subproblems, thereby transforming a PDE with infinitely many DoFs into a finite-dimensional system of algebraic equations. Specifically, FEM divides the entire computational domain into a set of interconnected subregions, or elements, each consisting of a finite number of nodes. Within each element, the unknown function is approximated using simple interpolation functions, whose coefficients are determined by the nodal values. In this way, the global solution of the problem can be represented as an assembly of local approximations. Subsequently, by applying variational principles or weighted residual methods, the PDEs are reformulated into algebraic equations at the element level. These local equations are then assembled into a global system of equations according to the element connectivity, which is solved numerically for the unknown coefficients. The essence of FEM lies in a "divide-and-conquer" strategy through local approximation and global assembly; it not only adapts effectively to complex geometries and boundary conditions but also reduces computational difficulty while maintaining accuracy.

Therefore, the procedure of a finite element analysis can be summarized as follows:

1. **Preprocessing and Discretizing:** Discretize the continuous physical domain into a finite number of interconnected elements to generate a mesh.

2. **Formulation and Assembly:** Select suitable interpolation functions to approximate the unknown field within each element. Formulate the characteristic

matrices and vectors (e.g., stiffness matrix and load vector) for each element by applying the variational principle or the weighted residual method, and then assemble them into a global system of equations.

3. **Solving and Postprocessing:** Apply boundary conditions to the global system and solve the resulting set of linear algebraic equations. The primary unknown solution at nodes can thus be obtained, and subsequently, quantities of interest can be computed and the results can be analyzed.

3.2.1.1 Mesh Generation

Mesh generation is a crucial step in the FEM, as the quality of the mesh strongly influences both the accuracy and efficiency of the solution. A poorly generated mesh may lead to inaccurate results or convergence problems, while a well-generated mesh enables efficient computations and reliable results.

As shown in Figure 3.9, in one-dimensional (1-D) problems, the computational domain reduces to a line segment, and the solution is a function of only one variable (usually the spatial coordinate x). In 2-D problems, the domain is represented by a plane or surface, and the solution depends on two variables, typically the spatial coordinates (usually the spatial coordinates x and y). In 3-D problems, the domain becomes a volume, and the solution is expressed as a function of three variables, usually the spatial coordinates x, y, and z. Compared with 1-D cases, mesh generation in 2-D and 3-D is considerably more complex.

In 2-D problems, FEM commonly employs triangular or quadrilateral elements, while in 3-D problems, tetrahedral or hexahedral elements are typically used. These subdomains are connected through nodes, and the process is generally

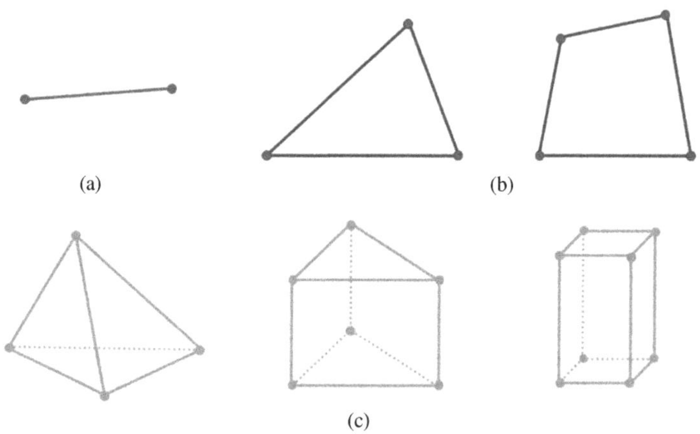

Figure 3.9 Element examples of FEM for (a) one dimension, (b) two dimension, and (c) three dimension.

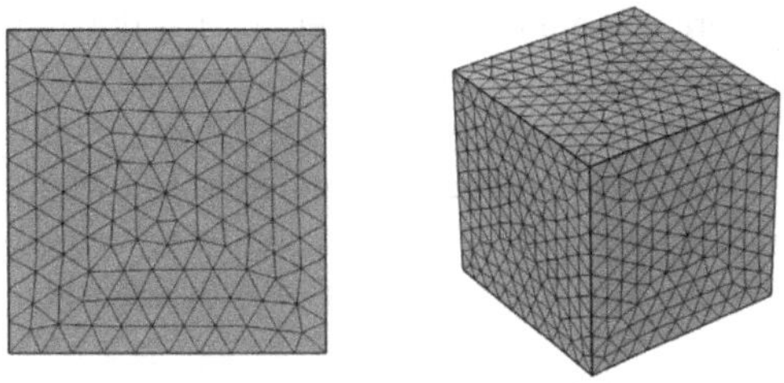

(a) (b)

Figure 3.10 Typical meshes of FEM. (a) 2-D triangle mesh and (b) 3-D tetrahedral mesh.

referred to as triangulation or tetrahedralization, which forms the foundation of mesh generation. A variety of techniques are available for generating high-quality meshes, such as Delaunay triangulation, advancing front methods, and octree approaches. In addition, mesh adaptation techniques can be applied to locally refine or coarsen the mesh in specific regions, thereby achieving a reasonable balance between computational efficiency and solution accuracy. Figure 3.10 illustrates examples of 2-D triangular meshes and 3-D tetrahedral meshes.

3.2.1.2 Finite Element Formulation

The Ritz method and the Galerkin method form the theoretical foundation of the FEM. The Ritz method originates from variational principles and is applicable to self-adjoint problems that can be expressed in terms of an energy functional or variational formulation [6, 7]. Its fundamental idea is to approximate the unknown function as a linear combination of basis functions and substitute this representation into the energy functional. The unknown coefficients are then determined by minimizing the functional. This approach is highly effective in structural and elasticity problems but is limited by the requirement that the problem must admit a variational form, which may not always exist. To overcome this limitation, the Russian mathematician Galerkin introduced the Galerkin method. This method belongs to the class of weighted residual methods, whose fundamental idea is to enforce the residual of the governing equations, obtained after substituting the approximate solution, to be orthogonal to a chosen space of weighting functions. In this way, approximate coefficients can be determined without relying on the existence of a variational functional. The FEM can be regarded as a numerical generalization of these two approaches, it is equivalent to the Ritz method when energy functional exists and corresponds to the Galerkin method in more general cases. Thus, the Ritz and Galerkin methods together constitute the theoretical cornerstone of modern FEM.

Suppose the solution of the following equation in the domain Ω:

$$\mathcal{L}(u) = f \tag{3.32}$$

where $\mathcal{L}$ is a differential operator, f is the excitation or forcing function, and u is the unknown function to be determined.

Following mesh generation, the computational domain Ω is partitioned into smaller finite elements, denoted as Ω^e, where the superscript e indexes the elements. Within a given element, let u^e represent the approximate solution. The corresponding local residual R^e is defined as

$$R^e = \mathcal{L}u^e - f^e \tag{3.33}$$

The unknown function u^e is then approximated by a weighted summation of basis functions, expressed as follows

$$u^e = \sum_{i=1}^{n} u_i^e N_i^e \tag{3.34}$$

where the coefficient u_i^e is the unknown constant corresponding to the value of u at the i-th node of the element (for nodal elements) and n is the number of DOFs in one element. The basis functions N_i^e have local support, being non-zero only within their associated elements. The type of basis function, vector (for edge elements) or scalar (for nodal elements), is selected based on the governing equation. See Figure 3.11 for an illustration of a tetrahedral edge element.

Increasing the order of polynomial shape functions is an effective way to boost simulation accuracy, particularly for problems with geometric complexity or material nonlinearity (Figure 3.12). However, the gains in precision come at the cost of higher computational overhead. Thus, selecting the appropriate shape function order necessitates a trade-off between desired accuracy and available computational resources.

In the Galerkin approach, the weighting functions ω_i are chosen from the same set of shape functions N_i. Using the weighted residual method leads to

$$\int_{\Omega^e} \omega_j^e \cdot (\mathcal{L}u^e - f^e) d\Omega^e = 0 \tag{3.35}$$

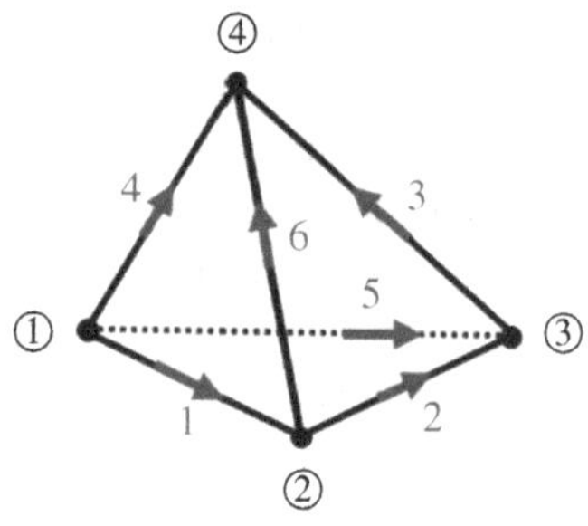

Figure 3.11 A tetrahedral edge element.

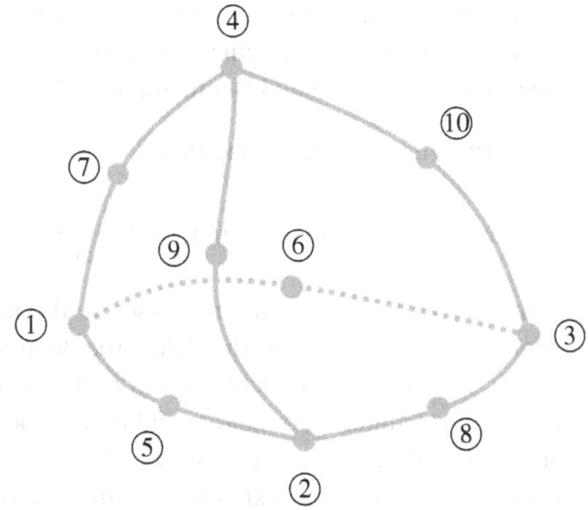

Figure 3.12 A quadratic tetrahedral element.

Substituting (3.34) into (3.35) and using the Galerkin approach where ($\omega_i = N_i$), (3.35) would then become

$$\sum_{i=1}^{n} u_i^e \int_{\Omega^e} \left(N_j^e \mathcal{L} \left(N_i^e \right) \right) d\Omega^e = \int_{\Omega^e} N_j^e f^e d\Omega^e \tag{3.36}$$

(3.36) can be written as the matrix equation

$$\mathbf{A}^e \mathbf{u}^e = \mathbf{b}^e \tag{3.37}$$

where $(\mathbf{A}^e)_{ji} = \int_{\Omega^e} N_j^e \mathcal{L} \left(N_i^e \right) d\Omega^e$ and $\mathbf{b}_j^e = \int_{\Omega^e} N_j^e f^e d\Omega^e$. The global system is constructed through the assembly of element matrices. This process involves summing the contributions from every finite element in the mesh. The assembled global matrix (Eq. 3.38) is characteristically large and sparse, representing the complete set of relationships between the unknown solution values at all nodes (or edges) within the mesh. Efficient global matrix assembly is paramount for large-scale simulations, driving the development of optimization strategies like parallelization and sparse storage formats. The resulting global system is the matrix

$$\mathbf{A}\mathbf{u} = \mathbf{b} \tag{3.38}$$

where N is the total number of unknowns, $\mathbf{A}$ is the $N \times N$ system matrix (e.g., combining stiffness and mass matrices), $\mathbf{u}$ is the $N \times 1$ unknown vector, and $\mathbf{b}$ is the $N \times 1$ known right-hand side vector.

3.2.1.3 Imposing the Boundary Conditions

Boundary conditions are essential to the FEM, prescribing the values of the solution or its derivatives on the domain boundaries. They are a fundamental component of the problem formulation, ensuring the solution is physically meaningful and the mathematical problem is well-posed. The two primary categories are Dirichlet and Neumann boundary conditions.

$$\text{Dirichlet boundary conditions:} \quad u|_{\Gamma_1} = u_0 \tag{3.39.a}$$

$$\text{Neumann boundary conditions:} \quad \left.\frac{\partial u}{\partial r}\right|_{\Gamma_2} = q \tag{3.39.b}$$

Dirichlet boundary conditions prescribe the solution's value on the domain boundary, effectively fixing the field variable at specified locations. In thermal analysis, for instance, a Dirichlet condition sets a known temperature at a surface. Conversely, Neumann boundary conditions stipulate the value of the solution's normal derivative at the boundary, which often corresponds to a flux condition. In the same thermal context, a Neumann condition would define the heat flux across a boundary.

Boundary conditions in FEM are implemented by modifying the global matrix **A** and right-hand side vector **b**. Dirichlet conditions fix solution values by adjusting rows of **A** and setting entries in **b**. Neumann conditions prescribe fluxes by updating **b** at the relevant nodes. For problems with both types, the modifications are combined to apply all constraints to the linear system.

3.2.1.4 Solving the Matrix Equations

Upon assembly of the matrix equation and imposition of boundary conditions, numerical methods specifically designed for matrix systems are applied to obtain the solution. The approaches for solving such matrix equations are generally divided into two primary categories, namely, direct methods and iterative methods.

Direct solvers leverage matrix factorization, such as LU or Cholesky decomposition, to transform the system into triangular forms that are solved efficiently via forward and backward substitution. Gaussian elimination is a foundational example of this class. Conversely, iterative solvers employ an iterative refinement procedure, generating a sequence of approximate solutions that converge to the exact result. These methods are particularly advantageous for large-scale systems where memory and computational cost are constraints, as seen in the conjugate gradient and Jacobi methods.

3.2.2 Strategies for Integrating GSTCs in FEM

Considering a 2-D electromagnetic problem within the XoZ-plane, a metasurface is positioned at the interface $z = 0$, oriented parallel to the XoY-plane. The electric field exhibits variation along the x- and z-directions while maintaining independence from the y-coordinate, that is $\frac{\partial}{\partial y} = 0$. For the case of a simple metasurface, transformation involves only χ_{ee}^{xx}, χ_{mm}^{yy}, χ_{ee}^{yy}, or χ_{mm}^{xx}. For scenarios involving non-zero field components, the governing scalar equation for the TM polarized wave can be expressed as [8, 9]

$$\frac{\partial}{\partial x}\left(\frac{1}{\varepsilon_r}\frac{\partial h_y}{\partial x}\right) + \frac{\partial}{\partial z}\left(\frac{1}{\varepsilon_r}\frac{\partial h_y}{\partial z}\right) + k_0^2\mu_r h_y = 0 \tag{3.40}$$

However, for more versatile and consequently more complex metasurface transformations, a complete description of all electric field components becomes necessary. The electric field vector can be decomposed into components lying within the XoZ-plane and a component orthogonal to it ($\boldsymbol{e}(x, z) = \boldsymbol{e}_t(x, z) + \hat{\mathbf{y}}e_y(x, z)$). Accordingly, the vectorial Helmholtz equation governing the electric field is formulated as

$$\nabla \times \boldsymbol{\mu}_r^{-1} \cdot \nabla \times \mathbf{e} - k_0^2\boldsymbol{\varepsilon}_r\mathbf{e} = 0 \tag{3.41}$$

and

$$\nabla_{\parallel} \times \mu_r^{-1}\nabla_{\parallel} \times \mathbf{e}_{\parallel} - k_0^2\varepsilon_r\mathbf{e}_{\parallel} = 0 \tag{3.42.a}$$

$$\nabla_{\parallel} \times \mu_r^{-1}\nabla_{\parallel} \times \hat{y}e_y - k_0^2\varepsilon_r\hat{y}e_y = 0 \tag{3.42.b}$$

where $\nabla_{\parallel} = \hat{\mathbf{x}}\frac{\partial}{\partial x} + \hat{\mathbf{z}}\frac{\partial}{\partial z}$; ε_r and μ_r represent the relative permittivity and permeability of the medium, respectively. Both the scalar formulation (3.40) and the full-vectorial formulation (3.41) are examined in this study. Therefore, different discretization methods should be adopted according to the unknown structure of the equation. By applying the Galerkin method and Green's theorem [10], the corresponding weak forms for the scalar equation and the full-vectorial Helmholtz equation are respectively derived as follows

$$\begin{aligned}&\left(\varepsilon_r^{-1}\frac{\partial h_y}{\partial x}, \frac{\partial q}{\partial x}\right) + \left(\varepsilon_r^{-1}\frac{\partial h_y}{\partial z}, \frac{\partial q}{\partial z}\right) - k_0^2(\mu_r h_y, q)\\ &= \left(\hat{\mathbf{n}} \cdot \varepsilon_r^{-1}\frac{\partial h_y}{\partial x}\hat{\mathbf{x}}, q\right)_L + \left(\hat{\mathbf{n}} \cdot \varepsilon_r^{-1}\frac{\partial h_y}{\partial z}\hat{\mathbf{z}}, q\right)_L\end{aligned} \tag{3.43}$$

$$\left(\mu_r^{-1}\nabla_{\parallel} \times \mathbf{e}_{\parallel}, \nabla_{\parallel} \times \mathbf{v}\right) + \left(\hat{\mathbf{n}} \times \mu_r^{-1}\nabla_{\parallel} \times \mathbf{e}_{\parallel}, \mathbf{v}\right)_L - \left(k_0^2\varepsilon_r\mathbf{e}_{\parallel}, \mathbf{v}\right) = 0 \tag{3.44}$$

$$\left(\mu_r^{-1}\nabla_{\parallel}e_y, \nabla_t q\right) - \left(\hat{\mathbf{n}} \cdot \mu_r^{-1}\nabla_{\parallel}e_y, q\right)_L - \left(k_0^2\varepsilon_r e_y, q\right) = 0 \tag{3.45}$$

where $\mathbf{v} \subset \mathbf{H}(\text{curl}; \Omega)$ and $q \subset \mathbf{H}^1(\Omega)$ are test functions. Subscript L represents the line integrals of the boundary and $\mathbf{n}$ represents the unit normal outward vector. In Sections 3.2 and 3.3, $(\cdot)$ represents the integration of two functions, namely, $(u, v) = \iint_{\Omega} u \cdot v \, d\Omega$.

3.2.2.1 GSTC Implementation in 2-D FEM

Applying Faraday's law, the line integrals in (3.44) can be reformulated as follows

$$\left(\hat{\mathbf{n}} \times \mu_r^{-1}\nabla_{\|} \times \mathbf{e}_{\|}, \mathbf{v}\right)_L = -(\hat{\mathbf{n}} \times j\omega\mu_0\hat{y}h_y, \mathbf{v})_L \tag{3.46}$$

$$\left(\hat{\mathbf{n}} \cdot \mu_r^{-1}\nabla_{\|}e_y, q\right)_L = (\hat{\mathbf{n}} \times j\omega\mu_0\hat{\mathbf{z}}h_x, q)_L \tag{3.47}$$

The line integrals in (3.47) initially involve the term $j\omega\mu_0\hat{\mathbf{x}}h_z$. However, according to the GSTCs, field discontinuities are exclusively associated with tangential components adjacent to the metasurface, implying that $h_z = 0$ along contour L. Owing to the discontinuities introduced by the equivalent GSTC relations, nodes located on the GSTC interface L are assigned distinct DoFs on the dual quadrilateral elements, as illustrated in Figure 3.13. Assuming the metasurface is positioned at z=0, the line integrals are decomposed into contributions from the two adjoining parts of the dual elements sharing the interface L, i.e., L_1 for the position on $z = 0^+$ and L_2 for the position on $z = 0^-$. The line integrals in (3.46) and (3.47) include the upper L_1 and the lower L_2 terms

$$j\omega\mu_0(\hat{\mathbf{n}} \times \hat{\mathbf{y}}h_y, \mathbf{v})_L = j\omega\mu_0(\hat{\mathbf{n}}_1 \times \hat{\mathbf{y}}h_{y1}, \mathbf{v})_{L_1} + j\omega\mu_0(\hat{\mathbf{n}}_2 \times \hat{\mathbf{y}}h_{y2}, \mathbf{v})_{L_2} \tag{3.48}$$

$$j\omega\mu_0(\hat{\mathbf{n}} \cdot \hat{\mathbf{z}}h_x, q)_L = j\omega\mu_0(\hat{\mathbf{n}}_1 \cdot \hat{\mathbf{z}}h_{x1}, q)_{L_1} + j\omega\mu_0(\hat{\mathbf{n}}_2 \cdot \hat{\mathbf{z}}h_{x2}, q)_{L2} \tag{3.49}$$

where $\mathbf{n}_2 = -\mathbf{n}_1$ is used in these equations. Once the susceptibility tensors are derived from the target electromagnetic response, the magnetic field discontinuities across the GSTC boundary L can be explicitly formulated. These expressions then enable the precise integration of the corresponding line integral terms into the weak form, denoted as (3.49). For conciseness, this discussion focuses solely on the GSTC implementation within the full-vectorial Helmholtz framework; an analogous procedure is applicable to the scalar formulation (3.43).

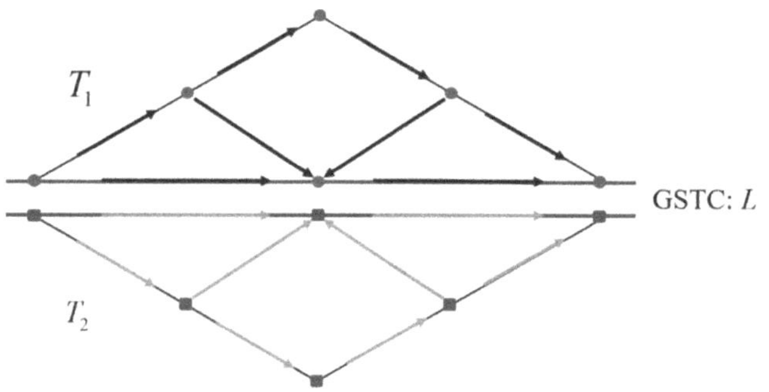

Figure 3.13 Distinct DoFs are allocated to the nodes residing on the GSTC interface line L. In the schematic, arrows denote the mixed-order vector basis functions, while circular dots and square markers respectively represent the nodal basis functions associated with the two dual elements.

3.2.2.2 Discretization of FEM-GSTC

The linear-tangential/quadratic-normal (LT/QN) edge elements are used to expand the tangential electric field $\mathbf{e}_{\parallel}$, and the second-order nodal basis functions are used to expand the components h_y and e_y.

$$h_y(x,z) = \sum_{m=1}^{N^d} h_{y,m}\psi_m^e(x,z) \tag{3.50.a}$$

$$\mathbf{e}_{\parallel}(x,z) = \sum_{i=1}^{N^e} \mathbf{e}_{\parallel,i}\mathbf{\Phi}_i^e(x,z) \tag{3.50.b}$$

$$e_y(x,z) = \sum_{j=1}^{N^d} e_{y,j}\psi_j^e(x,z) \tag{3.50.c}$$

where $h_{y,m}$ is the m-th unknown of the scalar field h_y for TM polarization wave, $\mathbf{e}_{\parallel,i}$ is the tangential field along i-th edge, and $e_{y,j}$ is the j-th unknown of the field. The term $\mathbf{\Phi}_i^e$ denotes the vector basis function, and $N^e = 24$ is the number of edge-based basis functions in one element. ψ_m^e and ψ_j^e are the nodal basis function, and $N^d = 6$ is the number of second-order nodal basis functions in one element.

Substituting (3.50) into (3.43) and (3.44) leads to

$$\mathbf{K}\mathbf{h}_y = \mathbf{b} \tag{3.51}$$

$$\begin{bmatrix} \mathbf{K}_{\parallel} & \mathbf{K}_{\parallel y} \\ \mathbf{K}_{y\parallel} & \mathbf{K}_{yy} \end{bmatrix} \begin{Bmatrix} \mathbf{e}_{\parallel} \\ \mathbf{e}_y \end{Bmatrix} = \begin{Bmatrix} \mathbf{b}_{\parallel} \\ \mathbf{b}_y \end{Bmatrix} \tag{3.52}$$

The matrix $\mathbf{K}$ in (3.51) corresponds to the scalar formulation and incorporates the global stiffness matrix, mass matrix, as well as contributions arising from the surface susceptibility terms. The matrices $\mathbf{K}_{\parallel}$ and $\mathbf{K}_{yy}$ denote the assembled elemental stiffness and mass matrices from (3.46) and (3.47), respectively. These matrices inherently account for the discontinuous field terms associated with the bianisotropic medium. Furthermore, matrices $\mathbf{K}_{\parallel y}$ and $\mathbf{K}_{y\parallel}$ represent the contributions of the discontinuity terms introduced via the line integrals specified by the GSTCs. Taking $\mathbf{K}_{\parallel y}$ as an example, the subscripts t and y indicate that this matrix is constructed from the integrals involving the vectorial test functions $\mathbf{\Phi}$ and the scalar field component $\mathbf{e}_y$.

3.2.3 Simulation Examples

Following the derivation of the FEM-GSTC framework, its numerical implementation is demonstrated through two characteristic examples. A beam deflector with monoanisotropic uniaxial properties and a polarization beam splitter (PBS) employing a complex tensor design are presented herein. The simulation of both cases using the proposed 2-D FEM-GSTC approach confirms the method's accuracy and efficiency.

3.2.3.1 Deflecting Metasurface

In conventional media, the relationship between the angles of incidence, reflection, and transmission is governed by the media's permittivity and permeability. In this work, the method for synthesizing metasurfaces via GSTCs to achieve the desired wave deflection is employed. Specifically, a deflecting metasurface capable of deflecting an incident wave into transmitted and reflected waves at independently specified angles is presented.

First of all, consider a metasurface that deflects a normally incident plane wave by 45° to the right, achieving perfect transmission without reflection, as illustrated in Figure 3.14. The computational domain is defined as a square region with dimensions of $26\lambda \times 26\lambda$ in the XoZ-plane, and the metasurface is located at the center of the region with a length of 20 λ paralleling x-axis. Specify a 5-GHz TM-polarized wave with Gaussian variation along with the x-axis normal incidence to metasurface from below, and the transmitted wave is designed to be the same polarized with a 45° transmission angle.

The expressions of incident electric and magnetic fields are

$$\mathbf{E}^i = \hat{\mathbf{x}} e^{-jk_0 z} \tag{3.53.a}$$

$$\mathbf{H}^i = \hat{\mathbf{y}} \frac{1}{Z_0} e^{-jk_0 z} \tag{3.53.b}$$

where Z_0 is the wave impedance in vacuum. The transmitted fields are expressed as

$$\mathbf{E}^t = \hat{\mathbf{x}} \frac{\sqrt{2}}{2} e^{-jk_0(\sqrt{2}/2)x} \tag{3.54.a}$$

$$\mathbf{H}^t = \hat{\mathbf{y}} \frac{\sqrt{2}}{2Z_0} e^{-jk_0(\sqrt{2}/2)x} \tag{3.54.b}$$

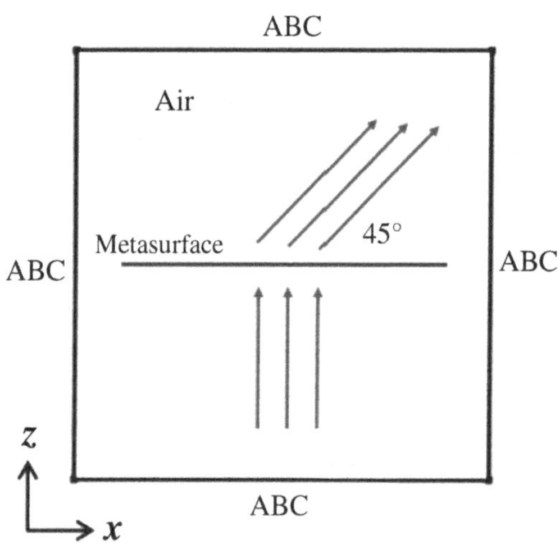

Figure 3.14 Schematic diagram of the deflecting without reflection metasurface.

Since the intended functionality of this metasurface is to deflect a normally incident wave by 45° to the right with perfect transmission and without polarization conversion, the magnetoelectric coupling tensors and the off-diagonal elements of the susceptibility tensors are set to zero. Consequently, the required surface susceptibilities, synthesized based on rigorous boundary conditions for the tangential electric and magnetic fields, are derived as follows:

$$\chi_{ee}^{xx} = \frac{4j}{k_0} \frac{e^{-jk_0\sqrt{2}/2x} - 1}{2 + \sqrt{2}e^{-jk_0\sqrt{2}/2x}} \tag{3.55.a}$$

$$\chi_{mm}^{yy} = \frac{j}{k_0} \frac{\sqrt{2}e^{-jk_0\sqrt{2}/2x} - 2}{1 + e^{-jk_0\sqrt{2}/2x}} \tag{3.55.b}$$

It's worth noting that, the value of χ_{ee}^{xx} and χ_{mm}^{yy} are only related to the x-axis in 2-D case. χ_{ee}^{xx} and χ_{mm}^{yy} are shown in Figure 3.15. The determined susceptibility tensors are combined with FEM-GSTC derived in previous section. The simulated space is divided into triangular meshes satisfied the 10 points per wavelength (PPW) sampling density, shown as Figure 3.16(a). The metasurface is modeled as a zero-thickness sheet, which can be conceptually treated as a line discontinuity. To validate the efficiency and accuracy of the proposed method, the same example

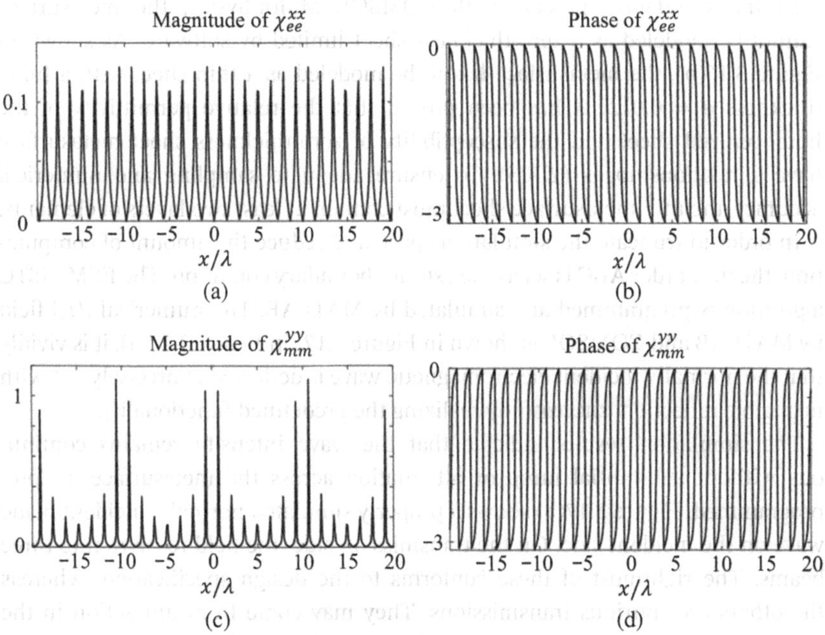

Figure 3.15 (a) Magnitude and (b) phase of χ_{ee}^{xx} and (c) magnitude and (d) phase of χ_{mm}^{yy}.

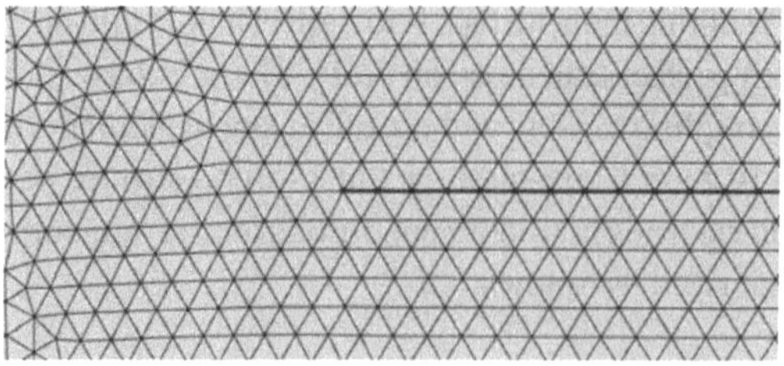

(a)

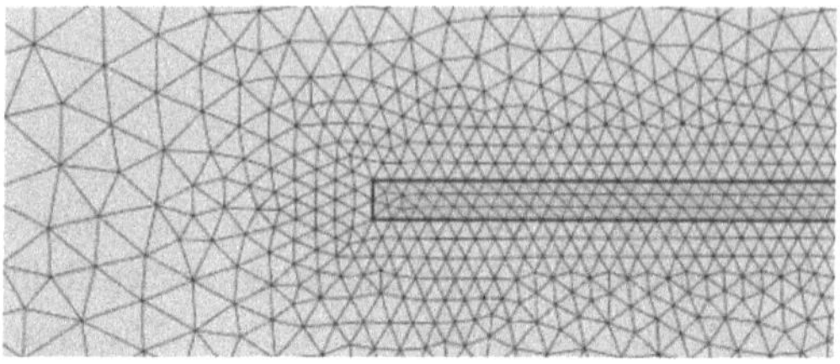

(b)

Figure 3.16 Comparison of mesh configurations between the COMSOL simulation and the proposed FEM-GSTC method. (a) discretization scheme in the FEM-GSTC approach for modeling the metasurface as a zero-thickness interface and (b) Mesh discretization in COMSOL for modeling the metasurface with a finite thickness of $d = 0.01\lambda$.

with identical parameters is simulated using the commercial software COMSOL Multiphysics for comparative analysis.

In the simulation process with COMSOL Multiphysics, the metasurface cannot be modeled as a zero-thickness sheet limited by software. As shown in Figure 3.16(b), the metasurface has to be modeled as a thin sheet with a finite thickness $d = 0.01\lambda$. It has been proved that the relative permittivity of the homogenized sheet and the susceptibility of zero-thickness sheet metasurface have a relationship as (2.105). To ensure adequate sampling and numerical accuracy, the thin metasurface sheet must contain at least two layers of elements.

In order to truncate the simulation space and reduce the amount of computation, the first-order ABC is set as the exterior boundary condition. The FEM-GSTC algorithm is programmed and simulated by MATLAB. The numerical $|H_y|$ field by MATLAB and COMSOL is shown in Figure 3.17. In Figure 3.17(a), it is vividly that the normally incident electromagnetic wave is deflected at precisely 45° with negligible reflection, successfully realizing the predefined functionality.

The simulation results indicate that the wave intensity remains continuous without substantial jump or attenuation across the metasurface, as preprogrammed. Figure 3.17(b) shows a properly simulated normally incident plane wave on the incident side. On the transmission side, the field resolves into three beams. The rightmost of these conforms to the design specification, whereas the others are spurious transmissions. They may come from diffraction in the metasurface when replacing the zero-thickness sheet with the thin metasurface

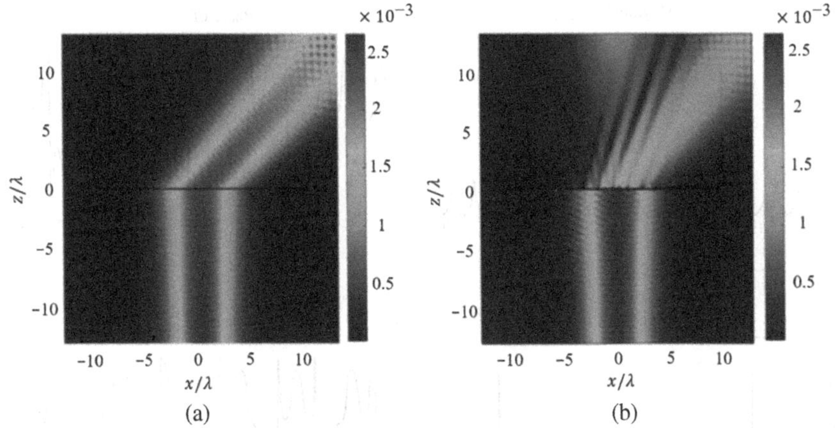

Figure 3.17 Comparison of the simulation results $|H_y|$ by the FEM-GSTC (a) and COMSOL (b) for the metasurface refracting in Figure 3.14. *Source:* Cai et al. [11] / with permission of IEEE.

sheet. The wave intensity substantially attenuates when crossing the metasurface, which is also caused by unexpected waves.

Since the limited accuracy of the first-order ABCs, standing waves are formed in both the MATLAB and COMSOL results, as the reflected waves are not fully absorbed at the domain boundaries. Fortunately, these spurious modes do not affect the propagation of the desired wave and can be effectively suppressed by employing PMLs or high-order ABCs. By comparing FEM-GSTC results with COMSOL, it is clear that the FEM-GSTC method has incomparable accuracy in zero-thickness sheet metasurface synthesis and can achieve the preset function perfectly.

A more general validation of the FEM-GSTC method for beam deflection is performed by synthesizing a metasurface that also generates a specified reflection. The designed structure splits a normally incident Gaussian beam equally into an obliquely reflected wave ($\theta_r = \pi/9$) and an obliquely transmitted wave ($\theta_t = \pi/6$), maintaining the original polarization. The electric and magnetic fields for the incident (i), reflected (r), and transmitted (t) waves can be directly formulated. Due to the presence of the reflected wave, both the transmitted and reflected wave amplitudes are set to half that of the incident wave, as given by

$$\mathbf{E}^i = \hat{\mathbf{x}} \cos\theta_i e^{-jk_0 \sin\theta_i x} \tag{3.56.a}$$

$$\mathbf{E}^t = \hat{\mathbf{x}} 0.5 \cos\theta_t e^{-jk_0 \sin\theta_t x} \tag{3.56.b}$$

$$\mathbf{E}^r = \hat{\mathbf{x}} 0.5 \cos\theta_r e^{-jk_0 \sin\theta_r x} \tag{3.56.c}$$

$$\mathbf{H}^i = \hat{\mathbf{y}} \frac{1}{Z_0} e^{-jk_0 \sin\theta_i x} \tag{3.56.d}$$

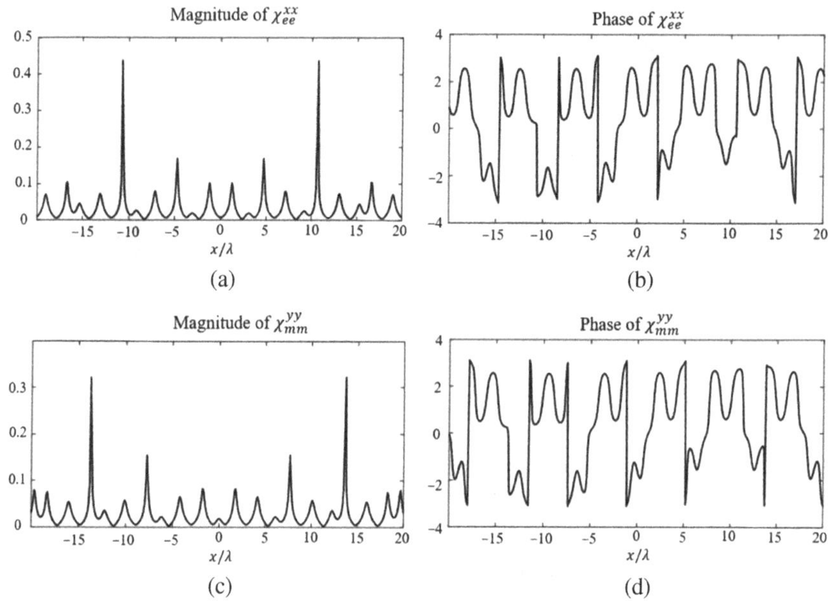

Figure 3.18 (a) Magnitude and (b) phase of χ_{ee}^{xx} and (c) magnitude and (d) phase of χ_{mm}^{yy}.

$$\mathbf{H}^t = \hat{\mathbf{y}}\frac{1}{2Z_0}e^{-jk_0 \sin\theta_t x} \tag{3.56.e}$$

$$\mathbf{H}^r = -\hat{\mathbf{y}}\frac{1}{2Z_0}e^{-jk_0 \sin\theta_r x} \tag{3.56.f}$$

The susceptibilities χ_{ee}^{xx} and χ_{mm}^{yy} are synthesized according to (2.97), which are shown in Figure 3.18, and it can be found that the susceptibility tensor varies sharply along the x-direction, necessitating a sufficiently dense mesh to accurately resolve its spatial features.

The obtained susceptibility parameters are substituted into the FEM-GSTC framework for simulation. Since the corresponding COMSOL solution is not accurate for reliable comparison, the assessment is confined to the FEM-GSTC output, which is displayed on a decibel scale in Figure 3.19(a). The figure shows that the reflected and transmitted beams propagate in their prescribed directions. Interference fringes form on the incident side between the incident and reflected waves. The amplitude of the transmitted beam is half that of the incident one, confirming the intended equal-energy-splitting design.

Without loss of generality, FEM-GSTC can also achieve negative reflection. In Figure 3.19(b), the metasurface with incident ($\theta_i = 0$), reflected ($\theta_r = -\pi/4$), and transmitted ($\theta_t = \pi/3$) waves is successfully simulated. Through GSTC theory, it is possible to generate reflected and transmitted waves at arbitrary angles from a

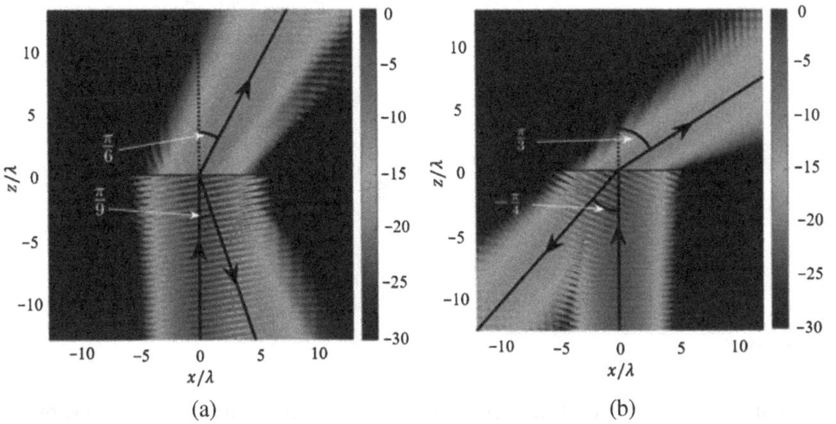

Figure 3.19 FEM-GSTC results of $|H_y|$ in dB. (a) A metasurface splitting a normally incident Gaussian wave into an obliquely reflected ($\theta_r = \pi/9$) and a transmitted ($\theta_t = \pi/6$) Gaussian wave. (b) Another metasurface with $\theta_r = -\pi/4$ and $\theta_t = \pi/3$. *Source:* Cai et al. [11] / with permission of IEEE.

plane wave incident at an arbitrary angle, which provides a new idea and method for wavefront manipulation.

3.2.3.2 PBS with Adjustable Transmission Angle

Birefringence refers to the phenomenon where a light beam incident on an anisotropic crystal split into two refracted beams propagating along different directions. Building upon this principle, birefringent metasurfaces are designed to independently manipulate the amplitude, phase, and polarization state of two orthogonal electromagnetic wave components, offering substantial application prospects across both optical and microwave frequency ranges. Transparent calcite is a kind of traditional birefringent crystal, whose birefringence is determined by the material properties. The angle of refraction of the two kinds of light is also fixed. Here, a PBS is designed to independently manipulate orthogonally polarized beams, refracting each into distinct directions [12, 13]. As shown in Figure 3.20, a normal incident beam with two polarizations is split into two beams. The x-polarized wave transmitted through the metasurface is deflected at an angle θ_1 toward the negative x-direction, while the y-polarized wave is deflected at an angle θ_2 toward the positive x-direction.

An incident field with quadrature polarization with amplitude $\sqrt{2}$ V/m is of the form

$$\mathbf{E}^i = \hat{\mathbf{x}}e^{-jk_0z} + \hat{\mathbf{y}}e^{-jk_0z} \tag{3.57.a}$$

$$\mathbf{H}^i = \frac{1}{Z_0}\left(\hat{\mathbf{y}}e^{-jk_0z} - \hat{\mathbf{x}}e^{-jk_0z}\right) \tag{3.57.b}$$

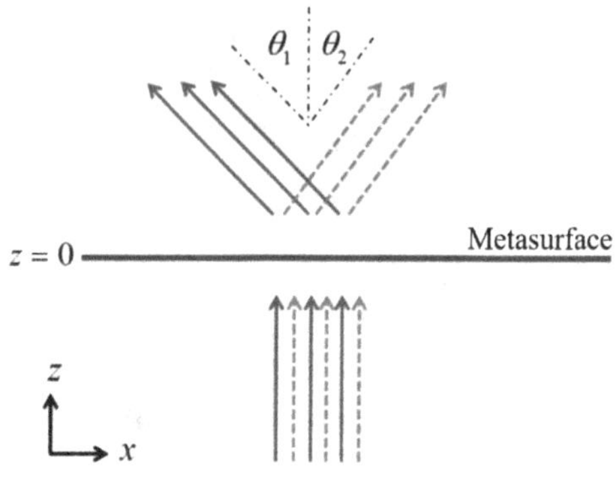

Figure 3.20 Functional schematic of the polarization beam splitter.

Meanwhile, the form of an x-polarized wave deflected to the left region by an angle $\theta_1 = 30°$ and a y-polarized wave deflected to the right region by an angle $\theta_2 = 45°$

$$\mathbf{E}_1^t = \hat{\mathbf{x}}\cos(-\theta_1)e^{-jk_0 \sin(-\theta_1)x} \tag{3.58.a}$$

$$\mathbf{E}_2^t = \hat{\mathbf{y}}\cos(\theta_2)e^{-jk_0 \sin(\theta_2)x} \tag{3.58.b}$$

$$\mathbf{H}_1^t = \hat{\mathbf{y}}\frac{1}{Z_0}e^{-jk_0 \sin(-\theta_1)x} \tag{3.58.c}$$

$$\mathbf{H}_2^t = \hat{x}\frac{1}{Z_0}e^{-jk_0 \sin(\theta_2)x} \tag{3.58.d}$$

Only birefringence is considered here, and polarization transformation is not included, thus four susceptibilities are needed, without electromagnetic coupling parameters. As noted in reference [14], the susceptibility tensors for a birefringent metasurface exhibit the following structural forms

$$\begin{pmatrix} -\Delta H_y \\ \Delta H_x \end{pmatrix} = j\omega\varepsilon \begin{pmatrix} \chi_{ee}^{xx} & 0 \\ 0 & \chi_{ee}^{yy} \end{pmatrix} \begin{pmatrix} E_x \\ E_y \end{pmatrix} \tag{3.59}$$

$$\begin{pmatrix} \Delta E_y \\ -\Delta E_x \end{pmatrix} = j\omega\mu \begin{pmatrix} \chi_{mm}^{xx} & 0 \\ 0 & \chi_{mm}^{yy} \end{pmatrix} \begin{pmatrix} H_x \\ H_y \end{pmatrix} \tag{3.60}$$

The susceptibilities can be determined by (2.97)

$$\chi_{ee}^{xx} = \frac{2j}{k_0}\frac{e^{jk_0 \sin\theta_1 x} - 1}{e^{jk_0 \sin\theta_1 x}\cos\theta_1 + 1} \tag{3.61.a}$$

$$\chi_{ee}^{yy} = \frac{2j}{k_0}\frac{e^{-jk_0 \sin\theta_2 x}\cos\theta_2 - 1}{e^{-jk_0 \sin\theta_2 x} + 1} \tag{3.61.b}$$

$$\chi_{mm}^{xx} = \frac{2j}{k_0}\frac{e^{-jk_0 \sin\theta_2 x} - 1}{e^{-jk_0 \sin\theta_2 x}\cos\theta_2 + 1} \tag{3.61.c}$$

$$\chi_{mm}^{yy} = \frac{2j}{k_0}\frac{e^{jk_0 \sin\theta_1 x}\cos\theta_1 - 1}{e^{jk_0 \sin\theta_1 x} + 1} \tag{3.61.d}$$

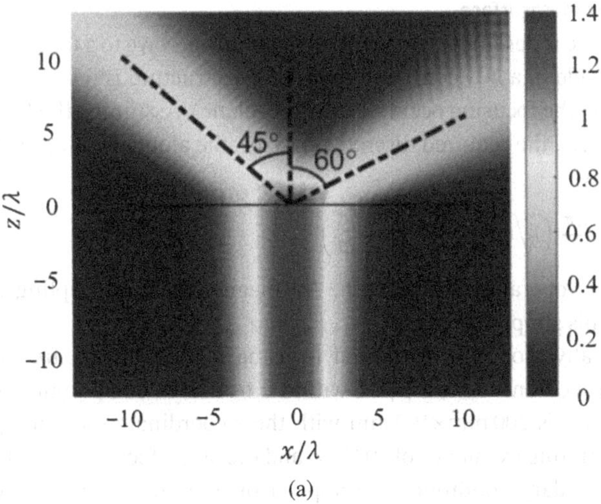

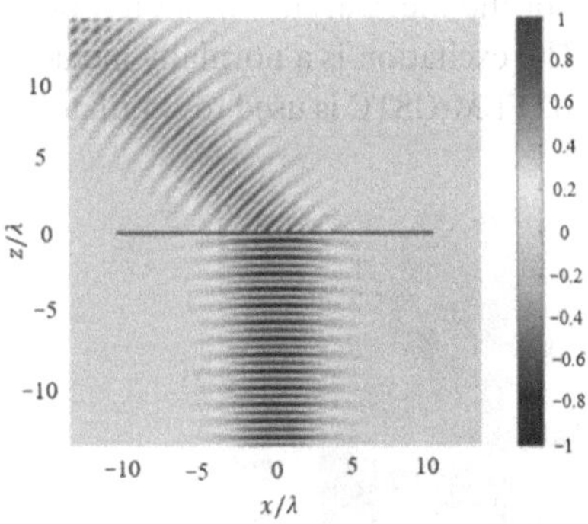

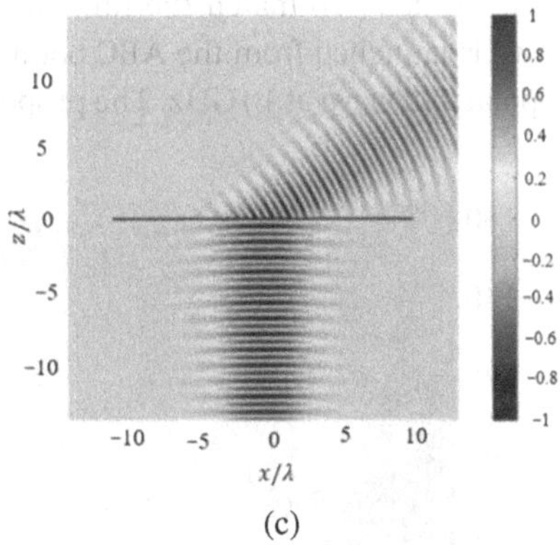

Figure 3.21 FEM-GSTC results of PBS. (a) Electric field intensity distribution $|e_t|$ of PBS, (b) the real part of E_x, and (c) the real part of E_y.

The dimensions of the square computational domain in vacuum are $26\lambda \times 26\lambda$ in the *XoZ*-plane with the first-order ABC. The simulation result obtained by FEM-GSTC is shown in Figure 3.21. Figure 3.21(a) shows the electric field intensity of PBS, and Figure 3.21(b,c) is the real part of E_x and E_y, respectively. It can be seen that after passing through the metasurface, the beams are split into left and right beams as the desired angles, and E_x polarization propagates to the left of the normal, while E_y polarization to the right of normal as Figure 3.21(b,c). In addition, the magnitudes of the resultant field intensities are approximately $\sqrt{2}/2$, corresponding to half the peak amplitude of the incident field.

3.2.3.3 Focus Metasurface

The metasurface is designed to focus an incident plane wave to a focal point on the transmission side in a 2-D configuration (i.e., the structure is invariant along the y-direction and the focusing occurs in the XoZ-plane). Assuming the focal point is located at (x_f, z_f), the required discrete phase profile along the metasurface aperture is given by

$$\phi(x) = \frac{2\pi}{\lambda}\left(\sqrt{(x - x_f)^2 + z_f^2} - z_f\right) \tag{3.62}$$

where λ is the operating wavelength. All magnetoelectric coupling terms are neglected in this simplified design, i.e., $\chi_{em} = \chi_{me} = 0$.

To numerically model a finite-sized focusing metasurface in free space, the structure is placed on the XoY-plane within a truncated computational domain. The domain size is 200 mm×195 mm with the z-coordinate spanning [−30,165] mm under operating frequency of 10 GHz, and the target focus is set at (0, 60) mm. Absorbing boundary conditions are applied on the bottom boundary to both emulate the open-space truncation and serve as the excitation boundary for the incident field. To match the finite aperture of the metasurface, a flat-top plane wave is launched from the ABC boundary. The excitation is a normally incident, x-polarized wave at 10 GHz. The proposed 2-D FEM-GSTC is used to simulate the

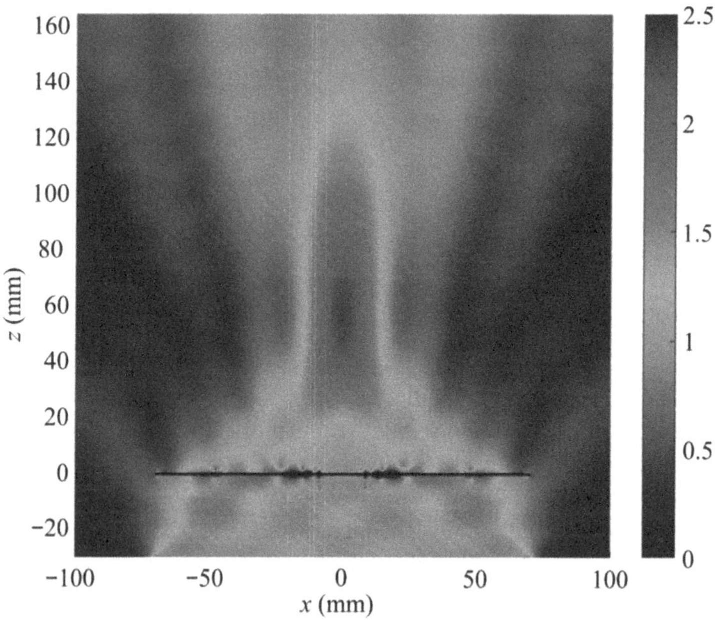

Figure 3.22 Electric field intensity distribution $|e_t|$ of the focus metasurface.

focus metasurface, whose equivalent susceptibility tensors are calculated from the transmission and reflection coefficients. Figure 3.22 presents the $|e_t|$ in the XoZ-plane. The results show that the transmitted field is tightly concentrated at a single focus after traversing the metasurface, and the peak intensity occurs at the intended target location.

3.3 SEM with GSTCs

3.3.1 Introduction of SEM

The SEM is a high-order numerical technique used for solving PDEs. It combines features of both the FEM and spectral methods, offering a powerful framework for simulating complex physical phenomena. Initially introduced in the 1980s by Polish mathematician Anthony Patera and researchers from the Massachusetts Institute of Technology (MIT) [15], SEM combines the high accuracy and convergence properties of spectral methods [16] with the localized discretization capabilities of FEM [17].

In the SEM, basis functions within each element are constructed using the roots of Chebyshev polynomials, Legendre polynomials, or other orthogonal polynomials. These interpolation points are often non-uniformly spaced and include boundary points, effectively mitigating the Runge phenomenon encountered in FEMs and the Gibbs phenomenon associated with spectral methods. The earliest applications of SEM focused primarily on solving fluid dynamics problems within complex geometries, including the numerical solution of the Navier–Stokes equations. During the 1990s, SEM underwent further development and found widespread application across various fields of engineering and science, including electromagnetics, fluid mechanics [18], and solid mechanics [19].

In the field of electromagnetics, Q. H. Liu's team has made significant contributions to the application of the SEM method. Lee adopted Gauss–Lobatto–Legendre (GLL) polynomial-based basis functions in SEM to analyze waveguide eigenvalue problems [20]. Later, the SEM was extended to the time domain for the analysis of transient electromagnetic fields [21–24].

Here's the general workflow for applying SEM to electromagnetic problems

1) Problem definition and formulation of Helmholtz equations
2) Domain discretization and mesh generation
3) Interpolation and selection of basis functions
4) System discretization and matrices assembly
5) Solving the discretized system
6) Implementing boundary conditions and source terms
7) Postprocessing and result analysis

3.3.1.1 Governing Equations and Boundary Conditions

The approach to solving the electric field wave equation using the SEM and the FEM follows a unified framework. Both methods apply the Galerkin method to derive the weak form of the solution and discretize the computational domain into a finite number of elements, approximating the field variables within each element using basis functions. Here, considering a 3-D anisotropic medium, Maxwell's equations can be reformulated as a vector Helmholtz equation

$$\nabla \times \left(\overline{\overline{\mu}}_r^{-1} \nabla \times \mathbf{E} \right) - k_0^2 \overline{\overline{\varepsilon}}_r \mathbf{E} = f \tag{3.64}$$

where $\overline{\overline{\varepsilon}}_r$ and $\overline{\overline{\mu}}_r$ are the relative permittivity and permeability tensors, and k_0 is the vacuum wave number. The Galerkin method is applied to (3.64) by testing function $\mathbf{v}$

$$\int_V \mathbf{v} \cdot \nabla \times \left(\overline{\overline{\mu}}_r^{-1} \nabla \times \mathbf{E} \right) dV - \int_V \mathbf{v} \cdot k_0^2 \overline{\overline{\varepsilon}}_r \mathbf{E} dV = \int_V \mathbf{v} \cdot \mathbf{f} dV \tag{3.65}$$

Using the vector identity $\nabla \cdot (\mathbf{A} \times \mathbf{B}) = \mathbf{B} \cdot \nabla \times \mathbf{A} - \mathbf{A} \cdot \nabla \times \mathbf{B}$, and identifying $\mathbf{A} = \overline{\overline{\mu}}_r^{-1} (\nabla \times \mathbf{E})$ and $\mathbf{B} = \mathbf{v}$, the weak form is

$$\int_V \overline{\overline{\mu}}_r^{-1} \nabla \times \mathbf{v} \cdot (\nabla \times \mathbf{E}) dV - \int_V k_0^2 \overline{\overline{\varepsilon}}_r \mathbf{v} \cdot \mathbf{E} dV = \oint_S \mathbf{v} \cdot \mathbf{n} \times \overline{\overline{\mu}}_r^{-1} (\nabla \times \mathbf{E}) dS + \int_V \mathbf{v} \cdot f dV \tag{3.66}$$

where the subscript S is the boundary of volume V. The weak-form equations can be solved together with the appropriate outer boundary conditions.

Depending on different applications, various boundary conditions can be selected. For example, the following are some boundary conditions that might be used in this chapter

1) Dirichlet boundary conditions for the electric field

$$\mathbf{n} \times \mathbf{E} = \mathbf{g}, \quad \mathbf{r} \in S \tag{3.67}$$

Dirichlet boundary conditions specify the value of the solution at the boundary. When $\mathbf{g} = 0$, it is known as PEC.

2) Absorbing boundary conditions (ABC) [25]

$$\mathbf{n} \times (\nabla \times \mathbf{E}) + jk_0 \mathbf{n} \times (\mathbf{n} \times \mathbf{E}) = \mathbf{n} \times (\nabla \times \mathbf{E}^i) + jk_0 \mathbf{n} \times (\mathbf{n} \times \mathbf{E}^i) \tag{3.68}$$

ABCs are designed to absorb waves incident at the boundary, aiming to minimize non-physical reflections. ABCs are regarded as a mixed boundary condition, where $\mathbf{E}$ represents the total field, and $\mathbf{E}^i$ represents the incident field. By using ABCs, the excitation source can be introduced into the SEM.

3) Periodic boundary conditions
   These conditions are applied when the structure or problem exhibits periodicity. PBC assumes that the fields on one side of the boundary are identical to those on the opposite side, allowing for the simulation of an infinite periodic structure by modeling only a single unit cell. According to Floquet PBCs theory [26], only one phase factor $C = e^{-jkL}$ differs between corresponding boundaries.

3.3.1.2 Mesh and Vector Basis Function

The SEM is a specialized high-order variant of the FEM, blending the geometric flexibility of FEM with the exponential convergence of spectral methods. SEM provides several advantages, including high accuracy, low dispersion, and computational efficiency. In 3-D space, SEM employs hexahedral mesh elements. Generally, the hexahedral elements in physical space are irregular, and performing calculations directly on these physical elements would increase the computational burden. Therefore, a coordinate transformation is required to map them onto the reference element $(\xi, \eta, \zeta) \in [-1,1] \times [-1,1] \times [-1,1]$, as shown in Figure 3.23. The mapping from the physical element to the reference element satisfies the following relationship

$$\mathbf{\Phi}(x, y, z) = \mathbf{J}^{-1}\hat{\mathbf{\Phi}}(\xi, \eta, \zeta) \tag{3.69.a}$$

$$\nabla \times \mathbf{\Phi}(x, y, z) = \frac{1}{|\mathbf{J}|}\mathbf{J}^T \nabla \times \hat{\mathbf{\Phi}}(\xi, \eta, \zeta) \tag{3.69.b}$$

where $\mathbf{\Phi}$ denotes the covariant components in the reference domain, and $\mathbf{J}$ denotes the Jacobian matrix as

$$\mathbf{J} = \begin{bmatrix} \frac{\partial x}{\partial \xi} & \frac{\partial y}{\partial \xi} & \frac{\partial z}{\partial \xi} \\ \frac{\partial x}{\partial \eta} & \frac{\partial y}{\partial \eta} & \frac{\partial z}{\partial \eta} \\ \frac{\partial x}{\partial \zeta} & \frac{\partial y}{\partial \zeta} & \frac{\partial z}{\partial \zeta} \end{bmatrix} \tag{3.70}$$

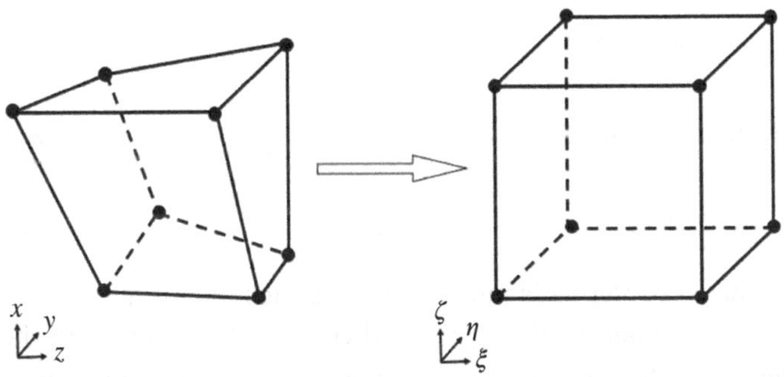

Figure 3.23 Hexahedral mesh. Mapping from physical elements to reference element.

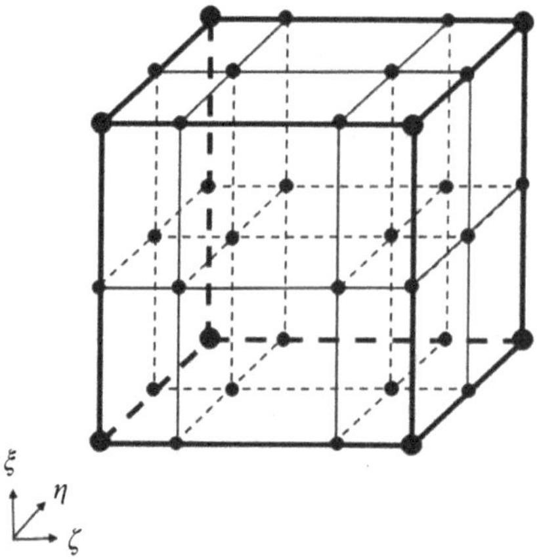

Figure 3.24 Distribution of GLL points in the reference element, when $N_\xi = 3$, $N_\eta = 2$, and $N_\zeta = 2$.

In this book, the SEM selects GLL points as interpolation points and employs Lagrange functions to construct the vectorial basis functions, resulting in high-order vector basis functions [20], and then an N-th order basis function requires $N+1$ GLL points. The basis functions can have different orders in different directions, for example, Figure 3.24 shows the distribution of GLL points when $N_\xi = 3$, $N_\eta = 2$, and $N_\zeta = 2$, where N_ξ, N_η, and N_ζ are interpolation orders of the reference domain along the parametric coordinates, respectively.

The Nth-order GLL basis functions in 1-D reference element $\xi \in [-1, 1]$ are

$$\phi_j^N(\xi) = \frac{-1}{N(N+1)L_N(\xi_j)} \frac{(1-\xi^2)L'_N(\xi)}{(\xi - \xi_j)} \tag{3.71}$$

For $j = 0, \cdots, N$, where $L_N(\xi)$ is the Nth-order Legendre polynomial. The variables ξ_j are the GLL points. The 3-D curl conforming vector basis functions in SEM have been utilized, which can be written as

$$\hat{\mathbf{\Phi}}_{ijk}^{\xi} = \hat{\boldsymbol{\xi}}\phi_i^{N_\xi - 1}(\xi)\phi_j^{N_\eta}(\eta)\phi_k^{N_\zeta}(\zeta) \tag{3.72.a}$$

$$\hat{\mathbf{\Phi}}_{ijk}^{\eta} = \hat{\boldsymbol{\eta}}\phi_i^{N_\xi}(\xi)\phi_j^{N_\eta - 1}(\eta)\phi_k^{N_\zeta}(\zeta) \tag{3.72.b}$$

$$\hat{\mathbf{\Phi}}_{ijk}^{\zeta} = \hat{\boldsymbol{\zeta}}\phi_i^{N_\xi}(\xi)\phi_j^{N_\eta}(\eta)\phi_k^{N_\zeta - 1}(\zeta) \tag{3.72.c}$$

An example of the 2nd-order basis functions is presented to provide a clearer illustration of the distribution in Figure 3.25, where the arrows represent the different components of the vectorial basis function $\rightarrow \hat{\mathbf{\Phi}}_{ijk}^{\xi}$, $\nearrow \hat{\mathbf{\Phi}}_{ijk}^{\eta}$, and $\uparrow \hat{\mathbf{\Phi}}_{ijk}^{\zeta}$, respectively.

Figure 3.25 Curl conforming vector basis functions in the reference element where $N_\xi = 2$, $N_\eta = 2$, and $N_\zeta = 2$.

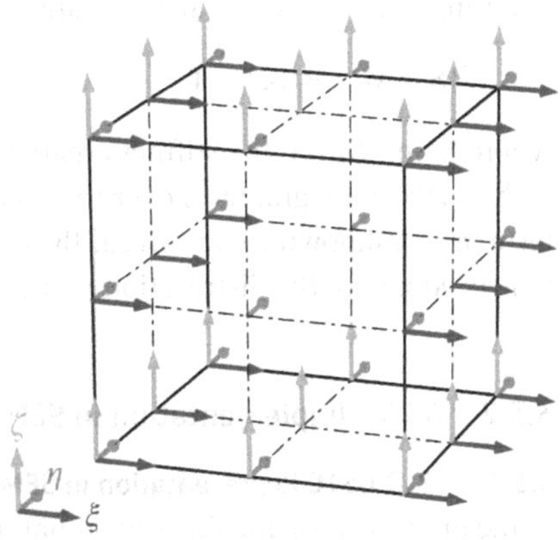

3.3.1.3 Discrete Equations

Once the basis functions are constructed, the unknown quantities are approximated in terms of these functions. The electric field on the reference element is expressed through an expansion of the vector basis functions as follows

$$\mathbf{E}(\xi,\eta,\zeta) = \sum_{m=1}^{N_e} e_m \hat{\mathbf{\Phi}}_m(\xi,\eta,\zeta) \tag{3.73}$$

where e_m is the m-th unknown of the electric field, $N_e = 3N(N+1)^2$ is the DoFs of $\mathbf{E}$ in one element with N-th order, and $\hat{\mathbf{\Phi}}_m$ is the curl-conforming basis function along the m-th edge. By substituting (3.73) into the weak form of the governing Eq. (3.66), the discrete equation of SEM on the reference element becomes

$$\sum_{m=1}^{N_e} \left[\mathbf{S}^e_{im} - \mathbf{M}^e_{im} - \mathbf{B}^e_{im}\right] \mathbf{e}^e_m = f^e_i \tag{3.74}$$

where

$$\mathbf{S}^e_{im} = \int_{V_e} \overline{\overline{\mu}}_r^{-1} \nabla \times \hat{\mathbf{\Phi}}_i \cdot \nabla \times \hat{\mathbf{\Phi}}_m dV \tag{3.75.a}$$

$$\mathbf{M}^e_{im} = \int_V k_0^2 \overline{\overline{\varepsilon}}_r \hat{\mathbf{\Phi}}_i \cdot \hat{\mathbf{\Phi}}_m dV \tag{3.75.b}$$

$$\mathbf{B}^e_{im} = \oint_S \hat{\mathbf{\Phi}}_i \cdot \mathbf{n} \times \overline{\overline{\mu}}_r^{-1} \nabla \times \hat{\mathbf{\Phi}}_m \tag{3.75.c}$$

$$f^e_i = \int_{V_e} \hat{\mathbf{\Phi}}_i \cdot f dV \tag{3.75.d}$$

$\hat{\mathbf{\Phi}}_i$ is the i-th basis function in the e-th element. The element matrices are assembled into the global matrix based on the global numbering of the basis functions, resulting in the system matrix equation to be solved

$$[\mathbf{S} - \mathbf{M} - \mathbf{B}]\{\mathbf{e}\} = \mathbf{f} \tag{3.76}$$

where **S** and **M** are the stiffness matrix and mass matrix, respectively. Matrix **B** is the surface-integral term over the boundary, and vector **f** represents the source term. **e** is unknown to be solved. By selecting an appropriate solver to solve the matrix equation, the distribution of the electric field can be finally obtained.

3.3.2 GSTCs Implementation in SEM

3.3.2.1 3-D GSTC Representation in SEM

In the process of solving the wave equation for the electric field using SEM, there is only one unknown (either **E** or **H**). Therefore, when combining GSTC with SEM, it is necessary to derive the substitute relationships between the electric and magnetic fields on both sides of the equivalent metasurface through GSTC. Consider a GSTC equivalent metasurface placed in a Cartesian coordinate system, as shown in Figure 3.26. Ignoring the normal components of the susceptibilities, the GSTC is simplified to

$$\hat{\mathbf{n}} \times \Delta\mathbf{H} = j\omega\mathbf{P}_{\|} \tag{3.77.a}$$

$$\hat{\mathbf{n}} \times \Delta\mathbf{E} = -j\omega\mu\mathbf{M}_{\|} \tag{3.77.b}$$

where **n** is the unit outward vector normal to the metasurface.

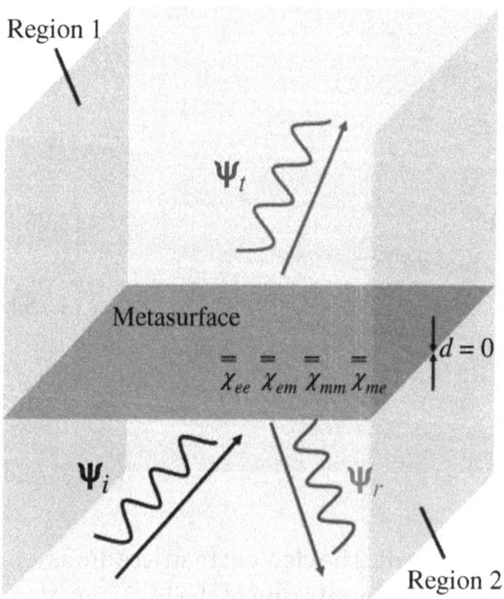

Figure 3.26 Schematic diagram of a 3-D GSTC metasurface.

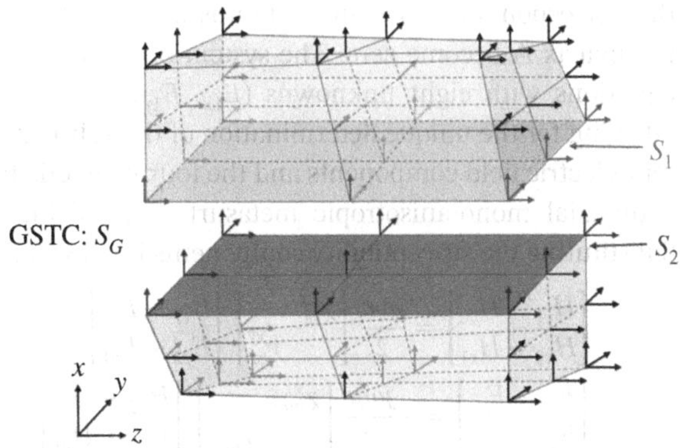

Figure 3.27 Two adjacent hexahedron elements with the same GSTC surface S_G. S_G is virtually divided into two surfaces S_1 and S_2, and different DoFs are assigned on them during the discretization of SEM. The arrows represent the second-order basis function.

Due to the field discontinuity on the GSTC, the GSTC surface (represented as S_G) is virtually divided into two zero-thickness surfaces, S_1 and S_2. Different DoFs are assigned on S_1 and S_2, as shown in Figure 3.27. Using the fields on both sides of S_G to represent the surface polarizations

$$\mathbf{P}_{\|} = \frac{1}{2}\overline{\overline{\chi}}_{ee}(\mathbf{E}_1 + \mathbf{E}_2) + \frac{1}{2}\overline{\overline{\chi}}_{em}(\mathbf{H}_1 + \mathbf{H}_2) \tag{3.78.a}$$

$$\mathbf{M}_{\|} = \frac{1}{2}\overline{\overline{\chi}}_{me}(\mathbf{E}_1 + \mathbf{E}_2) + \frac{1}{2}\overline{\overline{\chi}}_{mm}(\mathbf{H}_1 + \mathbf{H}_2) \tag{3.78.b}$$

Substituting (3.78) into (3.77) leads to

$$\hat{\mathbf{n}} \times (\mathbf{H}_2 - \mathbf{H}_1) = j\omega\hat{\mathbf{n}} \times \hat{\mathbf{n}} \times \left(\varepsilon\overline{\overline{\chi}}_{ee}\frac{\mathbf{E}_1 + \mathbf{E}_2}{2} + \overline{\overline{\chi}}_{em}\sqrt{\varepsilon\mu}\frac{\mathbf{H}_1 + \mathbf{H}_2}{2} \right) \tag{3.79.a}$$

$$\mathbf{n} \times (\mathbf{E}_2 - \mathbf{E}_1) = -j\omega\mu\hat{\mathbf{n}} \times \hat{\mathbf{n}} \times \left(\overline{\overline{\chi}}_{mm}\frac{\mathbf{H}_1 + \mathbf{H}_2}{2} + \overline{\overline{\chi}}_{me}\sqrt{\frac{\varepsilon}{\mu}}\frac{\mathbf{E}_1 + \mathbf{E}_2}{2} \right) \tag{3.79.b}$$

where $\mathbf{E}_1 = [E_{x,1}, E_{y,1}, E_{z,1}]^T$, $\mathbf{E}_2 = [E_{x,2}, E_{y,2}, E_{z,2}]^T$, $\mathbf{H}_1 = [H_{x,1}, H_{y,1}, H_{z,1}]^T$, and $\mathbf{H}_2 = [H_{x,2}, H_{y,2}, H_{z,2}]^T$. By rearranging the system of equations in (3.79.a) and introducing a 6×6 matrix $\mathbf{A}$ to represent such substitute relationship between the electric and magnetic fields

$$\begin{pmatrix} H_{x1} \\ H_{y1} \\ H_{z1} \\ H_{x2} \\ H_{y2} \\ H_{z2} \end{pmatrix} = \frac{1}{j\omega\mu_0}\begin{bmatrix} A_{11} & A_{12} & \dots & A_{16} \\ A_{21} & A_{22} & \dots & A_{26} \\ \vdots & \vdots & \ddots & \vdots \\ A_{61} & A_{62} & \dots & A_{66} \end{bmatrix}\begin{pmatrix} E_{x1} \\ E_{y1} \\ E_{z1} \\ E_{x2} \\ E_{y2} \\ E_{z2} \end{pmatrix} \tag{3.80}$$

Since the normal susceptibility components are neglected, the normal components of the electric and magnetic fields in (3.78.a) are set to zero. Consequently, the corresponding rows and columns associated with the normal direction in the matrix **A** become zero. The system of equations in (3.79.a) consists of four equations with eight unknowns (E_{1x}, E_{1y}, E_{2x}, E_{2y}, H_{1x}, H_{1y}, H_{2x}, and H_{2y}), allowing for the unique determination of the substitute relationship between the four electric field components and the four magnetic field components. Consider a uniaxial mono-anisotropic metasurface involving χ_{ee}^{xx}, χ_{ee}^{yy}, χ_{mm}^{xx}, and χ_{mm}^{yy}. Substituting the susceptibility components into (3.79.a) leads to

$$\begin{bmatrix} H_{y1} - H_{y2} \\ H_{x2} - H_{x1} \end{bmatrix} = \frac{j\omega\varepsilon}{2} \begin{bmatrix} \chi_{ee}^{xx} & \\ & \chi_{ee}^{yy} \end{bmatrix} \begin{bmatrix} E_{x2} + E_{x1} \\ E_{y2} + E_{y1} \end{bmatrix} \tag{3.81.a}$$

$$\begin{bmatrix} E_{y1} - E_{y2} \\ E_{x2} - E_{x1} \end{bmatrix} = -\frac{j\omega\mu}{2} \begin{bmatrix} \chi_{mm}^{xx} & \\ & \chi_{mm}^{yy} \end{bmatrix} \begin{bmatrix} H_{x2} + H_{x1} \\ H_{y2} + H_{y1} \end{bmatrix} \tag{3.81.b}$$

Solving the system of equations (3.81)

$$\mathbf{A} = \begin{pmatrix} 0 & A_4 & 0 & 0 & A_3 & 0 \\ -A_2 & 0 & 0 & -A_1 & 0 & 0 \\ 0 & 0 & 0 & 0 & 0 & 0 \\ 0 & -A_3 & 0 & 0 & -A_4 & 0 \\ A_1 & 0 & 0 & A_2 & 0 & 0 \\ 0 & 0 & 0 & 0 & 0 & 0 \end{pmatrix} \tag{3.82}$$

where

$$A_1 = \frac{k_0^2 \chi_{mm}^{yy} \chi_{ee}^{xx} + 4}{4\mu_r \chi_{mm}^{yy}} \tag{3.83.a}$$

$$A_2 = \frac{k_0^2 \chi_{mm}^{yy} \chi_{ee}^{xx} - 4}{4\mu_r \chi_{mm}^{yy}} \tag{3.83.b}$$

$$A_3 = \frac{k_0^2 \chi_{mm}^{xx} \chi_{ee}^{yy} + 4}{4\mu_r \chi_{mm}^{xx}} \tag{3.83.c}$$

$$A_4 = \frac{k_0^2 \chi_{mm}^{xx} \chi_{ee}^{yy} - 4}{4\mu_r \chi_{mm}^{xx}} \tag{3.83.d}$$

Then, the GSTC is expressed by the matrix **A**, and the surface integral only depends on **E**.

3.3.2.2 Combination of SEM with GSTC

GSTC is introduced into SEM through the surface integral term, i.e., $\mathbf{B}_{im}^{e}$ in (3.74). To eliminate normal component of **E**, the Faraday's law is applied to the surface integral term in the reference element. For distinguishing from traditional boundary conditions, the surface integral is represented by $\mathbf{G}^e$

$$\mathbf{G}^e = \int_{S_G} \mathbf{v} \cdot \mathbf{n} \times \overline{\overline{\mu}}_r^{-1} \nabla \times \mathbf{E}\, dS = \int_{S_G} \mathbf{v} \cdot \mathbf{n} \times (-j\omega\mu_0 \mathbf{H})\, dS \tag{3.84}$$

Substituting the relation **A** in (3.80) into (3.84) transforms the unknowns into **E**

$$\begin{aligned}\mathbf{G}^e &= \int_{S_1} \mathbf{v} \cdot \mathbf{n} \times (-j\omega\mu_0 \mathbf{H}_1)\, dS + \int_{S_2} \mathbf{v} \cdot \mathbf{n} \times (-j\omega\mu_0 \mathbf{H}_2)\, dS \\ &= \int_{S_1} \mathbf{v} \cdot \mathbf{n} \times (-\mathbf{A}_{1:3,1:3}\mathbf{E}_1 - \mathbf{A}_{1:3,4:6}\mathbf{E}_2)\, dS \\ &\quad + \int_{S_2} \mathbf{v} \cdot \mathbf{n} \times (-\mathbf{A}_{4:6,1:3}\mathbf{E}_1 - \mathbf{A}_{4:6,4:6}\mathbf{E}_2)\, dS\end{aligned} \tag{3.85}$$

Plug (3.73) into (3.85), the surface integral **G** is discretized as

$$\mathbf{G}^e_{im} = \sum_{m=1}^{N_e} e_m \begin{bmatrix} \int_{S_{1e}} \hat{\mathbf{\Phi}}_i \cdot \mathbf{n} \times (A_{1:3,1:3}\hat{\mathbf{\Phi}}_{m1} + A_{1:3,4:6}\hat{\mathbf{\Phi}}_{m2}) dS \\ + \int_{S_{2e}} \hat{\mathbf{\Phi}}_i \cdot \mathbf{n} \times (A_{4:6,1:3}\hat{\mathbf{\Phi}}_{m1} + A_{4:6,4:6}\hat{\mathbf{\Phi}}_{m2}) dS \end{bmatrix} \tag{3.86}$$

By incorporating this term into the SEM, the discrete element matrix equation becomes

$$\sum_{m=1}^{N_e} \left[\mathbf{S}^e_{im} - \mathbf{M}^e_{im} - \mathbf{B}^e_{im} + \mathbf{G}^e_{im}\right] \mathbf{e}^e_m = f^e_i \tag{3.87}$$

After assembling the element matrices into the global system matrix, the final system of equations to be solved is obtained

$$[\mathbf{S} - \mathbf{M} - \mathbf{B} + \mathbf{G}]\{\mathbf{e}\} = \mathbf{f} \tag{3.88}$$

Here, **G** represents the surface-integral term originating from the GSTCs. It should be noted that if the GSTC and ABC surfaces intersect, only the ABC condition is applied to the DoFs on the shared boundary, thereby allowing the computational domain to be truncated. The GSTC formulation is thus successfully integrated into the SEM framework. Solving the resulting system of equations yields the electric field distribution within the computational domain.

3.3.3 Simulation Examples

In the previous sections of this chapter, the SEM and its combination with GSTC are introduced. The effectiveness and accuracy of SEM-GSTC through several classic metasurface examples are demonstrated. This subsection presents some examples. The first one simulates a dual-polarization conversion metasurface using the 2-D SEM. The second example features a hybrid modulation metasurface, where the polarization rate is controlled by scattering parameters, enabling simultaneous phase and amplitude modulation. To balance accuracy and computational cost, the basis function order of $N = 5$ is chosen for the 2-D SEM-GSTC, and an order of $N = 3$ is chosen for the 3-D SEM-GSTC.

3.3.3.1 Polarization Conversion Metasurface

Traditional metasurfaces manipulate the polarization properties of electromagnetic waves through the phase and amplitude of anisotropic materials, as well as the electromagnetic coupling between individual unit cells. In contrast, GSTC-based metasurfaces enable polarization conversion by controlling susceptibility tensors. Moreover, a polarization conversion metasurface that generates distinct responses to incident fields with varying polarizations is presented.

A metasurface that performs two conversions is designed [11]: (1) converting x-polarized incident fields into y-polarized transmitted fields and (2) fully absorbing y-polarized incident fields. Both conversions should not generate any reflected waves. Considering that the metasurface only controls the fields in the x and y polarization directions, a 2-D vector SEM is sufficient for simulation. For computational efficiency, this case will only use the 2-D SEM-GSTC. The dimensions of the computational space are set as $26\lambda \times 26\lambda$ in the XoZ-plane. A metasurface of 20λ is placed at the center, parallel to the x-axis. A 1 GHz Gaussian beam with x-polarization or y-polarization is normal incident in the metasurface. The expressions for incident waves and transmitted waves are

$$\mathbf{E}^i = \hat{\mathbf{x}} e^{-jk_0 z} \tag{3.89.a}$$

$$\mathbf{H}^i = \hat{\mathbf{y}} \frac{1}{Z_0} e^{-jk_0 z} \tag{3.89.b}$$

$$\mathbf{E}^t = \hat{\mathbf{y}} e^{-jk_0 z} \tag{3.89.c}$$

$$\mathbf{H}^t = -\hat{\mathbf{x}} \frac{1}{Z_0} e^{-jk_0 z} \tag{3.89.d}$$

and

$$\mathbf{E}^i = \hat{\mathbf{y}} e^{-jk_0 z} \tag{3.90.a}$$

$$\mathbf{H}^i = -\hat{\mathbf{x}} \frac{1}{Z_0} e^{-jk_0 z} \tag{3.90.b}$$

$$\mathbf{E}^t = \mathbf{H}^t = 0 \tag{3.90.c}$$

Since the metasurface incorporates two conversion functions, eight susceptibility components are required to achieve the desired behavior. In this case, the diagonal components of the susceptibility tensors and the cross-polarization terms are non-zero. Based on the GSTC framework, the following system of equations is obtained

$$-\Delta E_x = j\omega\mu \chi_{mm}^{yy} H_y + j\omega\sqrt{\mu\varepsilon}\chi_{me}^{yy} E_y \tag{3.91.a}$$

$$-\Delta H_y = j\omega\mu \chi_{ee}^{xx} E_x + j\omega\sqrt{\mu\varepsilon}\chi_{em}^{xx} H_x \tag{3.91.b}$$

$$\Delta E_y = j\omega\mu \chi_{mm}^{xx} H_x + j\omega\sqrt{\mu\varepsilon}\chi_{me}^{xx} E_x \tag{3.91.c}$$

$$\Delta H_x = j\omega\varepsilon \chi_{ee}^{yy} E_y + j\omega\sqrt{\mu\varepsilon}\chi_{em}^{yy} H_y \tag{3.91.d}$$

Substituting the specified incident and transmitted fields into (3.91) leads to

$$\chi_{ee}^{xx} = \chi_{ee}^{yy} = \chi_{mm}^{xx} = \chi_{mm}^{yy} = -2j/k_0 \tag{3.92.a}$$

$$\chi_{em}^{xx} = \chi_{me}^{yy} = 0, \quad \chi_{em}^{yy} = 4j/k_0, \quad \chi_{me}^{xx} = -4j/k_0 \tag{3.92.b}$$

In this case, (3.80) is specialized into a substitution relation consisting of eight susceptibilities components.

$$A = \frac{1}{j\omega\mu_0}\begin{bmatrix} A_6 & A_8 & 0 & A_5 & A_7 & 0 \\ -A_2 & -A_4 & 0 & -A_1 & -A_3 & 0 \\ 0 & 0 & 0 & 0 & 0 & 0 \\ A_5 & -A_7 & 0 & A_6 & -A_8 & 0 \\ A_1 & -A_3 & 0 & A_2 & -A_4 & 0 \\ 0 & 0 & 0 & 0 & 0 & 0 \end{bmatrix} \tag{3.93}$$

where

$$A_1 = \frac{k^2 \chi_{mm}^{yy}\left(\chi_{ee}^{xx}\chi_{mm}^{xx} - \chi_{em}^{xx}\chi_{me}^{xx}\right) + 4\chi_{mm}^{xx}}{4\mu_r \chi_{mm}^{xx}\chi_{mm}^{yy}} \tag{3.94.a}$$

$$A_2 = \frac{k^2 \chi_{mm}^{yy}\left(\chi_{ee}^{xx}\chi_{mm}^{xx} - \chi_{em}^{xx}\chi_{me}^{xx}\right) - 4\chi_{mm}^{xx}}{4\mu_r \chi_{mm}^{xx}\chi_{mm}^{yy}} \tag{3.94.b}$$

$$A_3 = -\frac{jk_0}{2}\sqrt{\frac{\varepsilon_r}{\mu_r}}\left(\frac{\chi_{em}^{xx}\chi_{mm}^{yy} - \chi_{me}^{yy}\chi_{mm}^{xx}}{\chi_{mm}^{xx}\chi_{mm}^{yy}}\right) \tag{3.94.c}$$

$$A_4 = \frac{jk_0}{2}\sqrt{\frac{\varepsilon_r}{\mu_r}}\left(\frac{\chi_{em}^{xx}\chi_{mm}^{yy} + \chi_{me}^{yy}\chi_{mm}^{xx}}{\chi_{mm}^{xx}\chi_{mm}^{yy}}\right) \tag{3.94.d}$$

$$A_5 = \frac{jk_0}{2}\sqrt{\frac{\varepsilon_r}{\mu_r}}\left(\frac{\chi_{em}^{xx}\chi_{mm}^{xx} - \chi_{me}^{xx}\chi_{mm}^{yy}}{\chi_{mm}^{xx}\chi_{mm}^{yy}}\right) \tag{3.94.e}$$

$$A_6 = -\frac{jk_0}{2}\sqrt{\frac{\varepsilon_r}{\mu_r}}\left(\frac{\chi_{em}^{xx}\chi_{mm}^{xx} + \chi_{me}^{xx}\chi_{mm}^{yy}}{\chi_{mm}^{xx}\chi_{mm}^{yy}}\right) \tag{3.94.f}$$

$$A_7 = \frac{-k^2 \chi_{mm}^{xx}\left(\chi_{em}^{yy}\chi_{me}^{yy} - \chi_{ee}^{yy}\chi_{mm}^{yy}\right) - 4\chi_{mm}^{yy}}{4\mu_r \chi_{mm}^{xx}\chi_{mm}^{yy}} \tag{3.94.g}$$

$$A_8 = \frac{-k^2 \chi_{mm}^{xx}\left(\chi_{em}^{yy}\chi_{me}^{yy} - \chi_{ee}^{yy}\chi_{mm}^{yy}\right) + 4\chi_{mm}^{yy}}{4\mu_r \chi_{mm}^{xx}\chi_{mm}^{yy}} \tag{3.94.h}$$

The simulation results of the 2-D SEM-GSTC are shown in Figures 3.28 and 3.29. To facilitate the observation of the results, the metasurface is excited separately using x- and y-polarized incident fields. Figure 3.28 shows the example where only the x-polarized wave is incident. Figure 3.28(a) represents the real part of E_x, while Figure 3.28(b) represents the real part of the E_y. It can be seen that since

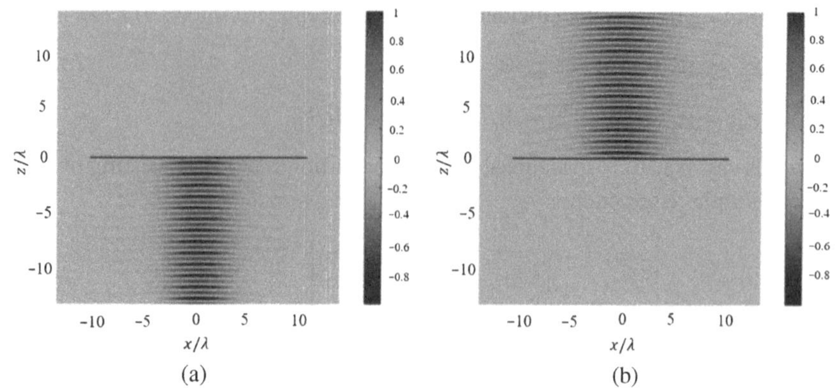

Figure 3.28 The results for the dual-polarization conversion metasurface with only the x-polarized wave incident. The real part of (a) x-component and (b) y-component of electric fields.

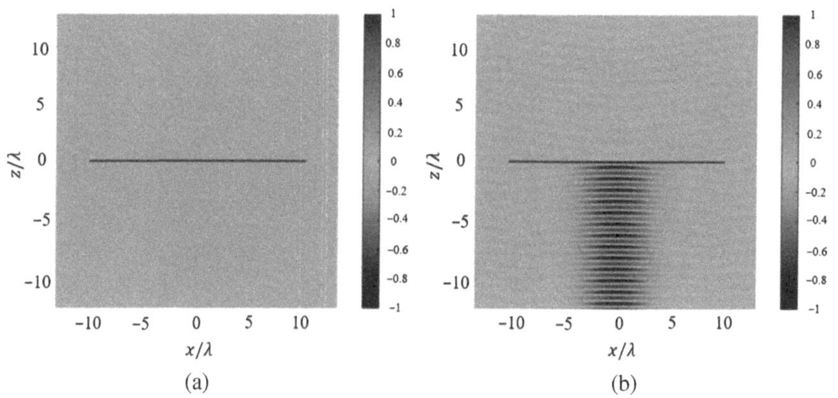

Figure 3.29 The results for the dual-polarization conversion metasurface with only the y-polarized wave incident. The real part of (a) x-component and (b) y-component of electric fields.

the x-polarized incident wave is completely converted into the y-polarized transmitted wave after passing through the metasurface, E_x only exists on the incident side ($z < 0$), while E_y only exists on the transmission side ($z > 0$). The conversion effect of the metasurface on the y-polarized incident wave is shown in Figure 3.29. Figure 3.29(b) illustrates that the y-polarized wave disappears after it is incident on the metasurface, while Figure 3.29(a) shows that there is no x-polarized wave present in the computational domain. Therefore, the y-polarized wave is not converted into an x-polarized wave but is instead absorbed by the metasurface.

The simulation results confirm that the metasurface successfully achieves the desired polarization conversion functionality. This example highlights the effectiveness of the full-vector SEM-GSTC in 2-D scenarios, broadening its applicability to a wider range of 2-D problems. Furthermore, by specifying eight polarization rate components, the metasurface can implement two distinct transformation functions.

3.3.3.2 Hybrid Modulation Metasurface

Here, the transmission and reflection coefficients are used to synthesize susceptibility tensors. The field distribution of the metasurface or scattering cell can be determined through measurable scattering parameters. As a result, the hybrid modulation metasurface constructed with this approach integrates both amplitude and phase modulation capabilities, providing promising opportunities for a wide range of applications. For a y-polarized wave propagating along the x-direction, the response of the metasurface is defined by the reflection coefficient R_y and transmission coefficient T_y, where $\mathbf{E}^r = R_y\mathbf{E}^i$ and $\mathbf{E}^t = T_y\mathbf{E}^i$. The relationship between the scattering parameters and the susceptibilities can be expressed as [27]

$$\chi_{ee}^{yy} = \frac{2j(T_y + R_y - 1)}{k_0(T_y + R_y + 1)} \tag{3.95.a}$$

$$\chi_{mm}^{zz} = \frac{2j(T_y - R_y - 1)}{k_0(T_y - R_y + 1)} \tag{3.95.b}$$

Consider a hybrid modulation metasurface that controls the amplitude and phase of both the reflected and transmitted fields [28]. The transmitted field amplitude is 0.3 times that of the incident field, with a phase delay of $\pi/3$; the reflected field amplitude is 0.7 times with a phase advance of $\pi/4$. That is, $R_y = 0.7e^{-j\frac{\pi}{4}}$ and $T_y = 0.3e^{j\frac{\pi}{3}}$. The dimension of the computational domain is $6\lambda \times 2\lambda \times 1\lambda$. The metasurface is located at $x = 0$ and parallel to the YoZ-plane. A 13.2 GHz y-polarized plane incident wave propagate along the positive x-direction. Analytically, the electric field on the incident side and the transmitted side are

$$\mathbf{E}^i + \mathbf{E}^r = \hat{y}e^{-jk_0(x+3\lambda)} + \hat{y}R_y e^{jk_0x} = \hat{y}\left(1 + 0.7 \times e^{j2k_0x - j\frac{\pi}{4}}\right)e^{-jk_0x} \tag{3.96.a}$$

$$\mathbf{E}^t = \hat{y}T_y e^{-jk_0x} = \hat{y}0.3e^{-jk_0x + j\frac{\pi}{3}} \tag{3.96.b}$$

Figure 3.30 presents a 3-D propagation model of the real part of E_y in the simulation result of hybrid modulation metasurface. It shows that the wave maintains the plane waveform throughout its propagation. After modulation by the metasurface, a clear difference in amplitude and a phase shift are evident, successfully achieving the intended functionalities.

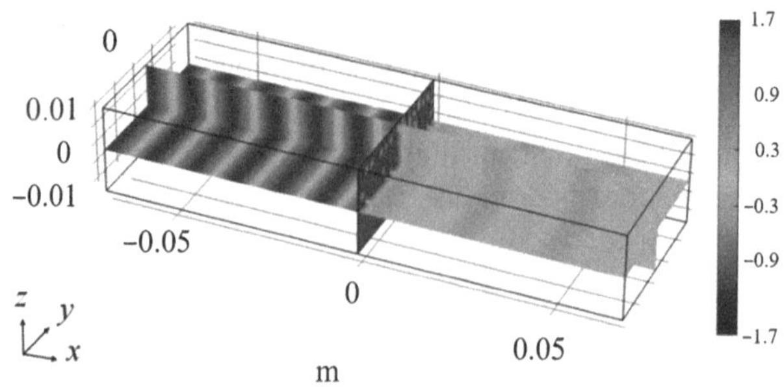

Figure 3.30 The 3-D propagation model of the real part of y-component of electric field through hybrid modulation metasurface calculated using the 3-D SEM-GSTC method.

To further analyze the effectiveness and accuracy of the SEM-GSTC simulation for the 3-D model, the field values along the x-axis are selected to plot. This distribution is then compared with the theoretical results derived from (3.96.a). The amplitude of E_y is shown in Figure 3.31(a). On the side of region $x < 0$, due to the superposition of the incident and reflected fields, the amplitude exhibits a sinusoidal fluctuation, with a frequency twice that of the incident field. The peak is approximately 1.7, which aligns with the calculations from the field superposition formula. On the side of region $x > 0$, the electric field amplitude stabilizes at 0.3 with no fluctuations. Figure 3.31(b) shows the phase distribution of E_y. It is visually evident that there is a phase jump of $\pi/3$ at $x = 0$. These characteristics match the predefined functionalities. Additionally, the comparison with the analytical solution further confirms this conclusion, with a relative error of only 0.041% between the 3-D SEM-GSTC and the analytical value, perfectly meeting the requirements of engineering computations.

3.4 IE Method with GSTCs

3.4.1 Introduction of IE Method

The original concepts of the IE method were developed in the early 20th century and applied to electromagnetic problems in the 1960s. In 1968, R. F. Harrington first systematically explained the IE method [29]. Afterward, the IE method underwent decades of development and gradually matured. In 1990, Rokhlin proposed the fast multipole method (FMM) [30], which tremendously improves the computational efficiency of the IE method. Nowadays, the IE method has been unquestionably regarded as one of the most widespread numerical methods in electromagnetic simulation.

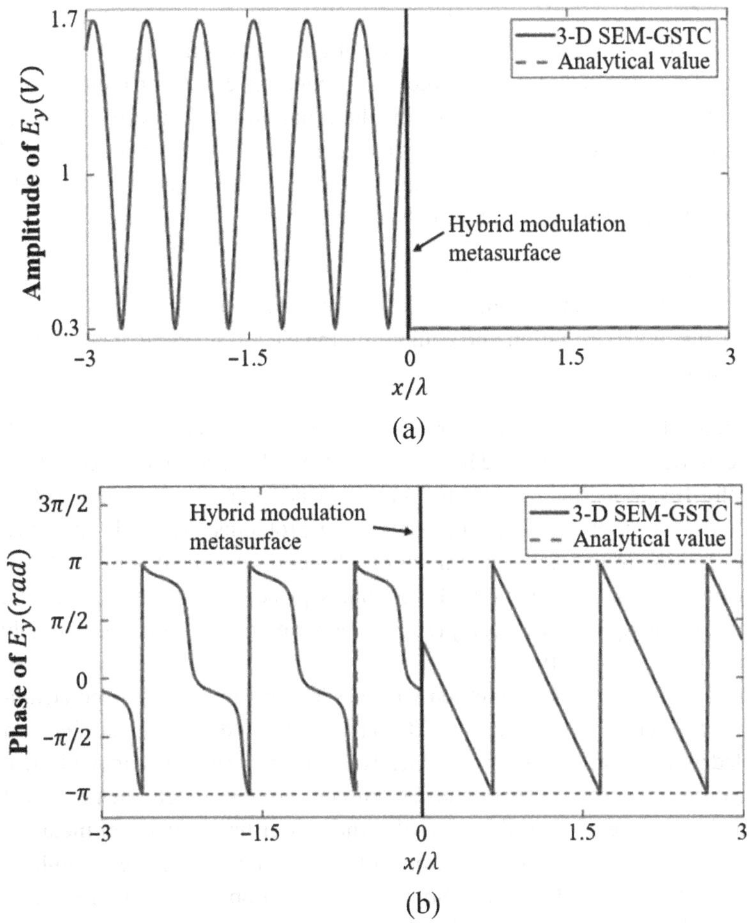

Figure 3.31 Comparison of 3-D SEM-GSTC and analytical value of hybrid modulation metasurface. (a) Amplitude of E_y. (b) Phase of E_y.

Generally, the prerequisite of the IE method is combining the boundary conditions of the electromagnetic interface with the free space radiation equations to obtain the IEs of the specific problems. Subsequently, the IEs should be discretized, including discretizing the equivalent currents with basis functions, testing the IEs with a group of trial functions, and assembling the matrix equation. It's noteworthy that the mesh generation is performed only on the interfaces between different materials where the equivalent currents exist. When computing moments of the coefficient matrix through numerical integration, the singularity needs to be specially processed. The scheme to deal with singularity and the selection of basis and testing functions in discretization have a direct effect on the accuracy and complexity of the IE method [29]. Besides, solving a matrix with large DoFs is also a difficulty. Denoting the DoF as N, the requirement of computer memory and operations is $O(N^2)$ and $O(N^3)$ when solving the equations with the direct method. This is unbearable when N becomes very large. In this situation, more advanced strategy needs to be imposed on the IE method to solve the equations effectively. In summary, the procedure of IE methods involves

1) Establishing the electromagnetic IEs,
2) Discretizing the IEs and assembling the matrix equation, and
3) Solving the matrix equation and computing the electromagnetic fields within the simulation domain.

Compared with the numerical electromagnetic simulation method based on differential equations, the IE method has two advantages. First, as mentioned before, because the IEs inherently satisfy the radiation boundary conditions (RBCs), the IE method is more suitable for open-space problems, which is the most common situation in metasurface simulations. Second, combined with the BEM, mesh partitioning of the IE method only takes place on the electromagnetic boundaries, reducing the space complexity by one dimension, which remarkably spares the time and memory cost.

The incorporation of the IE method with GSTC further enhances the advantages of the IE method. By applying GSTC to the IE method, the metasurface can be modeled as a zero-thickness sheet which is no different from a general electromagnetic interface except its special surface susceptibilities model. The process of the IE method can be easily utilized to solve the scattering problems of metasurfaces just with a tiny adjustment. Therefore, lots of work has been done to combine GSTC with the IE method. In 2019, Smy's team first proposed a scattering field solution based on BEM-GSTC for 2-D interconnected regions [31]. The IE-GSTC method was then applied to develop a general design procedure to synthesize a metasurface hologram in 2020 by Smy's team [32]. Also in 2020, Caloz's team utilized the IE-GSTC method to design the metasurface cloak [33]. Afterward, Smy's team improved the IE-GSTC method by involving normal polarizabilities in the IE-GSTC solver in 2022 [34]. In 2023, Smy's team further extended the GSTCs model into spatially dispersive metasurfaces and incorporated it with the standard IE method [35–37]. In summary, the IE-GSTC method has proved its great advantages in solving scattering problems involved with metasurfaces.

3.4.1.1 General Scattering Equations on the Electromagnetic Interface

The IEs of electromagnetic fields are established by a set of equivalent current sources, i.e., electric current $\mathbf{J}$ and magnetic current $\mathbf{K}$, as shown in Figure 3.32.

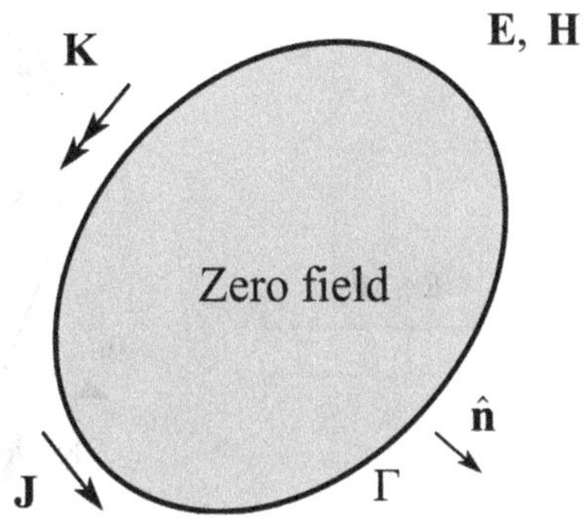

Figure 3.32 Illustration of the equivalent principle.

Γ is a hypothetical closed surface. According to the equivalent principle, the fields outside Γ can be regarded as the scattering fields excited by the set of currents on the boundary. This set of currents is given by

$$\mathbf{K} = \mathbf{E} \times \hat{\mathbf{n}} \tag{3.97}$$

$$\mathbf{J} = \hat{\mathbf{n}} \times \mathbf{H} \tag{3.98}$$

where $\hat{\mathbf{n}}$ is the unit normal outward vector. Solving Maxwell's equations in free space, the scattering field motivated by currents in free space can be represented by the equivalent currents as

$$\mathbf{E}^s = Z\mathcal{L}(\mathbf{J}) - \mathcal{K}(\mathbf{K}) \tag{3.99}$$

$$\mathbf{H}^s = \frac{1}{Z}\mathcal{L}(\mathbf{K}) + \mathcal{K}(\mathbf{J}) \tag{3.100}$$

where $Z = \sqrt{\mu/\varepsilon}$ is the wave impedance, and the linear operators $\mathcal{L}$ and $\mathcal{K}$ are given by

$$\mathcal{L}(\mathbf{X}) = -\mathrm{j}k \int \left[\mathbf{X} + \frac{1}{k^2}\nabla_{r_p}(\nabla_{r_s} \cdot \mathbf{X})\right] G d\tau_s \tag{3.101}$$

$$\mathcal{K}(\mathbf{X}) = -\int \mathbf{X} \times \nabla_{r_p} G d\tau_s \tag{3.102}$$

where τ represents the region of definition of testing functions, and subscripts p and s indicate the field quantity and the source quantity, respectively. G stands for the Green's function. It's worth noting that, in (3.101) and (3.102), ∇_{r_p} denotes the spatial differential operator operating on the field point, and ∇_{r_s} denotes operating on the source point. Consider a general interface dividing the computational domain into two regions, as shown in Figure 3.33. Denoted by Ω_1 and Ω_2, both regions are half-open and filled with homogeneous medium.

For general electromagnetic interfaces, both tangential electric and magnetic currents exist on the interface. That is, $\mathbf{J}_1$ and $\mathbf{K}_1$ on side 1, and $\mathbf{J}_2$ and $\mathbf{K}_2$ on side 2.

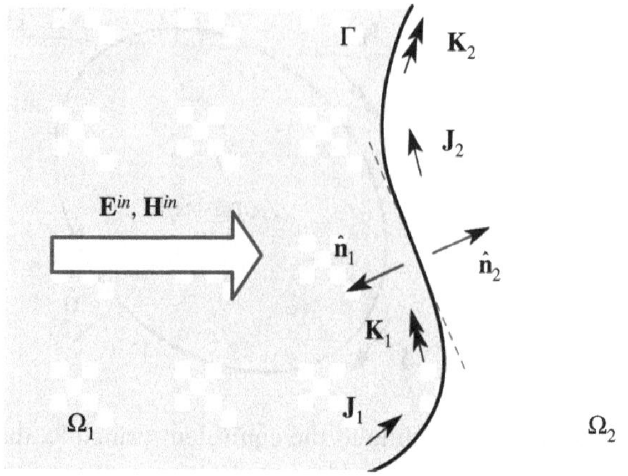

Figure 3.33 Electromagnetic interface in the free space.

Assume that the incident fields are induced in Ω_1, the electromagnetic fields on side 1 are given by

$$\mathbf{E}_1 = \mathbf{E}^{in} + \mathbf{E}_1^s = \mathbf{E}^{in} + Z_1\mathcal{L}(\mathbf{J}_1) - \mathcal{K}(\mathbf{K}_1) \tag{3.103}$$

$$\mathbf{H}_1 = \mathbf{H}^{in} + \mathbf{H}_1^s = \mathbf{H}^{in} + \frac{1}{Z_1}\mathcal{L}(\mathbf{K}_1) + \mathcal{K}(\mathbf{J}_1) \tag{3.104}$$

In Ω_2, only scattering electromagnetic fields exist which writes

$$\mathbf{E}_2 = \mathbf{E}_2^s = Z_2\mathcal{L}(\mathbf{J}_2) - \mathcal{K}(\mathbf{K}_2) \tag{3.105}$$

$$\mathbf{H}_2 = \mathbf{H}_2^s = \frac{1}{Z_2}\mathcal{L}(\mathbf{K}_2) + \mathcal{K}(\mathbf{J}_2) \tag{3.106}$$

According to (3.97) and (3.98), the tangential fields on the two sides of the interface satisfy

$$\mathbf{K}_1 = \mathbf{E}_1 \times \hat{\mathbf{n}}_1 \tag{3.107.a}$$

$$\mathbf{J}_1 = \hat{\mathbf{n}}_1 \times \mathbf{H}_1 \tag{3.107.b}$$

$$\mathbf{K}_2 = \mathbf{E}_2 \times \hat{\mathbf{n}}_2 \tag{3.107.c}$$

$$\mathbf{J}_2 = \hat{\mathbf{n}}_2 \times \mathbf{H}_2 \tag{3.107.d}$$

Substituting (3.107.a) into (3.103)–(3.106) leads to

$$-\mathbf{K}_1 - \hat{\mathbf{n}}_1 \times [Z_1\mathcal{L}_1(\mathbf{J}_1) - \mathcal{K}_1(\mathbf{K}_1)] = \hat{\mathbf{n}}_1 \times \mathbf{E}^{in} \tag{3.108}$$

$$J_1 - \hat{\mathbf{n}}_1 \times \left[\frac{1}{Z_1}\mathcal{L}_1(\mathbf{K}_1) + \mathcal{K}_1(\mathbf{J}_1)\right] = \hat{\mathbf{n}}_1 \times \mathbf{H}^{in} \tag{3.109}$$

$$\mathbf{K}_2 + \hat{\mathbf{n}}_2 \times [Z_2\mathcal{L}_2(\mathbf{J}_2) - \mathcal{K}_2(\mathbf{K}_2)] = 0 \tag{3.110}$$

$$-\mathbf{J}_2 + \hat{\mathbf{n}}_2 \times \left[\frac{1}{Z_2}\mathcal{L}_2(\mathbf{K}_2) + \mathcal{K}_2(\mathbf{J}_2)\right] = 0 \tag{3.111}$$

Equations (3.108) and (3.109) on side 1 are equivalent as well as (3.110) and (3.111) on side 2. Thus, the linear combination of them can assemble the scattering equations of the general electromagnetic interface. The IEs for the two sides are

$$\mathbf{K}_1 + \mathbf{K}_2 - \hat{\mathbf{n}} \times [Z_1\mathcal{L}_1(\mathbf{J}_1) - \mathcal{K}_1(\mathbf{K}_1) - Z_2\mathcal{L}_2(\mathbf{J}_2) + \mathcal{K}_2(\mathbf{K}_2)] = \hat{\mathbf{n}} \times \mathbf{E}^{in} \tag{3.112.a}$$

$$-\mathbf{J}_1 - \mathbf{J}_2 - \hat{\mathbf{n}} \times \left[\frac{1}{Z_1}\mathcal{L}_1(\mathbf{K}_1) + \mathcal{K}_1(\mathbf{J}_1) - \frac{1}{Z_2}\mathcal{L}_2(\mathbf{K}_2) - \mathcal{K}_2(\mathbf{J}_2)\right] = \hat{\mathbf{n}} \times \mathbf{H}^{in} \tag{3.112.b}$$

where $\hat{\mathbf{n}} = \hat{\mathbf{n}}_1 = -\hat{\mathbf{n}}_2$. The system of equations (3.112.a) is the famous PMCHW equation. Previous researches show its system matrix has better condition number, thus bearing better numerical characteristics [38].

3.4.1.2 Discretization of IEs

All operations on $\mathbf{J}_1$, $\mathbf{K}_1$, $\mathbf{J}_2$, and $\mathbf{K}_2$ in the Eq. (3.112.a) are linear. Thus, it can be rewritten with a linear operator $\mathcal{L}$

$$\mathcal{L}\mathbf{s} = \mathbf{b} \tag{3.113}$$

where $\mathbf{s}$ refers to the unknown equivalent surface currents. Express the unknown currents as the linear combination of a group of basis functions $\{\mathbf{g}_1, \mathbf{g}_2, \cdots\}$

$$\mathbf{s} = \sum_{i=1}^{N} s_i \mathbf{g}_i \tag{3.114}$$

where s_i is the unknown expansion coefficient. In fact, s_i corresponds to the value on the sampling nodes of each meshing grid. The basis functions $\{\mathbf{g}_1, \mathbf{g}_2, \cdots\}$ are specifically designed. Once the discrete interpolations $\{s_1, s_2, \cdots\}$ are solved, the value of $\mathbf{S}$ on the interface can be determined with basis functions.

Substituting (3.114) into (3.113) and testing both sides of (3.113) with testing functions $\{\mathbf{h}_1, \mathbf{h}_2, \cdots\}$ lead to

$$\sum_{j=0}^{N} \left[s_j \int (\mathbf{h}_i \cdot \mathcal{L}\mathbf{g}_j) d\tau\right] = \int \mathbf{h}_i \cdot \mathbf{b} d\tau \tag{3.115}$$

This group of equations can be expressed in the matrix form as

$$\mathbf{Z}\mathbf{x} = \mathbf{q} \tag{3.116}$$

The elements of matrices in (3.116) are

$$[Z]_{i,j} = \int (\mathbf{h}_i \cdot \mathcal{L}\mathbf{g}_j) d\tau \tag{3.117}$$

The elements of column vectors $\mathbf{x}$ and $\mathbf{q}$ are

$$x_i = s_i \tag{3.118.a}$$

$$q_i = \int \mathbf{h}_i \cdot \mathbf{b} d\tau \tag{3.118.b}$$

The vector of unknowns $\mathbf{x}$ can be solved by

$$\mathbf{x} = \mathbf{Z}^{-1}\mathbf{q} \tag{3.119}$$

3.4.1.3 Solving the Matrix Equations

As introduced in Section 3.1.1.2, traditional methods to solve the matrix equations contain direct methods and iterative methods. In the IE method, the matrix is dense, unlike the matrices in FDFD and FEM. Many acceleration techniques for the IE method have been developed since the 1980s, such as the adaptive cross approximation (ACA), conjugate gradient-fast Fourier transformation (CG-FFT), and adaptive integral method (AIM). Among them, the ACA has wide adaptability and simple implementation, but the acceleration effect is average. The CG-FFT and AIM have an optimal computational complexity of $O(N \log N)$ but require a certain arrangement of basis functions to achieve optimal acceleration.

Compared with the mentioned methods above, the FMM is one of the most commonly used and efficient acceleration techniques. The calculation of the matrix-vector multiplication in the iterative method is considered to be the interaction of equivalent currents. Separate the currents into several subgroups. The interactions between the current groups are in two manners far interactions and near interactions. The far interactions can be decomposed into three steps: (1) aggregate the currents corresponding to the unknowns in the group into the center, (2) transfer the functions to the center of the other group, and (3) distribute the functions to the currents associated with the unknowns in the new group. Mathematically, the process is performed by the multipole expansion of the Green's function. Since the interactions are processed by groups, the FMM requires fewer operations to perform the matrix-vector multiplication. The acceleration effect is obvious. The minimum number of operations that the FMM requires is $O(N^{3/2})$ [39].

3.4.2 Strategies for Integrating GSTCs in IE

3.4.2.1 Scattering Equations of Metasurfaces Based on GSTCs

GSTCs give the transition conditions of electromagnetic fields on the two sides of the metasurface as

$$\Delta\mathbf{E} = \hat{\mathbf{n}} \times j\omega(\mu\overline{\overline{\chi}}_{mm}\overline{\mathbf{H}} + \sqrt{\mu\varepsilon}\overline{\overline{\chi}}_{me}\overline{\mathbf{E}}) - \nabla_{\parallel}\left[\left(\varepsilon\overline{\overline{\chi}}_{ee}\overline{\mathbf{E}} + \sqrt{\frac{\mu}{\varepsilon}}\overline{\overline{\chi}}_{em}\overline{\mathbf{H}}\right)_n\right] \tag{3.120.a}$$

$$\Delta\mathbf{H} = -\hat{\mathbf{n}} \times j\omega(\varepsilon\overline{\overline{\chi}}_{ee}\overline{\mathbf{E}} + \sqrt{\mu\varepsilon}\overline{\overline{\chi}}_{em}\overline{\mathbf{H}}) - \nabla_{\parallel}\left[\left(\overline{\overline{\chi}}_{mm}\overline{\mathbf{H}} + \sqrt{\frac{\varepsilon}{\mu}}\overline{\overline{\chi}}_{me}\overline{\mathbf{E}}\right)_n\right] \tag{3.120.b}$$

where $\Delta\boldsymbol{\Psi} = \boldsymbol{\Psi}_2 - \boldsymbol{\Psi}_1$ and $\overline{\boldsymbol{\Psi}} = \frac{1}{2}(\boldsymbol{\Psi}_1 + \boldsymbol{\Psi}_2)$, $\boldsymbol{\Psi}$ referring to **E** or **H**. The second term on the right-hand side of (3.120.a) is related to the normal electric and magnetic polarization density P_n and M_n. In many instances, especially under the condition of normal incidence or small incident angles, the contribution of the normal electromagnetic polarization densities can be ignored. Consequently, (3.120.a) can be further simplified to

$$\Delta\mathbf{E} = \hat{\mathbf{n}} \times j\omega(\mu\overline{\overline{\chi}}_{mm}\mathbf{H} + \sqrt{\mu\varepsilon}\overline{\overline{\chi}}_{me}\mathbf{E}) \tag{3.121.a}$$

$$\Delta\mathbf{H} = -\hat{\mathbf{n}} \times j\omega(\varepsilon\overline{\overline{\chi}}_{ee}\mathbf{E} + \sqrt{\mu\varepsilon}\overline{\overline{\chi}}_{em}\mathbf{H}) \tag{3.121.b}$$

In order to obtain the scattering equations for the metasurface, the GSTCs are imposed on the electromagnetic interface integral scattering equations. Assume a set of equivalent electromagnetic currents **J**, **K** lie on both sides of the metasurface. According to the discussion in Section 3.4.1.1, they are related to the tangential electromagnetic fields on both sides of the metasurface by (3.107.a). Taking the cross-product with $\hat{\mathbf{n}}$ on both sides of (3.121.a), and considering (3.107.a), the GSTCs can be rewritten as

$$\mathbf{K}_1 + \mathbf{K}_2 + \hat{\mathbf{n}} \times j\omega\left(-\mu\overline{\overline{\chi}}_{mm}\frac{\mathbf{J}_1 - \mathbf{J}_2}{2} + \sqrt{\mu\varepsilon}\overline{\overline{\chi}}_{me}\frac{\mathbf{K}_1 - \mathbf{K}_2}{2}\right) = 0 \tag{3.122.a}$$

$$\mathbf{J}_1 + \mathbf{J}_2 + \hat{\mathbf{n}} \times j\omega\left(\varepsilon\overline{\overline{\chi}}_{ee}\frac{\mathbf{K}_1 - \mathbf{K}_2}{2} - \sqrt{\mu\varepsilon}\overline{\overline{\chi}}_{em}\frac{\mathbf{J}_1 - \mathbf{J}_2}{2}\right) = 0 \tag{3.122.b}$$

Combine (3.122.a) with (3.112.a) to get the integral scattering equations based on the surface susceptibilities tensors. The complete system of equations are

$$\mathbf{K}_1 + \mathbf{K}_2 - \hat{\mathbf{n}} \times [\eta_1\mathcal{L}_1(\mathbf{J}_1) - \mathcal{K}_1(\mathbf{K}_1) - \eta_2\mathcal{L}_2(\mathbf{J}_2) + \mathcal{K}_2(\mathbf{K}_2)] = \hat{\mathbf{n}} \times \mathbf{E}^{in} \tag{3.123.a}$$

$$-\mathbf{J}_1 - \mathbf{J}_2 - \hat{\mathbf{n}} \times \left[\frac{1}{\eta_1}\mathcal{L}_1(\mathbf{K}_1) + \mathcal{K}_1(\mathbf{J}_1) - \frac{1}{\eta_2}\mathcal{L}_2(\mathbf{K}_2) - \mathcal{K}_2(\mathbf{J}_2)\right] = \hat{\mathbf{n}} \times \mathbf{H}^{in} \tag{3.123.b}$$

$$\mathbf{K}_1 + \mathbf{K}_2 + \hat{\mathbf{n}} \times j\omega\left(-\mu\overline{\overline{\chi}}_{mm}\frac{\mathbf{J}_1 - \mathbf{J}_2}{2} + \sqrt{\mu\varepsilon}\overline{\overline{\chi}}_{me}\frac{\mathbf{K}_1 - \mathbf{K}_2}{2}\right) = 0 \tag{3.123.c}$$

$$\mathbf{J}_1 + \mathbf{J}_2 + \hat{\mathbf{n}} \times j\omega\left(\varepsilon\overline{\overline{\chi}}_{ee}\frac{\mathbf{K}_1 - \mathbf{K}_2}{2} - \sqrt{\mu\varepsilon}\overline{\overline{\chi}}_{em}\frac{\mathbf{J}_1 - \mathbf{J}_2}{2}\right) = 0 \tag{3.123.d}$$

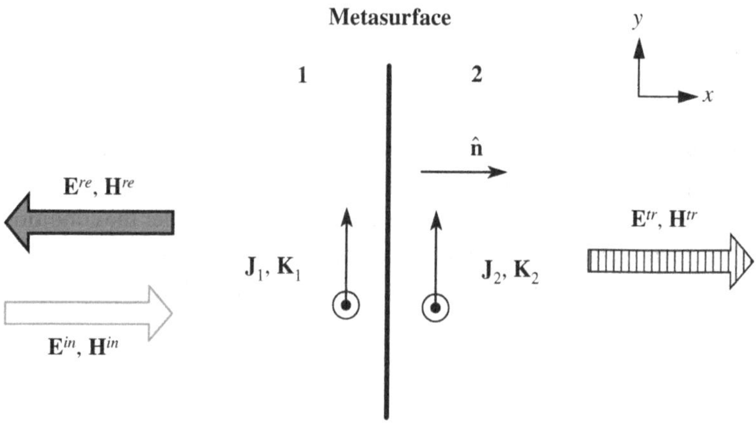

Figure 3.34 Equivalent currents on the metasurface.

3.4.2.2 Discrete Scalar Equations in 2-D IE-GSTC

Without loss of generality, consider the 2-D electromagnetic problems in the *XoY*-plane, an infinite metasurface is placed at plane $x=0$ parallel to the *YoZ*-plane. The electromagnetic fields are independent of z, that is, $\frac{\partial}{\partial z} = 0$. Assume that the metasurface is isotropic, so the susceptibilities are scalar, that is, $\overline{\overline{\chi}} = \chi$. The computational domain filled with homogeneous medium is divided into two open regions by the metasurface, as shown in Figure 3.34. Note that the metasurface here denotes the generalized sheet with equivalent surface susceptibilities, not the actual metasurface with physical electrically small structures.

It is obvious that the equivalent currents on the metasurface are also z-independent. Generally, the electromagnetic currents only distribute along the metasurface. Therefore, the electromagnetic currents only exist along the y-axis, that is,

$$\mathbf{J} = \mathbf{J}(y) \tag{3.124}$$

$$\mathbf{K} = \mathbf{K}(y) \tag{3.125}$$

Furthermore, currents without the normal component (x-component) exist since the metasurface is modeled as a zero-thickness sheet. So, the electromagnetic currents can be decomposed into

$$\mathbf{J} = \mathbf{J}_y(y) + \mathbf{J}_z(y) \tag{3.126}$$

$$\mathbf{K} = \mathbf{K}_y(y) + \mathbf{K}_z(y) \tag{3.127}$$

Hence, the divergence of **J** and **K** can be written as

$$\nabla \cdot \mathbf{J} = \frac{\partial J_y}{\partial y} \tag{3.128}$$

$$\nabla \cdot \mathbf{K} = \frac{\partial K_y}{\partial y} \tag{3.129}$$

Therefore, operators $\mathcal{L}$ and $\mathcal{K}$ in the vector equations (3.123.a)–(3.123.d) can be expanded as

$$\begin{aligned}
\mathcal{L}(\mathbf{X}) &= -jk\int \left[\mathbf{X} + \frac{1}{k^2}\nabla_{r_p}(\nabla_{r_s}\cdot\mathbf{X})G\right] d\tau_s \\
&= -jk\int \left[\mathbf{X}G + \frac{1}{k^2}\frac{\partial X_y}{\partial y_s}\left(\frac{\partial G}{\partial x_p}\cdot\hat{\mathbf{i}}_x + \frac{\partial G}{\partial y_p}\cdot\hat{\mathbf{i}}_y\right)\right] d\tau_s \\
&= -\frac{j}{k}\int \frac{\partial X_y}{\partial y_s}\frac{\partial G}{\partial x_p} d\tau_s \cdot \hat{\mathbf{i}}_x - jk\int\left(X_y G + \frac{1}{k^2}\frac{\partial X_y}{\partial y_s}\frac{\partial G}{\partial y_p}\right) d\tau_s \cdot \hat{\mathbf{i}}_y \\
&\quad - jk\int X_z G d\tau_s \cdot \hat{\mathbf{i}}_z
\end{aligned} \tag{3.130}$$

$$\begin{aligned}
\mathcal{K}(\mathbf{X}) &= -\int \mathbf{X}\times\nabla_{r_p} G d\tau_s \\
&= -\int \mathbf{X}\times\left(\frac{\partial G}{\partial x_p}\cdot\hat{\mathbf{i}}_x + \frac{\partial G}{\partial y_p}\cdot\hat{\mathbf{i}}_y\right) d\tau_s \\
&= \int X_z\frac{\partial G}{\partial y_p} d\tau_s\cdot\hat{\mathbf{i}}_x - \int X_z\frac{\partial G}{\partial x_p} d\tau_s\cdot\hat{\mathbf{i}}_y + \int X_y\frac{\partial G}{\partial x_p} d\tau_s\cdot\hat{\mathbf{i}}_z
\end{aligned} \tag{3.131}$$

where **X** refers to the currents **J** and **K**. Based on the discussion above, the vector equations (3.123.a)–(3.123.d) can be decomposed into the following scalar equations

$$-Z^{JE}_{yy,1}J_{y,1} - \left(Z^{KE}_{yz,1} - 1\right)K_{z,1} + Z^{JE}_{yy,2}J_{y,2} - \left(Z^{KE}_{yz,2} + 1\right)K_{z,2} = E^{in}_y \tag{3.132.a}$$

$$-Z^{JE}_{zz,1}J_{z,1} - \left(Z^{KE}_{zy,1} + 1\right)K_{y,1} + Z^{JE}_{zz,2}J_{z,2} + \left(Z^{KE}_{zy,2} - 1\right)K_{y,2} = E^{in}_z \tag{3.132.b}$$

$$-\left(Z^{JH}_{yz,1} + 1\right)J_{z,1} - Z^{KH}_{yy,1}K_{y,1} + \left(Z^{JH}_{yz,2} - 1\right)J_{z,2} + Z^{KH}_{yy,2}K_{y,2} = H^{in}_y \tag{3.132.c}$$

$$-\left(Z^{JH}_{zy,1} - 1\right)J_{y,1} - Z^{KH}_{zz,1}K_{z,1} + \left(Z^{JH}_{zy,2} + 1\right)J_{y,2} + Z^{KH}_{zz,2}K_{z,2} = H^{in}_z \tag{3.132.d}$$

$$-X_{mm}J_{y,1} + X_{me}K_{y,1} + K_{z,1} + X_{mm}J_{y,2} - X_{me}K_{y,2} + K_{z,2} = 0 \tag{3.132.e}$$

$$X_{mm}J_{z,1} + K_{y,1} - X_{me}K_{z,1} - X_{mm}J_{z,2} + K_{y,2} + X_{me}K_{z,2} = 0 \tag{3.132.f}$$

$$J_{y,1} + X_{em} J_{z,1} - X_{ee} K_{z,1} + J_{y,2} - X_{em} J_{z,2} + X_{ee} K_{z,2} = 0 \tag{3.132.g}$$

$$-X_{em} J_{y,1} + J_{z,1} + X_{ee} K_{y,1} + X_{em} J_{y,2} + J_{z,2} - X_{ee} K_{y,2} = 0 \tag{3.132.h}$$

where

$$Z^{JE}_{yy,1} = Z^{KH}_{yy,1} = -jZ_1 k_1 \int \left(G + \frac{1}{k_1{}^2} \frac{\partial G}{\partial y_p} \frac{\partial}{\partial y_s} \right) d\tau_s \tag{3.133.a}$$

$$Z^{JE}_{zz,1} = Z^{MH}_{zz,1} = -jZ_1 k_1 \int G d\tau_s \tag{3.133.b}$$

$$Z^{JH}_{zy,1} = -Z^{ME}_{zy,1} = -Z^{JH}_{yz,1} = Z^{ME}_{yz,1} = \int \frac{\partial G}{\partial x_p} d\tau_s \tag{3.133.c}$$

$$Z^{JE}_{yy,2} = Z^{MH}_{yy,2} = -jZ_2 k_2 \int \left(G + \frac{1}{k_2{}^2} \frac{\partial G}{\partial y_p} \frac{\partial}{\partial y_s} \right) d\tau_s \tag{3.133.d}$$

$$Z^{JE}_{zz,2} = Z^{MH}_{zz,2} = -jZ_2 k_2 \int G d\tau_s \tag{3.133.e}$$

$$Z^{JH}_{zy,2} = -Z^{ME}_{zy,2} = Z^{JH}_{yz,2} = -Z^{ME}_{yz,2} = -\int \frac{\partial G}{\partial x_p} d\tau_s \tag{3.133.f}$$

$$X_{ee} = \frac{1}{2} \mathrm{j}\omega\varepsilon\chi_{ee}, \quad X_{mm} = \frac{1}{2} \mathrm{j}\omega\mu\chi_{mm} \tag{3.133.g}$$

$$X_{me} = \frac{1}{2} \mathrm{j}\omega\sqrt{\mu\varepsilon}\chi_{me}, \quad X_{em} = \frac{1}{2} \mathrm{j}\omega\sqrt{\mu\varepsilon}\chi_{em} \tag{3.133.h}$$

The Eqs. (3.132.a)–(3.132.h) are the scalar IE system for 2-D IE-GSTC.

To solve Eq. (3.132.a) by numerical methods, the first step is to mesh the metasurface and choose the proper basis functions and testing functions.

The geometric model of the 2-D metasurface is a 1-D line, as presented in Figure 3.35. Dividing the metasurface into several segments and interpolating at the middle point of each segment, the electromagnetic currents can be expressed by a series of basis functions

$$J = \sum_{j=1}^{N} J_j g_j \tag{3.134}$$

$$K = \sum_{j=1}^{N} K_j g_j \tag{3.135}$$

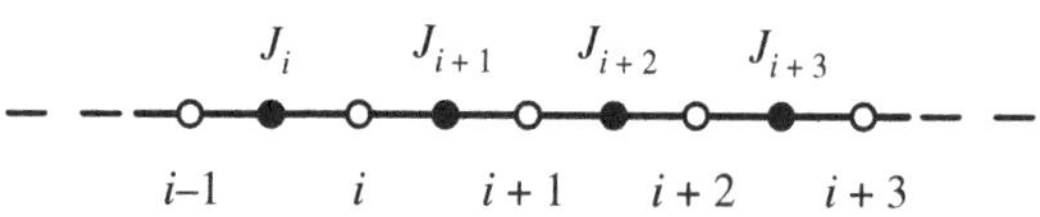

Figure 3.35 Interpolation of currents on the metasurface.

Figure 3.36 The first-order basis function.

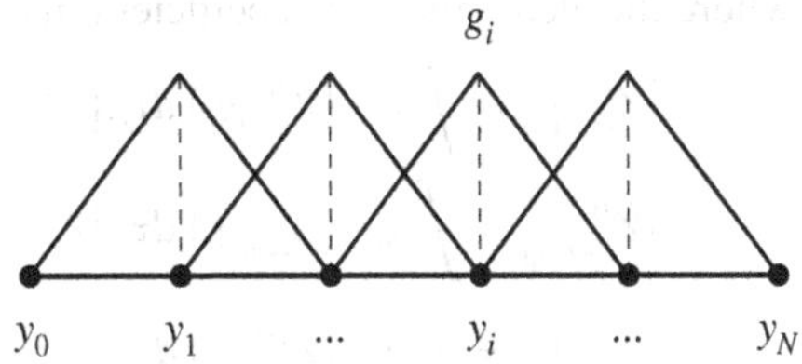

The unknown currents can be expanded by the first-order basis function. Figure 3.36 presents the amplitude of the basis function.

$$g_i(y) = \begin{cases} \frac{y-y_{i-1}}{y_i-y_{i-1}}, & y_{i-1} \leq y \leq y_i \\ \frac{y_{i+1}-y}{y_{i+1}-y_i}, & y_i \leq y \leq y_{i+1} \\ 0, & others \end{cases} \tag{3.136}$$

The choices of the testing functions are numerous with different implementation difficulty and computational performance. Generally, point-matching (testing with Dirac delta functions) is the easiest one to perform. On the other hand, the Galerkin method (testing with the same ones as the basis functions) is the more stable and accurate testing strategy. In this case, the Galerkin's testing strategy is used. Discretizing the scalar IEs (3.132.a)–(3.132.h) with the basis and testing functions leads to

$$\begin{bmatrix} \mathbf{Z}_{yy,1}^{JE} & \emptyset & \emptyset & \mathbf{Z}_{yz,1}^{KE} & \mathbf{Z}_{yy,2}^{JE} & \emptyset & \emptyset & \mathbf{Z}_{yz,2}^{KE} \\ \emptyset & \mathbf{Z}_{zz,1}^{JE} & \mathbf{Z}_{zy,1}^{KE} & \emptyset & \emptyset & \mathbf{Z}_{zz,2}^{JE} & \mathbf{Z}_{zy,2}^{KE} & \emptyset \\ \emptyset & -\left(\mathbf{Z}_{yz,1}^{JH} + \mathbf{I}\right) & -\mathbf{Z}_{yy,1}^{KH} & \emptyset & \emptyset & \mathbf{Z}_{yz,2}^{JH} - \mathbf{I} & \mathbf{Z}_{yy,2}^{KH} & \emptyset \\ -\left(\mathbf{Z}_{zy,1}^{JH} - \mathbf{I}\right) & \emptyset & \emptyset & -\mathbf{Z}_{zz,1}^{KH} & \mathbf{Z}_{zy,2}^{JH} + \mathbf{I} & \emptyset & \emptyset & \mathbf{Z}_{zz,2}^{KH} \\ \emptyset & \mathbf{X}_{mm} & \mathbf{I} & -\mathbf{X}_{me} & \emptyset & -\mathbf{X}_{mm} & \mathbf{I} & \mathbf{X}_{me} \\ -\mathbf{X}_{mm} & \emptyset & \mathbf{X}_{me} & \mathbf{I} & \mathbf{X}_{mm} & \emptyset & \mathbf{X}_{me} & \mathbf{I} \\ \mathbf{I} & \mathbf{X}_{em} & \emptyset & -\mathbf{X}_{ee} & \mathbf{I} & -\mathbf{X}_{em} & \emptyset & \mathbf{X}_{ee} \\ -\mathbf{X}_{em} & \mathbf{I} & \mathbf{X}_{ee} & \emptyset & \mathbf{X}_{em} & \mathbf{I} & -\mathbf{X}_{ee} & \emptyset \end{bmatrix}$$

$$\times \begin{bmatrix} \mathbf{J}_{y,1} \\ \mathbf{J}_{z,1} \\ \mathbf{K}_{y,1} \\ \mathbf{K}_{z,1} \\ \mathbf{J}_{y,2} \\ \mathbf{J}_{z,2} \\ \mathbf{K}_{y,2} \\ \mathbf{K}_{z,2} \end{bmatrix} = \begin{bmatrix} \mathbf{E}_y^{in} \\ \mathbf{E}_z^{in} \\ \mathbf{H}_y^{in} \\ \mathbf{H}_z^{in} \\ \emptyset \\ \emptyset \\ \emptyset \\ \emptyset \end{bmatrix}. \tag{3.137}$$

where $\emptyset$ denotes the null matrix. Write (3.137) in a concise matrix format as

$$\mathbf{Zx} = \mathbf{b} \tag{3.138}$$

where the elements of the coefficient matrix are

$$\left[Z^{JE}_{yy,q}\right]_{i,j} = \int g_i \cdot \left(Z^{JE}_{yy,q} g_j\right) d\tau_s, \left[Z^{MH}_{yy,q}\right]_{i,j} = \int g_i \cdot \left(Z^{MH}_{yy,q} g_j\right) d\tau_s$$

$$\left[Z^{JE}_{zz,q}\right]_{i,j} = \int g_i \cdot \left(Z^{JE}_{zz,q} g_j\right) d\tau_s, \left[Z^{MH}_{zz,q}\right]_{i,j} = \int g_i \cdot \left(Z^{MH}_{zz,q} g_j\right) d\tau_s$$

$$\left[Z^{JH}_{yz,q}\right]_{i,j} = \int g_i \cdot \left(Z^{JH}_{yz,q} g_j\right) d\tau_s, \left[Z^{ME}_{yz,q}\right]_{i,j} = \int g_i \cdot \left(Z^{ME}_{yz,q} g_j\right) d\tau_s$$

$$\left[Z^{JH}_{zy,q}\right]_{i,j} = \int g_i \cdot \left(Z^{JH}_{zy,q} g_j\right) d\tau_s, \left[Z^{ME}_{zy,q}\right]_{i,j} = \int g_i \cdot \left(Z^{ME}_{zy,q} g_j\right) d\tau_s, \quad q = 1, 2 \tag{3.139.a}$$

$$[\chi_{\alpha\beta}]_{i,j} = \int g_i \cdot (\chi_{\alpha\beta} g_j) d\tau_s, \quad \alpha, \beta = e, m \tag{3.139.b}$$

$$[I]_{i,j} = \int g_i \cdot g_j d\tau_s \tag{3.139.c}$$

The elements in the source vector $\mathbf{b}$ are

$$\mathbf{E}^{in}_y = \int g_i \cdot E^{in}_y d\tau_s \tag{3.140.a}$$

$$\mathbf{E}^{in}_z = \int g_i \cdot E^{in}_z d\tau_s \tag{3.140.b}$$

$$\mathbf{H}^{in}_y = \int g_i \cdot H^{in}_y d\tau_s \tag{3.140.c}$$

$$\mathbf{H}^{in}_z = \int g_i \cdot H^{in}_z d\tau_s \tag{3.140.d}$$

and the elements in the unknown vector $\mathbf{x}$ are

$$[J] = [J_1, J_2, \cdots]^T, \quad [K] = [K_1, K_2, \cdots]^T \tag{3.141}$$

The integral operation can be obtained by the numerical integration method. For the integration with singularity on the integral path, the integral operation can be decomposed into two parts. One is the integration of the area without singularity, and the other is the integration near the singularity. Here, the singularity is introduced mainly by the Green's function. The integration of the 2-D Green's function and its partial derivative on the segment with singularity is

$$\int G d\tau_s = \frac{1}{4\mathrm{j}} \left[1 - j\frac{2}{\pi} \ln\left(\frac{k_0 \gamma l}{4e}\right)\right] \tag{3.142}$$

$$\int \frac{\partial G}{\partial y} d\tau_s = 0, \quad \int \frac{\partial G}{\partial x} d\tau_s = \frac{1}{2} \tag{3.143}$$

Having assembled the coefficient matrices, the unknown currents can be solved by

$$\mathbf{x} = \mathbf{Z}^{-1}\mathbf{b} \tag{3.144}$$

3.4.2.3 IE-GSTC for Scattering from Curved Metasurfaces

For the curved metasurface, the incident electromagnetic wave is no longer normal to the metasurface. Instead, the incident angle may vary within a larger range. In this situation, the influence of the normal susceptibilities is non-negligible. In [40], a metasurface composed of isotropic PEC scatterers is analyzed, revealing the contribution of normal susceptibilities to the scattering fields

$$R(\theta) = \frac{-jk\left(\chi_{ee}^{\parallel} + \chi_{mm}^{nn}\sin^2\theta\right)}{2\cos\theta + jk\left(\chi_{ee}^{\parallel} + \chi_{mm}^{nn}\sin^2\theta\right)} \tag{3.145.a}$$

$$T(\theta) = \frac{2\cos\theta}{2\cos\theta + jk\left(\chi_{ee}^{\parallel} + \chi_{mm}^{nn}\sin^2\theta\right)} \tag{3.145.b}$$

Equations (3.145.a) and (3.145.b) show that the influence of normal susceptibilities is strengthened under a large incident angle. Thus, the scattering equations of the curved metasurface should be reconsidered. The transition conditions given by GSTCs with normal susceptibilities are presented in (3.120.a). According to free space radiation Eqs. (3.99) and (3.100), the average electromagnetic fields on both sides of the metasurface can be expressed as

$$\mathbf{E} = \frac{\mathbf{E}^{in} + \mathbf{E}_1^s + \mathbf{E}_2^s}{2} = \frac{1}{2}(\mathbf{E}^{in} + Z_1\mathcal{L}(\mathbf{J}_1) - \mathcal{K}(\mathbf{K}_1) + Z_2\mathcal{L}(\mathbf{J}_2) - \mathcal{K}(\mathbf{K}_2)) \tag{3.146.a}$$

$$\mathbf{H} = \frac{\mathbf{H}^{in} + \mathbf{H}_1^s + \mathbf{H}_2^s}{2} = \frac{1}{2}\left(\mathbf{H}^{in} + \frac{1}{Z_1}\mathcal{L}(\mathbf{K}_1) + \mathcal{K}(\mathbf{J}_1) + \frac{1}{Z_2}\mathcal{L}(\mathbf{K}_2) + \mathcal{K}(\mathbf{J}_2)\right) \tag{3.146.b}$$

Substituting (3.146.a) into (3.120.a) leads to

$$\mathbf{K}_1 + \mathbf{K}_2 + \hat{\mathbf{n}} \times j\omega\left(-\mu\overline{\overline{\chi}}_{mm}\frac{\mathbf{J}_1 - \mathbf{J}_2}{2} + \sqrt{\mu\varepsilon}\,\overline{\overline{\chi}}_{me}\frac{\mathbf{K}_1 - \mathbf{K}_2}{2}\right) + \mathbf{C}_1 = \mathbf{D}_1 \tag{3.147.a}$$

$$\mathbf{J}_1 + \mathbf{J}_2 + \hat{\mathbf{n}} \times j\omega\left(\varepsilon\overline{\overline{\chi}}_{ee}\frac{\mathbf{K}_1 - \mathbf{K}_2}{2} - \sqrt{\mu\varepsilon}\,\overline{\overline{\chi}}_{em}\frac{\mathbf{J}_1 - \mathbf{J}_2}{2}\right) + \mathbf{C}_2 = \mathbf{D}_2 \tag{3.147.b}$$

where

$$\mathbf{C}_1 = \frac{1}{2}\nabla_{\parallel}[\overline{\overline{\chi}}_{ee}(Z_1\mathcal{L}(\mathbf{J}_1) - \mathcal{K}(\mathbf{K}_1) + Z_2\mathcal{L}(\mathbf{J}_2) - \mathcal{K}(\mathbf{K}_2))]_n + \frac{1}{2}\nabla_{\parallel}\left[\sqrt{\frac{\mu}{\varepsilon}}\overline{\overline{\chi}}_{em}\left(\frac{1}{Z_1}\mathcal{L}(\mathbf{K}_1) + \mathcal{K}(\mathbf{J}_1) + \frac{1}{Z_2}\mathcal{L}(\mathbf{K}_2) + \mathcal{K}(\mathbf{J}_2)\right)\right]_n \tag{3.148.a}$$

$$\mathbf{D}_1 = -\frac{1}{2}\nabla_{\|}\left(\overline{\overline{\chi}}_{ee}\mathbf{E}^{in} + \sqrt{\frac{\mu}{\varepsilon}}\overline{\overline{\chi}}_{em}\mathbf{H}^{in}\right)_n \tag{3.148.b}$$

$$\begin{aligned}\mathbf{C}_2 = &\frac{1}{2}\nabla_{\|}\left[\overline{\overline{\chi}}_{mm}\left(\frac{1}{Z_1}\mathcal{L}(\mathbf{K}_1) + \mathcal{K}(\mathbf{J}_1) + \frac{1}{Z_2}\mathcal{L}(\mathbf{K}_2) + \mathcal{K}(\mathbf{J}_2)\right)\right]_n \\ &+ \frac{1}{2}\nabla_{\|}\left[\sqrt{\frac{\varepsilon}{\mu}}\overline{\overline{\chi}}_{me}(Z_1\mathcal{L}(\mathbf{J}_1) - \mathcal{K}(\mathbf{K}_1) + Z_2\mathcal{L}(\mathbf{J}_2) - \mathcal{K}(\mathbf{K}_2))\right]_n\end{aligned} \tag{3.148.c}$$

$$\mathbf{D}_2 = -\frac{1}{2}\nabla_{\|}\left(\overline{\overline{\chi}}_{mm}\mathbf{H}^{in} + \sqrt{\frac{\varepsilon}{\mu}}\overline{\overline{\chi}}_{me}\mathbf{E}^{in}\right)_n \tag{3.148.d}$$

Combining (3.147.a) with (3.112.a) leads to the IEs for scattering from curved metasurface.

$$\mathbf{K}_1 + \mathbf{K}_2 - \hat{\mathbf{n}} \times [Z_1\mathcal{L}_1(\mathbf{J}_1) - \mathcal{K}_1(\mathbf{K}_1) - Z_2\mathcal{L}_2(\mathbf{J}_2) + \mathcal{K}_2(\mathbf{K}_2)] = \hat{\mathbf{n}} \times \mathbf{E}^{in} \tag{3.149.a}$$

$$-\mathbf{J}_1 - \mathbf{J}_2 - \hat{\mathbf{n}} \times \left[\frac{1}{Z_1}\mathcal{L}_1(\mathbf{K}_1) + \mathcal{K}_1(\mathbf{J}_1) - \frac{1}{Z_2}\mathcal{L}_2(\mathbf{K}_2) - \mathcal{K}_2(\mathbf{J}_2)\right] = \hat{\mathbf{n}} \times \mathbf{H}^{in} \tag{3.149.b}$$

$$\mathbf{K}_1 + \mathbf{K}_2 + \hat{\mathbf{n}} \times j\omega\left(-\mu\overline{\overline{\chi}}_{mm}\frac{\mathbf{J}_1 - \mathbf{J}_2}{2} + \sqrt{\mu\varepsilon}\overline{\overline{\chi}}_{me}\frac{\mathbf{K}_1 - \mathbf{K}_2}{2}\right) + \mathbf{C}_1 = \mathbf{D}_1 \tag{3.149.c}$$

$$\mathbf{J}_1 + \mathbf{J}_2 + \hat{\mathbf{n}} \times j\omega\left(\varepsilon\overline{\overline{\chi}}_{ee}\frac{\mathbf{K}_1 - \mathbf{K}_2}{2} - \sqrt{\mu\varepsilon}\overline{\overline{\chi}}_{em}\frac{\mathbf{J}_1 - \mathbf{J}_2}{2}\right) + \mathbf{C}_2 = \mathbf{D}_2 \tag{3.149.d}$$

The unknown currents can be solved by performing discretization to obtain the matrix equations.

3.4.3 Simulation Examples

In the previous section, the integral scattering equation system for metasurfaces and its discretization are derived. The IE-GSTC solver is used to give several simulation examples to verify the proposed IE-GSTC. The first example proves the accuracy of the IE-GSTC with a simple instance. The second example demonstrates the powerful and flexible electromagnetic control capability of the metasurface. The third and fourth examples apply the IE-GSTC method to solve scattering fields of practical metasurface structures. The third example further verifies the accuracy of the IE-GSTC method through the simulation of a practical anisotropic metasurface. The fourth example of a *metal–insulator–metal* capacitor metasurface shows the effect of normal susceptibilities visually.

3.4.3.1 Phase-Amplitude Control

Considering an infinite metasurface in the z-direction and parallel to the YoZ-plane, the metasurface is monoisotropic, that is, $\overline{\overline{\chi}}_{ee} = \chi_{ee}$, $\overline{\overline{\chi}}_{mm} = \chi_{mm}$, and $\overline{\overline{\chi}}_{em} = \overline{\overline{\chi}}_{me} = 0$. A TE-polarized Gaussian beam is normally incident to the metasurface from the left region. After passing through the metasurface, the transmitted wave intensity attenuates to half. Besides, a 90° phase lag occurs to the transmitted wave. No reflection takes place. These properties of the metasurface can be represented by the reflection coefficient R and transmission coefficient T as

$$R = 0 \tag{3.150.a}$$

$$T = 0.5j \tag{3.150.b}$$

According to (2.97), there is

$$\chi_{ee} = \chi_{mm} = -0.0013 - j0.0010 \tag{3.151}$$

The dimensions of the square computation domain are 0.05 mm × 0.05 mm in XoZ-plane. The metasurface is placed at $x = 0$ with the length of 0.1 m. The waist width and the amplitude of the incident Gaussian beam are 0.01 m and 1 V/m. The frequency of the incident wave is 60 GHz. The discretization of the metasurface is set to 40 PPW. The simulation area is filled with air, that is, $\eta = \eta_0$. Figure 3.37 shows the simulation results of IE-GSTC. The electric field intensity of transmission wave attenuates obviously in Figure 3.37(a,b). The phase lag of the transmission wave can be observed clearly at $x = 0$ in Figure 3.37(b). Figure 3.38 goes a step further to prove the accuracy of numerical simulation by showing the electric field along the x-axis. There is a 90° phase lag at $x = 0$ and half attenuation

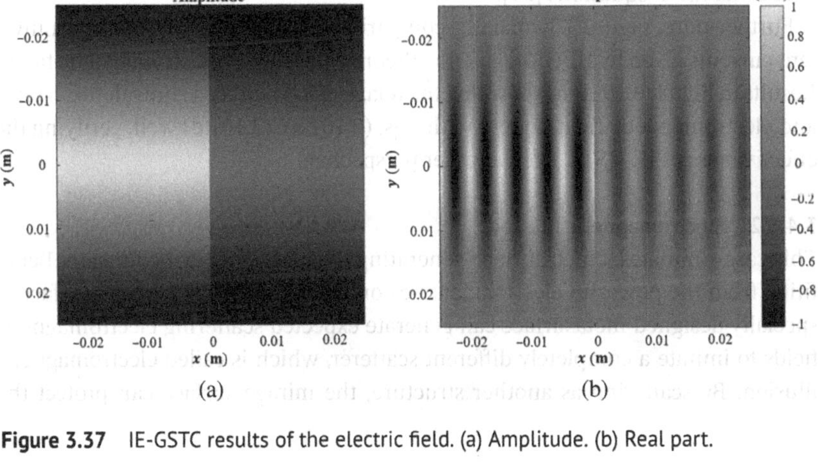

Figure 3.37 IE-GSTC results of the electric field. (a) Amplitude. (b) Real part.

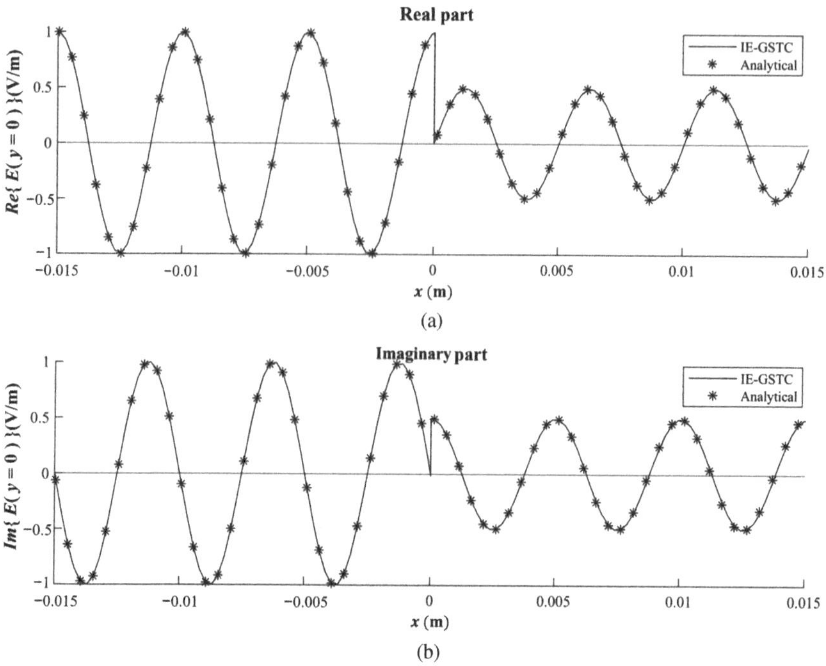

Figure 3.38 The electric field along the *x*-axis. (a) Real part. (b) Imaginary part.

of the wave intensity in the region of $x > 0$, both in Figure 3.38(a,b). The discussion above fully demonstrates that IE-GSTC simulates the metasurface precisely with results meeting expectations well. Compared with Figure 3.8, the results obtained by FDFD are caused by dissipation errors, which is one of the advantages of IEs over differential equations [31].

Furthermore, Figure 3.39 makes a comparison between the solutions of equivalent currents given by IE-GSTC and the theoretical value of electromagnetic fields. Results in Figure 3.39(a–h) show that the relationship between equivalent currents and electromagnetic fields agrees with Eqs. (3.107.a)–(3.107.d) well, verifying the effectiveness of IE-GSTC from another perspective.

3.4.3.2 Electromagnetic Illusion

This case simulates a metasurface generating the electromagnetic illusion. Benefiting from the powerful electromagnetic control capability of the metasurface, a specially designed metasurface can generate expected scattering electromagnetic fields to imitate a completely different scatterer, which is called electromagnetic illusion. By scattering as another structure, the mirage surface can protect the

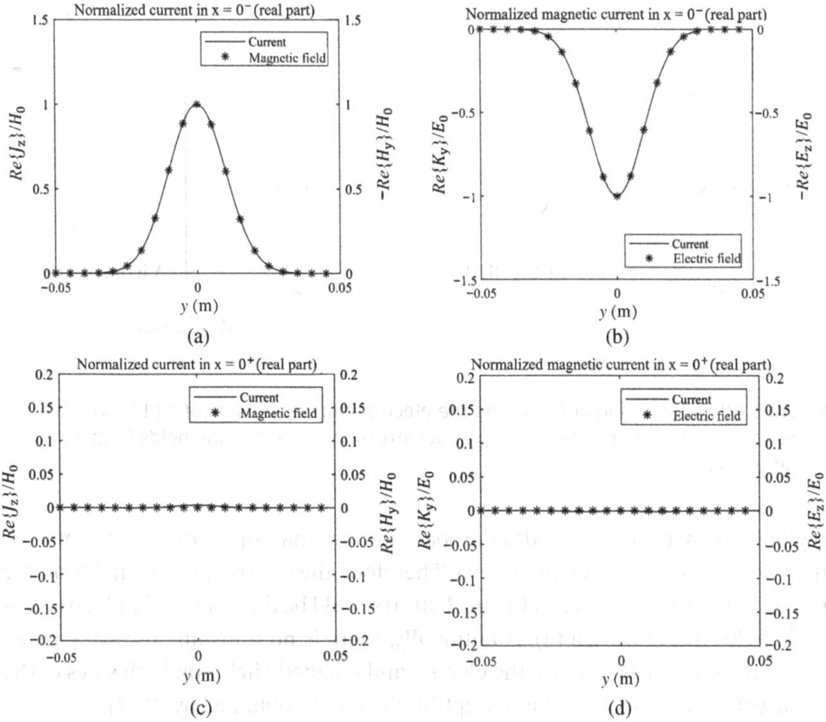

Figure 3.39 The comparison between the solutions of equivalent currents and tangential electromagnetic fields. (a)–(b) Normalized electric and magnetic currents and electromagnetic fields on the left side of the metasurface. (c)–(d) Normalized electric and magnetic currents and electromagnetic fields on the right side of the metasurface.

object covered by it from being detected by scattering as a common target which the observer is not interested in, or confuse the observer into misjudging that a target exists at a certain position, which actually is not.

Here, a metasurface faking the scattering fields from a 2-D PEC cylinder is simulated. As shown in Figure 3.40, a PEC cylinder leads to a specific distribution of electromagnetic fields in the observation area. Then the PEC cylinder is removed, and the metasurface generates similar scattering fields instead. As a result, the observer mistakenly believes a PEC cylinder is at the specified location.

To produce the expected electromagnetic illusion, the susceptibilities of the metasurface are synthesized by GSTCs. According to the uniqueness theorem, the electromagnetic fields in the observation area can be uniquely specified as long as the tangential electromagnetic fields are determined on the boundary of the

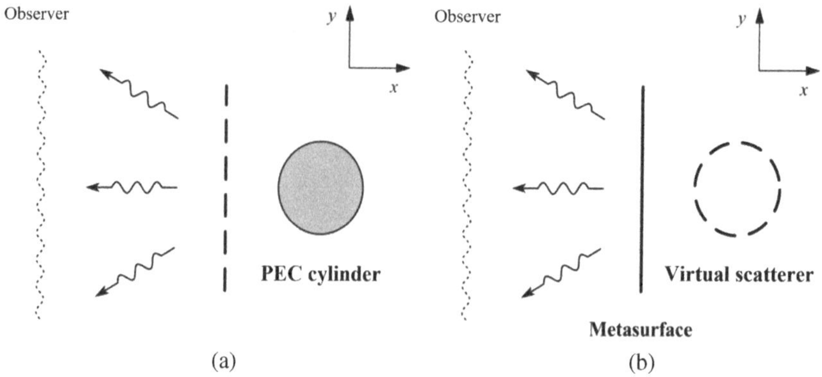

Figure 3.40 Illustration of producing the electromagnetic illusion of a PEC Cylinder. (a) Scattering fields from a true PEC cylinder structure. (b) Scattering fields from a metasurface.

region. The metasurface divides the space into two half-open areas and composes the boundary of the observation area. Therefore, the electromagnetic fields on the left side of the metasurface in Figure 3.40(b) should be the same as the electromagnetic fields in Figure 3.40(a). Additionally, there is no transmitted wave passing across the metasurface. Once the electric and magnetic fields on both sides of the metasurface are specified, the susceptibilities can be obtained by (2.97).

Assume that the metasurface is monoisotropic and the normal susceptibilities are zero for the sake of simplicity. The incident wave is a 10 GHz plane wave with $E_z = e^{-jk_0x}$ and $H_y = -e^{-jk_0x}/\eta_0$. The metasurface is placed at $x = 0$ with the length of $L = 0.6$ m. The center of the PEC cylinder is at $x = 0.09$ m with radius $r = 0.06$ m. The dimensions of the computation domain are 0.4 m × 0.4 m. The discretization of the metasurface and electromagnetic currents is 40 PPW. Calculating the electric and magnetic fields at $x = 0$, the synthesized susceptibilities of the metasurface are shown in Figure 3.41.

Applying IE-GSTC with the extracted susceptibilities, the simulation results are shown in Figure 3.42. Compared with Figure 3.42(a) with (b), the specially designed metasurface imitates the scattering of the PEC cylinder well in the observation region ($x < 0$). Figure 3.42(c) further shows the camouflage effect of the metasurface. The distribution of the electric field produced by the metasurface agrees with that of the PEC cylinder well.

This case of electromagnetic illusion fully illustrates the powerful electromagnetic control capability of metasurfaces. At the same time, it shows the effectiveness of susceptibility synthesis methods and demonstrates the promising potential of susceptibility synthesis methods in guiding the practical design of metasurfaces.

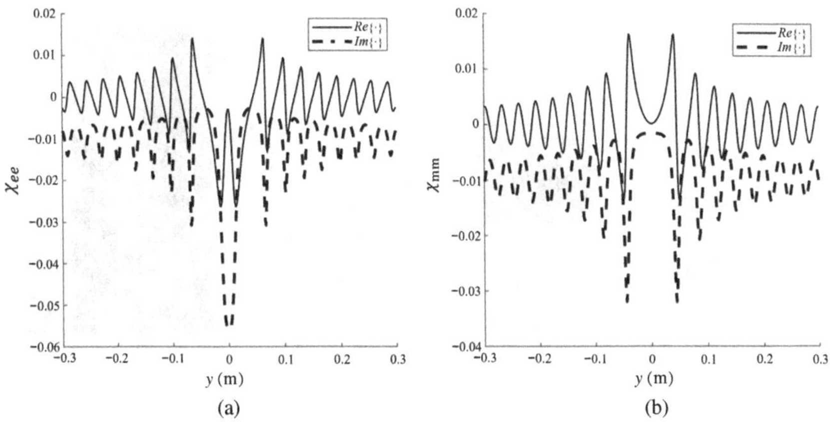

Figure 3.41 Extracted susceptibilities. (a) $\widetilde{\chi}_{ee}$ and (b) $\widetilde{\chi}_{mm}$.

3.4.3.3 A Reflective Metasurface Manipulating the Orthogonally Polarized Waves

In this case, the susceptibilities of the ten types of unit cells in an anisotropic planar metasurface presented in [41] are extracted. Then, the scattering of a metasurface composed of this type of unit cells is simulated to test the effectiveness of the IE-GSTC. This metasurface is capable of manipulating the reflection of orthogonal-polarized electromagnetic waves independently.

The unit cell of this metasurface is composed of an orthogonally I-shaped metal patch and a metal-grounded plane separated by an F4B dielectric substrate with the relative permittivity of $\varepsilon_r = 2.65$. The lengths of l_x and l_y in each cell varied periodically along the x-axis and y-axis, respectively, leading to the desired reflection phase of x- and y-polarized electromagnetic waves independently [41]. The metasurface and the unit cells are shown in Figure 3.43.

The metasurface is placed at the XoY-plane and illuminated by 10 GHz x- and y-polarized plane waves propagating along the z-axis, respectively. The metasurface is regarded to be infinitely periodic along the y-axis. Therefore, the calculation domain can be reduced to the 2-D XoZ-plane. The dimension of the calculation domain is $[-30, 30]\text{mm} \times [-10, 90]\text{mm}$.

Here, susceptibilities are extracted by each unit cell. Considering the TEM waves are incident to the metasurface vertically, the normal polarizabilities will make no contribution to the scattering fields when all electromagnetic fields are tangential. Therefore, all normal components of susceptibility tensors can be ignored. By symmetry and reciprocity assumptions, susceptibility tensors of the metasurface can be reduced to χ_{ee}^{xx}, χ_{ee}^{yy}, χ_{mm}^{xx}, and χ_{mm}^{yy}. Due to the anisotropy of the metasurface, the reflection and transmission coefficients, i.e., $\overline{\overline{R}} = \begin{bmatrix} R^{xx} & R^{xy} \\ R^{yx} & R^{yy} \end{bmatrix}$ and $\overline{\overline{S}}_{12} = \begin{bmatrix} T^{xx} & T^{xy} \\ T^{yx} & T^{yy} \end{bmatrix}$

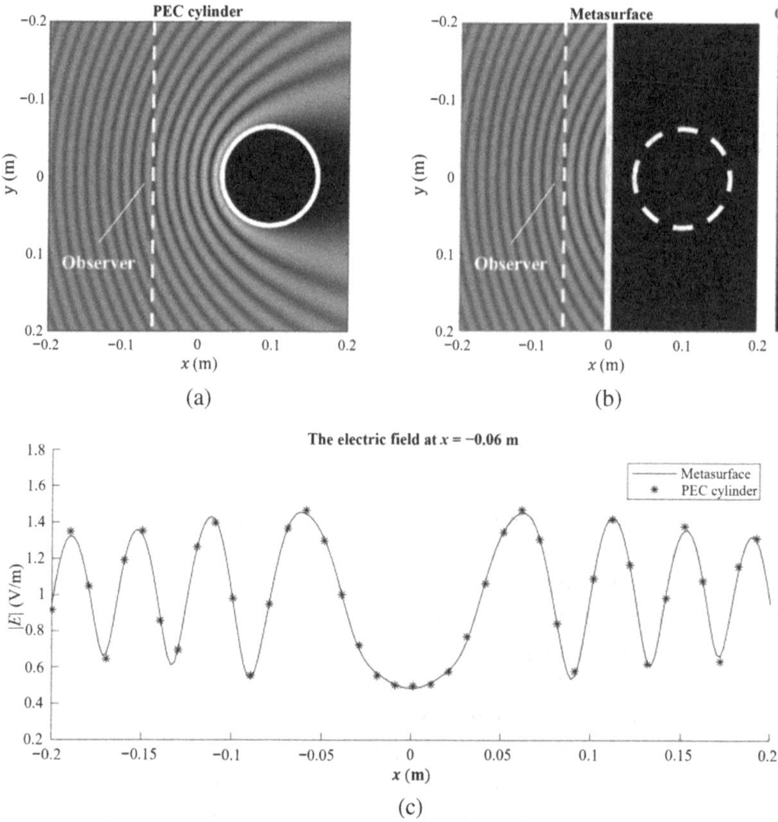

Figure 3.42 Simulation results. (a) Distribution of the electric fields generated by the metasurface. (b) Distribution of the electric fields generated by the PEC cylinder. (c) The electric field along $x = -2\lambda_0$.

are both tensors. The reflection fields and the transmission fields can be written by

$$\mathbf{E}^r = \overline{\overline{R}}\mathbf{E}^{in} \tag{3.153.a}$$

$$\mathbf{H}^r = \hat{\mathbf{z}} \times \overline{\overline{R}}\mathbf{E}^{in} \tag{3.153.b}$$

$$\mathbf{E}^t = \overline{\overline{T}}\mathbf{E}^{in} \tag{3.153.c}$$

$$\mathbf{H}^t = -\hat{\mathbf{z}} \times \overline{\overline{T}}\mathbf{E}^{in} \tag{3.153.d}$$

Substitute (3.153.a) into the GSTCs equations system (3.121.a) and then susceptibilities of each unit cell in the period array can be obtained. Extracted susceptibilities of each unit cell in the period shown in Figure 3.43(c) are listed in Table 3.1.

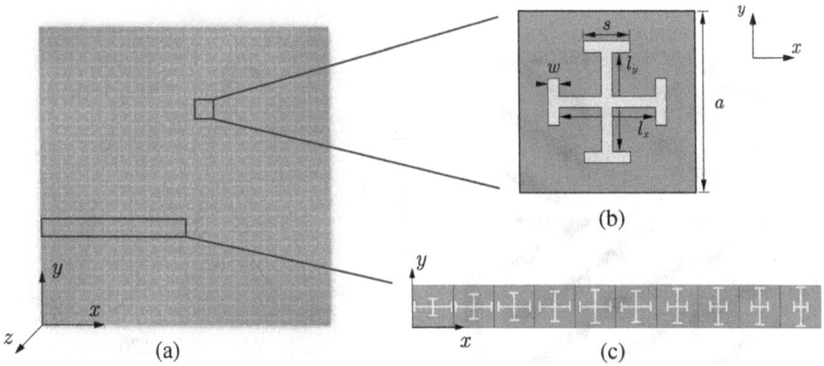

Figure 3.43 The schematic of the anisotropic metasurface and its unit cell. (a) The schematic of the metasurface. (b) A unit cell with $a = 6$ mm, $w = 0.4$ mm, $s = 2$ mm, and the thickness of the dielectric layer $d = 2$ mm. The thickness of metal patches is $d_m = 0.018$ mm. (c) The period array of unit cell composing the metasurface, where the lengths of l_x and l_y are increased and decreased gradually along the $+x$-axis ($l_x = 2$ mm, 3.8 mm, 4.2 mm, 4.4 mm, 4.5 mm, 4.65 mm, 4.75 mm, 4.9 mm, 5.05 mm, 5.15 mm; $l_y = 5.15$ mm, 5.05 mm, 4.9 mm, 4.75 mm, 4.65 mm, 4.5 mm, 4.4 mm, 4.2 mm, 3.8 mm, 2 mm), respectively. *Source:* Adapted from Ma et al. [41].

Table 3.1 The susceptibilities of each unit cell in the period array.

Cell	χ_{ee}^{xx}	χ_{ee}^{yy}	χ_{mm}^{xx}	χ_{mm}^{yy}
1	−0.016	−0.073	0.001	0.006
2	−0.009	0.159 − j0.001	−0.001	0.010
3	−0.005	0.030	−0.003	0.018
4	−0.002	0.010	−0.009	0.048 − j0.001
5	0	0.005	−0.020	−0.251 − j0.022
6	0.005	0	−0.402 − j0.056	−0.019
7	0.010	−0.002	0.046 − j0.001	−0.009
8	0.033	−0.005	0.018	−0.003
9	0.274 − j0.003	−0.009	0.010	0
10	−0.139	−0.016	0.006	0.001

Apply IE-GSTC to solve the reflection of the metasurface. As a reference, the anisotropic metasurface in Figure 3.43 is modeled with actual unit cell structures using the frequency-domain solver in commercial software ANSYS HFSS. The master-slave boundary is set to simulate the infinite periodic array in x- and

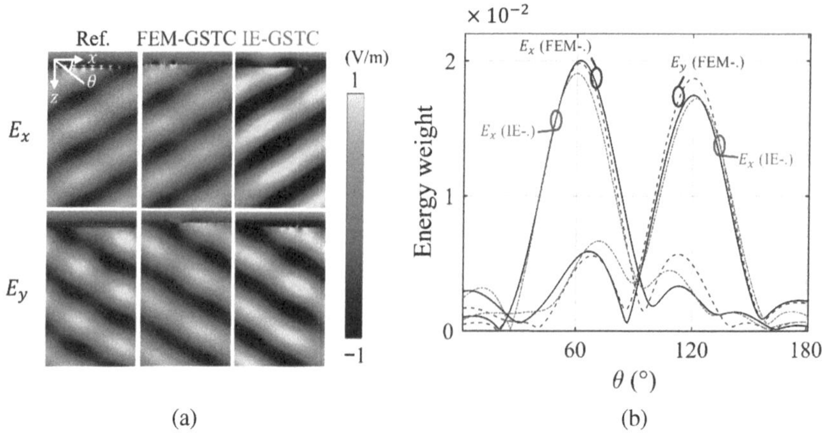

Figure 3.44 Results of the IE-GSTC method, FEM-GSTC method, and HFSS full-wave simulation (Ref. [41]). (a) The electric field distribution of reflected wave. (b) The energy weight curve at different angles.

y-directions. Besides, a FEM-GSTC simulation is also performed as a comparison. The simulation results are shown in Figure 3.44.

By artificially regulating the phase of reflection generated by each unit cell, the reflected wave with a specific deflection angle can be synthesized, as seen in Figure 3.44(a). To conduct a more in-depth analysis, the energy weight curve is plotted in Figure 3.44(b) to show the energy distribution of the reflected wave at different deflection angles. It can be seen that the *x*-polarized incident wave mainly reflects along the angle $\theta = 60°$ with respect to the $+x$-axis, and the *y*-polarized incident wave mainly reflects along the angle $\theta = 120°$. The relative error of IE-GSTC and FEM-GSTC with respect to HFSS is about −13 dB, the main reason for this error is that the surface susceptibilities are obtained from the periodic uniform array while applied in the non-uniform array. Even so, the GSTC-equivalence is still a relatively accurate reproduction of the independent control capacity of differently-polarized incidence of this metasurface. Considering that the array simulation time consumption of HFSS and IE-GSTC is 18 min and 23 s, respectively, the GSTC-equivalent is very efficient.

3.4.3.4 A Conformal Metasurface with Strong Normal Polarizabilities

In this case, the effect of normal susceptibilities is analyzed based on a conformal metasurface composed of the unit cell employed in [34]. First, the susceptibilities of the unit cell are extracted according to GSTCs. Then, the reflection and transmission rates of the unit cell from different incident angles are analyzed to illustrate the angle-dependence of scattering introduced by the normal susceptibilities. A conformal metasurface composed of the above-mentioned unit cell is simulated using the extracted susceptibilities.

The unit cell is composed of a split metallic loop on a polyimide (PI) film, loading with MIM unit cell [34]. Mainly, two types of electromagnetic interactions take place on the metasurface. One is the electric resonance induced by parallel electric currents in the MIM capacitor branches, which can be modeled by the z-polarized tangential electric susceptibility χ_{ee}^{zz}. The other is the strong x-polarized magnetic dipolar moment caused by the circulating current which is excited by a time dependent magnetic flux in the loops, which can be modeled by the x-polarized normal magnetic susceptibility χ_{mm}^{xx}. Other susceptibilities can be considered to be zero (Figure 3.45).

Following the same process introduced in Section 2.3.4.3, the susceptibilities of the MIM can be extracted, as shown in Figure 3.46.

Figure 3.47 makes a comparison between reflection rates R and transmission rates T calculated by the GSTCs and by the HFSS. The R and T obtained by both two methods agree well. It can be seen that, when taking consideration of the normal susceptibilities, the scattering properties of the unit cell are strongly angle-and-frequency dependent.

Here, a conformal metasurface composed of the above-mentioned unit cell is simulated by IE-GSTC solver to demonstrate the effect of normal susceptibilities intuitively. As shown in Figure 3.48, unit cells of the MIM are arranged in an

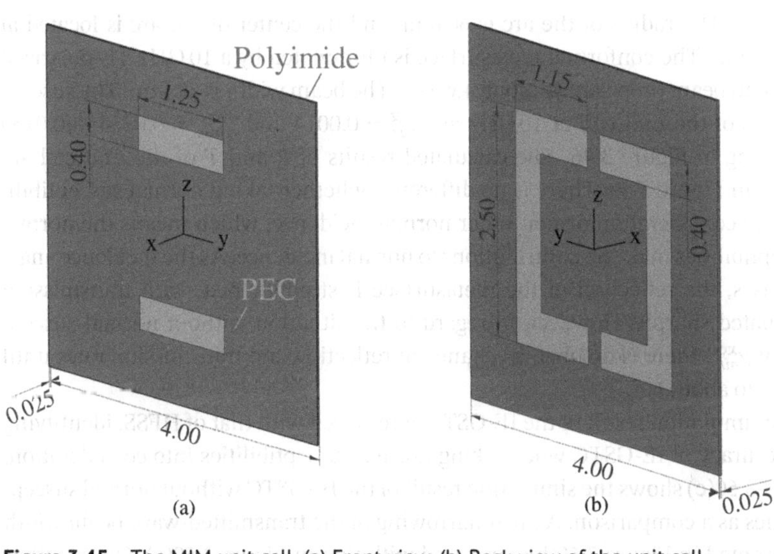

Figure 3.45 The MIM unit cell. (a) Front view. (b) Back view of the unit cell. *Source:* Adapted from Smy et al. [34].

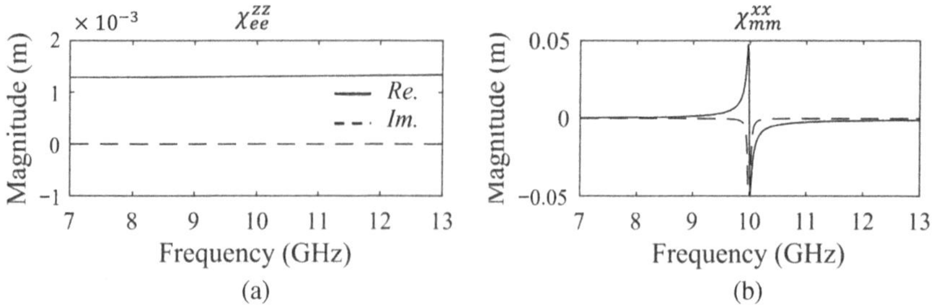

Figure 3.46 Extracted susceptibilities of the unit cell. (a) χ_{ee}^{zz} and (b) χ_{mm}^{xx}.

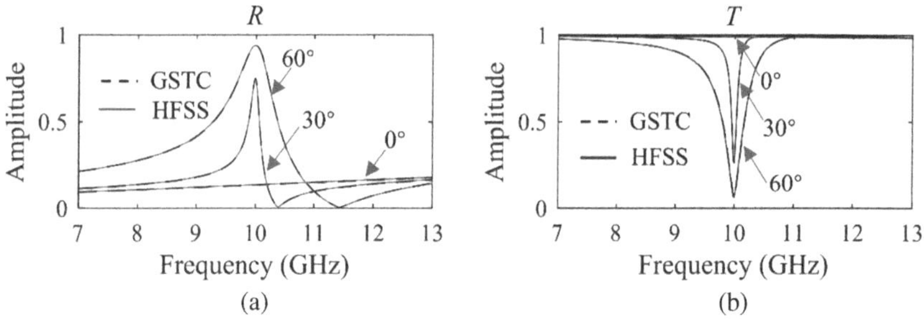

Figure 3.47 Reflection and transmission rates at different incident frequencies. (a) Reflection rate. (b) Transmission rate.

arc array. The radius of the arc is 60 mm, and the center of the arc is located at $x = 45$ mm. The conformal metasurface is illuminated by a 10 GHz TE-polarized Gaussian beam propagating along $+x$-axis. The beam width is 30 mm. The susceptibilities of the unit cell at 10 GHz are $\chi_{ee}^{zz} = 0.0014$ and $\chi_{mm}^{xx} = 0.0254 - j0.0159$ according to Figure 3.46. The calculated results of R and T of the unit cell are shown in Figure 3.49. There is no difference whether taking normal susceptibilities into consideration or not under normal incidence, which means the normal susceptibilities make no contributions to normal incidence. As the incidence angle increases, the reflection of the metasurface is strengthened, with transmission attenuated sharply. However, in regard to the situation without normal susceptibility χ_{mm}^{xx}, there is no obvious change in reflection and transmission rates until θ is up to about 60°.

The simulation result of the IE-GSTC agrees well with that of HFSS, identifying the accuracy of IE-GSTC when taking normal susceptibilities into consideration. Figure 3.50(c) shows the simulation result of the IE-GSTC without normal susceptibilities as a comparison. A great narrowing of the transmitted-wave beam width is presented in Figure 3.50(b) compared with that in Figure 3.50(c). On the other hand, reflection in Figure 3.50(b) is stronger in the area away from the central axis.

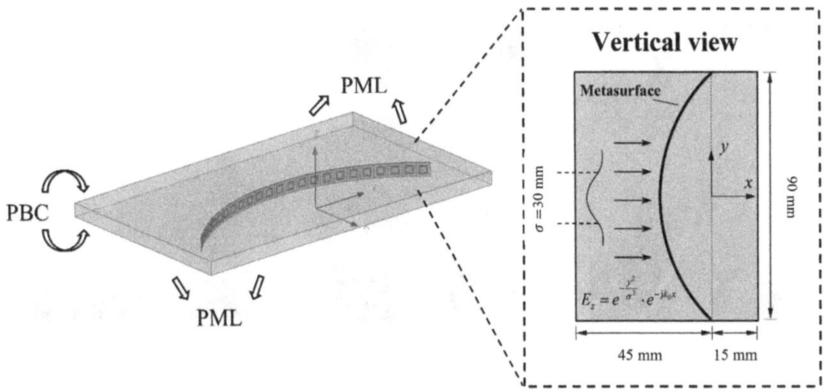

Figure 3.48 Arc-shaped metasurface composed of the MIM capacitors.

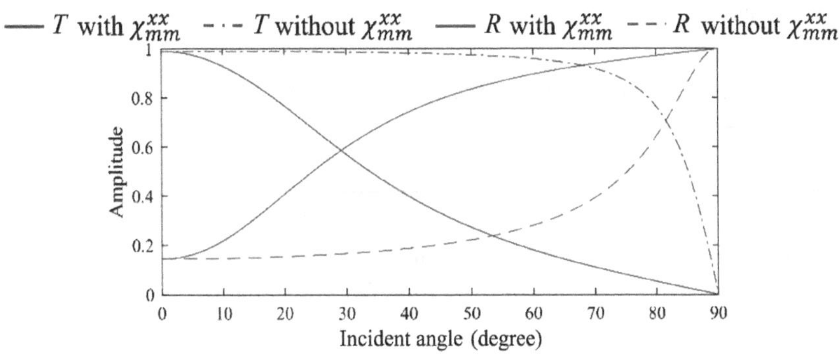

Figure 3.49 Reflection and transmission of the MIM capacitor illuminated by 10 GHz TE-polarized waves under different incidence angles.

This is because oblique incidence takes place on the curved metasurface except for the center point and at the location closer to the edge, the incidence angle is higher. As a result, the transmission gradually weakens while the reflection strengthens from the center to the sides, which agrees with the angle-dependent reflection and transmission shown in Figure 3.49. Apparently, IE-GSTC without normal susceptibilities in Figure 3.50(c) fails to simulate these phenomena correctly. Results in Figure 3.50 demonstrate the necessity to revise the IE-GSTC to take account of normal susceptibilities. Besides, the practical structure simulation consumption of HFSS and IE-GSTC is listed in Table 3.2, showing the outstanding calculation efficiency of the IE-GSTC again.

Besides, this example also demonstrates that the consideration of normal susceptibilities is indispensable for the analysis of metasurfaces with the strong angular scattering properties.

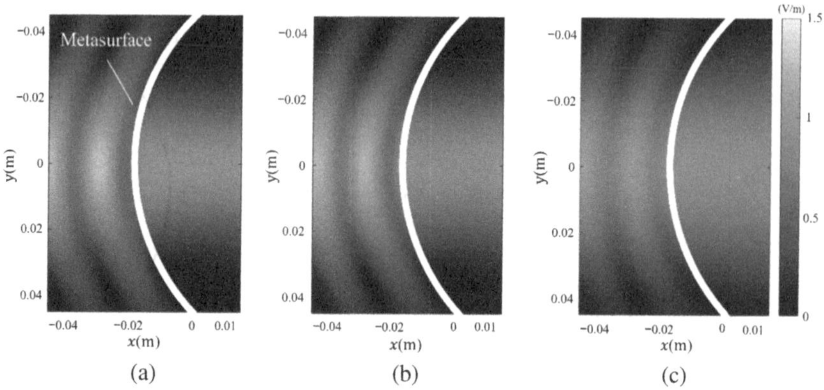

Figure 3.50 Total electric field distributions of the arc-shaped metasurface. (a) HFSS full-wave simulation. (b) IE-GSTC with normal susceptibilities. (c) IE-GSTC without normal susceptibilities.

Table 3.2 Computing resources and errors of example 3.4.3.4.

	Mesh (mm)	Time	Memory
IE-GSTC	1/4	102 s	2319 MB
HFSS	\	4 h 51 min	126.3 GB

3.5 Conclusion

In this chapter, how to use GSTCs in the frequency-domain method is presented. In Section 3.1, the traditional FDFD method and the matrix form of the calculation formula are introduced. Based on this, the strategies to couple the GSTCs into 2-D FDFD and 3-D FDFD methods are introduced, respectively. The 2-D FDFD-GSTC formula is simple and convenient to use, while the 3-D FDFD-GSTC more closely aligns with the practical scenarios of metasurfaces simulation. Two cases of metasurfaces are solved using the FDFD-GSTC solver, and the corresponding results are compared against actual-structure simulation results obtained from the commercial software COMSOL Multiphysics. While ensuring the effectiveness and accuracy of the proposed method, it also reflects the advantages of the proposed method in grid modeling and field value calculation accuracy, avoiding the problem of large computation caused by dense grids in subwavelength metasurface numerical simulation. Therefore, the FDFD-GSTC method is superior to the traditional FDFD method in subwavelength metasurface numerical simulation.

In Section 3.2, the basic theory of FEM-GSTC with derivation and simulation is presented. The basic concepts of FEM, including the basic idea, basis functions, mesh, and boundary conditions, are briefly reviewed. Then, the strategies to couple the GSTCs into are introduced. Two examples are presented to illustrate the effectiveness and accuracy of the FEM-GSTC algorithm. Also with comparison from commercial software COMSOL Multiphysics. The FEM-GSTC solver can simulate zero-thickness GSTC metasurfaces more efficiently and accurately by avoiding the extremely fine mesh and the resultant heavy computational burden when simulating the subwavelength metasurfaces without using thin sheet equivalence. Therefore, the FEM-GSTC is superior to the traditional method.

In Section 3.3, the incorporation of GSTCs and SEM is introduced. First, the basic concepts of the SEM are reviewed, including the governing equations, boundary conditions, mesh discretization, and the construction of vector basis functions based on GLL points. Then, the substitution relations for applying GSTC in SEM are derived, and GSTC is incorporated into SEM through surface integral terms. Finally, the solution equations containing GSTC terms, including the stiffness matrix, mass matrix, boundary matrix, and the matrix resulting from GSTC, are obtained. Based on the theory of SEM-GSTC, the effectiveness and accuracy of the algorithm are verified through three examples. The first is a polarization conversion metasurface, which uses eight susceptibilities to control the different responses of the metasurface to fields with different incident polarizations. Then a hybrid modulation metasurface, where susceptibilities are synthesized using scattering parameters to simultaneously control the amplitude and the phase of the transmitted and reflected fields. Finally, a non-diffracting vortex Bessel beam metasurface further demonstrates the broad applicability and reliability of SEM-GSTC through the analysis of fields with complex propagation characteristics.

In Section 3.4, the combination of the GSTCs and the IE method is introduced. Firstly, the general integral scattering equations on the electromagnetic interface based on the equivalent electric and magnetic current sources are introduced. Then, the strategies to integrate GSTCs into the IE method are present. Furthermore, the IE-GSTC method is extended to the curved metasurface problems by incorporating normal susceptibilities into scattering equations. Four interesting simulation cases are given at the end of this section. First, a simple case of a phase-amplitude metasurface is given to verify the accuracy of the IE-GSTC. The second case of electromagnetic illusion demonstrates the powerful capability of metasurfaces in electromagnetic control and gives the general procedure to design a metasurface with certain functions by susceptibility synthesis. The third case demonstrates the accuracy and calculation efficiency of the IE-GSTC by solving the reflection waves of an infinite anisotropic metasurface with practical unit cells. This case also shows the general procedure for applying the IE-GSTC to simulate a practical metasurface. The fourth case analyzes the effect of normal susceptibilities in-depth based on a conformal metasurface with strong normal polarizabilities. The IE-GSTC method is accelerated by FMM and performs satisfying calculation efficiency. Compared with numerical electromagnetic simulation methods based on differential equations, such as FDFD and FEM, the IE method enjoys unique advantages in scattering problems. The IEs satisfy the RBC inherently, so there is no need to set extra boundary conditions in the IE method. On the other hand, combined with the BEM, grid meshing in the IE method only takes place on the electromagnetic boundaries, reducing the space complexity by one dimension, which remarkably reduces the time and memory cost. Furthermore, the scattering equations of the IE method can be directly combined with the GSTCs equation system just after an uncomplicated modification of GSTCs without any extra processing for the mathematic or geometric model of the metasurface. Although all methods and cases introduced in this section are 2-D, the IE-GSTC framework can be extended directly to the 3-D model following a similar methodology using 3-D Green's function in the future [34].

References

1 K. Yee, "Numerical solution of initial boundary value problems involving Maxwell's equations in isotropic media," *IEEE Trans. Antennas Propag.*, vol. 14, no. 3, pp. 302–307, May 1966.

2 R. C. Rumpf, "Simple implementation of arbitrarily shaped total-field/scattered-field regions in finite-difference frequency-domain," *Prog. Electromagn. Res. B*, vol. 36, pp. 221–248, 2012.

3 W. Shin, "3D Finite-Difference Frequency-Domain Method for Plasmonics and Nanophotonics," Ph.D. dissertation, Stanford, CA, 2013.

4 Y. Vahabzadeh, K. Achouri, and C. Caloz, "Simulation of metasurfaces in finite difference techniques," *IEEE Trans. Antennas Propag.*, vol. 64, no. 11, pp. 4753–4759, Nov. 2016.

5 R. Courant, "Variational methods for the solution of problems of equilibrium and vibrations," *Bull. Am. Math. Soc.*, vol. 49, no. 1, pp. 1–24, Jan. 1943.

6 J.-M. Jin, *The Finite Element Method in Electromagnetics*. John Wiley & Sons, 2015.

7 J.-M. Jin, *Theory and Computation of Electromagnetic Fields*, 2nd ed. Hoboken, NJ: Wiley, 2015.

8 S. Sandeep, J.-M. Jin, and C. Caloz, "Finite-element modeling of metasurfaces with generalized sheet transition conditions," *IEEE Trans. Antennas Propag.*, vol. 65, no. 5, pp. 2413–2420, May 2017.

9 K. Wu, K. Shuai, and Q. Ren, "A DG-FEM-GSTC method for fast simulation of metasurfaces with strong normal polarization," *IEEE Trans. Antennas Propag.*, Early Access, 2025, doi: 10.1109/TAP.2025.3625208.

10 S. He, W. E. I. Sha, L. Jiang, W. C. H. Choy, W. C. Chew, and Z. Nie, "Finite-element-based generalized impedance boundary condition for modeling plasmonic nanostructures," *IEEE Trans. Nanotechnol.*, vol. 11, no. 2, pp. 336–345, Mar. 2012.

11 G. Cai, X. Liu, T. Shen, J. Liu, N. Liu, and Q. H. Liu, "A full-vectorial spectral element method with generalized sheet transition conditions for high-efficiency metasurface/metafilm simulation," *IEEE Trans. Antennas Propag.*, vol. 71, no. 3, pp. 2652–2660, Jan. 2023.

12 M. Farmahini-Farahani and H. Mosallaei, "Birefringent reflect array metasurface for beam engineering in infrared," *Opt. Lett.*, vol. 38, no. 4, pp. 462–464, 2013.

13 J. H. Lee, J. W. Yoon, M. J. Jung, J. K. Hong, S. H. Song, and R. Magnusson, "A semiconductor metasurface with multiple functionalities: A polarizing beam splitter with simultaneous focusing ability," *Appl. Phys. Lett.*, vol. 104, no. 23, pp. 1470–1474, 2014.

14 B. E. A. Saleh and M. C. Teich, *Fundamentals of Photonics*. John Wiley & Sons, 2019.

15 A. T. Patera, "A spectral element method for fluid dynamics: Laminar flow in a channel expansion," *J. Comput. Phys.*, vol. 54, no. 3, pp. 468–488, Jun. 1984.

16 M. Y. Hussaini and T. A. Zang, "Spectral methods in fluid dynamics," *Annu. Rev. Fluid Mech.*, vol. 19, no. 1, pp. 339–367, Jan. 1987.

17 K.-J. Bathe, "Finite element method," *Wiley Encycl. Comput. Sci. Eng.*, Jun. 2008.

18 R. D. Moser, J. Kim, and N. N. Mansour, "Direct numerical simulation of turbulent channel flow up to $Re\tau=5200$," *Phys. Fluids*, vol. 11, no. 4, pp. 943–945, Apr. 1999.

19 D. Komatitsch and J. Tromp, "Introduction to the spectral element method for three-dimensional seismic wave propagation," *Geophys. J. Int.*, vol. 139, no. 3, pp. 806–822, Dec. 1999.

20 J.-H. Lee, T. Xiao, and Q. H. Liu, "A 3-D spectral-element method using mixed-order curl conforming vector basis functions for electromagnetic fields," *IEEE Trans. Microw. Theory Tech.*, vol. 54, no. 1, pp. 437–444, Jan. 2006.

21 J.-H. Lee and Q. H. Liu, "A 3-D spectral-element time-domain method for electromagnetic simulation," *IEEE Trans. Microw. Theory Tech.*, vol. 55, no. 5, pp. 983–991, May 2007.

22 N. Liu, L. Tobón, Y. Tang, and Q. H. Liu, "Mixed spectral element method for 2D Maxwell eigenvalue problem," *Commun. Comput. Phys.*, vol. 17, no. 2, pp. 458–486, 2015.

23 N. Liu, L. E. Tobon, Y. Zhao, Y. Tang, and Q. H. Liu, "Mixed spectral-element method for 3-D Maxwell's eigenvalue problem," *IEEE Trans. Microw. Theory Tech.*, vol. 63, no. 2, pp. 317–325, Jan. 2015.

24 N. Liu, G. Cai, C. Zhu, Y. Tang, and Q. H. Liu, "The mixed spectral-element method for anisotropic, lossy, and open waveguides," *IEEE Trans. Microw. Theory Tech.*, vol. 63, no. 10, pp. 3094–3102, Oct. 2015.

25 Z. S. Sacks, D. M. Kingsland, R. Lee, and J.-F. Lee, "A perfectly matched anisotropic absorber for use as an absorbing boundary condition," *IEEE Trans. Antennas Propag.*, vol. 43, no. 12, pp. 1460–1463, 1995.

26 N.-W. Chen, M. Lu, F. Capolino, B. Shanker, and E. Michielssen, "Floquet wave-based analysis of transient scattering from doubly periodic, discretely planar, perfectly conducting structures," *Radio Sci.*, vol. 40, no. 4, pp. 1–21, Aug. 2005.

27 K. Achouri, M. A. Salem, and C. Caloz, "General metasurface synthesis based on susceptibility tensors," *IEEE Trans. Antennas Propag.*, vol. 63, no. 7, pp. 2977–2991, Jul. 2015.

28 Y.-Z. Yu and W.-B. Dou, "Vector analyses of nondiffracting Bessel beams," *Prog. Electromagn. Res. Lett*, vol. 5, pp. 57–71, 2008.

29 R. F. Harrington, *Field Computation by Moment Methods*. New York: Macmillan, 1968.

30 V. Rokhlin, "Rapid solution of integral equations of scattering theory in two dimensions," *J. Comput. Phys.*, vol. 86, no. 2, pp. 414–439, Feb. 1990.

31 S. A. Stewart, S. Moslemi-Tabrizi, T. J. Smy, and S. Gupta, "Scattering field solutions of metasurfaces based on the boundary element method for interconnected regions in 2-D," *IEEE Trans. Antennas Propag.*, vol. 67, no. 12, pp. 7487–7495, Dec. 2019.

32 T. J. Smy, S. A. Stewart, and S. Gupta, "Surface susceptibility synthesis of metasurface holograms for creating electromagnetic illusions," *IEEE Access*, vol. 8, pp. 93408–93425, 2020.

33 M. Dehmollaian and C. Caloz, "IE-GSTC analysis of metasurface cavities and application to redirection cloaking," 2020 Fourteenth International Congress on Artificial Materials for Novel Wave Phenomena (Metamaterials), New York, NY, USA, 2020, pp. 318-320.

34 T. J. Smy, V. Tiukuvaara, and S. Gupta, "IE-GSTC metasurface field solver using surface susceptibility tensors with normal polarizabilities," *IEEE Trans. Antennas Propag.*, vol. 70, no. 6, pp. 4752–4765, Jun. 2022.

35 J. G. N. Rahmeier, T. J. Smy, J. Dugan, and S. Gupta, "Zero thickness surface susceptibilities and extended GSTCs-part I spatially dispersive metasurfaces," *IEEE Trans. Antennas Propag.*, vol. 71, no. 7, pp. 5909–5919, Jul. 2023.

36 T. J. Smy, J. G. N. Rahmeier, J. Dugan, and S. Gupta, "Spatially dispersive metasurfaces-part II IE-GSTC-SD field solver with extended GSTCs," *IEEE Trans. Antennas Propag.*, vol. 71, no. 7, pp. 5920–5934, Jul. 2023.

37 J. Dugan, J. G. Nizer Rahmeier, T. J. Smy, and S. Gupta, "Part III – spatially dispersive metasurfaces zero thickness modeling of periodic and finite non-uniform surfaces," *IEEE Trans. Antennas Propag.*, vol. 71, no. 7, pp. 5935–5945, May 2023.

38 A. Poggio, *Computer Techniques for Electromagnetics*. Oxford: Pergamon Press, 1973.

39 J. Dugan, T. J. Smy, and S. Gupta, "Accelerated IE-GSTC solver for large-scale metasurface field scattering problems using fast multipole method (FMM)," *IEEE Trans. Antennas Propag.*, vol. 70, no. 10, pp. 9524–9533, Oct. 2022.

40 L. Xiao, F. Yang, M. Li, and S. Xu, "Generalized boundary conditions in surface electromagnetics: Fundamental theorems and surface characterizations," *Appl. Sci.*, vol. 9, no. 9, p. 1891, May 2019.

41 H. F. Ma, G. Z. Wang, G. S. Kong, and T. J. Cui, "Independent controls of differently-polarized reflected waves by anisotropic metasurfaces," *Sci. Rep.*, vol. 5, no. 9605, Apr. 2015.

4

Time-Domain Method for Metasurface Using GSTCs

In this chapter, the simulation of metasurfaces using GSTCs in the time-domain methods will be introduced. When engineers need to solve the broadband response of a metasurface, or when they need to simulate a time-varying or nonlinear metasurface, the time-domain method is a better choice. The time-domain method solves the Maxwell's equations in (1.1), and there will be some additional concepts in addition to those in the frequency-domain method in the previous chapter. The first is the time-stepping process. The time, t, as a new dimension here, is an important variable of the time-domain method that the simulation can be regarded as a process to approximate the exact solution $f(t)$ with a discrete numerical solution $\widetilde{f}(n\Delta t)$, where n denotes the index of the time step, and Δt is the time step interval. The corresponding computer programs appear to use the values of the unknowns at the past time steps to obtain the values of the unknowns at the current time step, and so on until the duration $N_T\Delta t$ is reached. The second concept is the spatial explicit method and stability. Different from the frequency-domain method to solve the large system matrix $\mathbf{Ax} = \mathbf{b}$, the time-domain method can avoid the inverse of the large matrix by spatial explicit algorithms, which makes the time complexity of the time-domain method linearly increase with the growth of unknowns. However, the time-domain methods need to consider the problem of stability when using an explicit algorithm. It will get a wrong result or even lead to divergence. The third concept is the parameter fitting process and auxiliary differential equations (ADEs). In the time-domain method, the electric and magnetic fields should be real numbers. Therefore, the imaginary number in the surface susceptibilities should be approximately replaced by a dispersion model within the specified frequency range, and become the time derivative term (or the time difference term) through the time-frequency domain transformation, resulting in new differential equations. These new differential equations are called ADEs.

The following sections include two time-domain methods and the strategies for their incorporation with GSTCs. Section 4.1 introduces the FDTD method, which is the most widely employed time-domain electromagnetics algorithm. Section 4.2

introduces the discontinuous Galerkin time-domain (DGTD) method, which is based on the unstructured grids and hence suitable for curved metasurfaces. The unique concepts in the time-domain method, including the broadband pulse excitation signals, stability criteria, parameter fitting methods, dispersion models, and ADEs for PMLs, will be covered in Section 4.1.

4.1 FDTD Method with GSTCs

4.1.1 Introduction of the FDTD Method

In 1966, K. S. Yee [1] proposed the famous Yee grid, which represents the beginning of the application of the FDTD method in solving Maxwell's equations, and the Yee grid has become the fundamental framework of the FDTD method. In 1975, A. Taflove and M. E. Brodwin used the FDTD method for the first time to analyze the electromagnetic scattering problem and the thermal problem of human tissue under electromagnetic radiation [2, 3]. In 1981, G. Mur first proposed the absorbing boundary condition, which is used to truncate the computation domain of infinity in the open-domain problem, and this technique later became an indispensable part of the FDTD method [4]. In 1982, K. Umashankar proposed the total-field/scatter-field method (TF/SF) technology to introduce plane wave excitation and near-far field conversion technology into the FDTD method to analyze scattering problems [5]. In 1985, the simulation software MAFIA based on the FDTD-like FIT (finite integration technique) was developed, which is the predecessor of the currently popular electromagnetic simulation product CST MWS (CST Microwave Studio). [6]. In 1994, J. P. Berenger proposed the PML method, which is another kind of ABC based on absorbing media, and also becomes one of the basic frameworks of the FDTD method [7].

In Section 3.1, the FDFD method is introduced to solve the time-harmonic Maxwell's equations. The FDTD method described in this chapter has the same mathematical principle as FDFD, i.e., the difference schematic under the Yee grid. The different point is that the governing equations are the time-domain Maxwell's equations as

$$\varepsilon\frac{\partial E_x}{\partial t} = \frac{\partial H_z}{\partial y} - \frac{\partial H_y}{\partial z} - J_x \tag{4.1.a}$$

$$\varepsilon\frac{\partial E_y}{\partial t} = \frac{\partial H_x}{\partial z} - \frac{\partial H_z}{\partial x} - J_y \tag{4.1.b}$$

$$\varepsilon\frac{\partial E_z}{\partial t} = \frac{\partial H_y}{\partial x} - \frac{\partial H_x}{\partial y} - J_z \tag{4.1.c}$$

$$\mu\frac{\partial H_x}{\partial t} = -\frac{\partial E_z}{\partial y} + \frac{\partial E_y}{\partial z} - M_x \tag{4.1.d}$$

$$\mu\frac{\partial H_y}{\partial t} = -\frac{\partial E_x}{\partial z} + \frac{\partial E_z}{\partial x} - M_y \tag{4.1.e}$$

$$\mu\frac{\partial H_z}{\partial t} = -\frac{\partial E_y}{\partial x} + \frac{\partial E_x}{\partial y} - M_z \tag{4.1.f}$$

where E, H, J, and M are the electric field, the magnetic field, the electric current density, and magnetic current density, respectively, and the subscripts x, y, and z denote the components along corresponding directions. Different from the equations in the frequency domain, time-domain methods need additional processing of the time partial derivative term $\frac{\partial}{\partial t}$.

4.1.1.1 The Yee Algorithm for FDTD

The Yee algorithm considers a 3-D uniformly spaced rectangular grids, as introduced in Section 3.1. Each grid cell has the dimensions of Δx, Δy, and Δz along each of the Cartesian axis. The discrete form of the coordinates of a certain node in a Yee grid can be expressed as $(i\Delta x, j\Delta y, k\Delta z) \to (x_i, y_j, z_k)$, where i, j, and k are all integers and can be regarded as the indices in a computer program storing the spatial information. Similarly, time is discretized as $t = n\Delta t$, and n is an integer standing for the step of time integration. Therefore, the exact solution $f(x, y, z, t)$ of the Maxwell's equations under specified boundary conditions can be approximated by the numerical solution $\widetilde{f}(x_i, y_j, z_k, t_n)$.

The field components E_x, E_y, E_z, H_x, H_y, and H_z are fixed at the edge center of the Yee grid, as shown in Figure 3.1. The index (i, j, k) denotes the spatial label of this cell in the whole mesh grid. The z-component of the electric field E_z, as shown in Figure 4.1(b), (4.1.c) can be approximated as

$$\varepsilon\frac{\partial}{\partial t}E_z(x_i, y_j, z_{k+1/2}) = \frac{H_y(x_{i+1/2}, y_j, z_{k+1/2}) - H_y(x_{i-1/2}, y_j, z_{k+1/2})}{\Delta x} - \frac{H_x(x_i, y_{j+1/2}, z_{k+1/2}) - H_x(x_i, y_{j-1/2}, z_{k+1/2})}{\Delta y} - J_z(x_i, y_j, z_{k+1/2}) \tag{4.2}$$

To implement FDTD in a computer program, arrays are used to store unknowns $E_{x/y/z}$ and $H_{x/y/z}$. Assuming that the computational domain is meshed by Yee grids, with N_x, N_y, and N_z cells in the x-, y-, and z-directions, an array with at least $N_x \times N_y \times N_z$ elements is needed to store one component of fields. In this book, indices are built in the way shown in Figure 3.1, where the point $E_z(x_i, y_j, z_{k+1/2})$ is represented as $E_z^{i,j,k}$ and stored at i-th row, j-th column, and k-th page of a 3-D array. For the sake of brevity, the array index is used instead of the actual spatial coordinates to abbreviate the field sampling points. As illustrated in Figure 4.1(a), the corresponding cell indices are used to rewrite (4.2) as

$$\varepsilon_z^{i,j,k}\frac{\partial E_z^{i,j,k}}{\partial t} = \frac{H_y^{i,j,k} - H_y^{i-1,j,k}}{\Delta x} - \frac{H_x^{i,j,k} - H_x^{i,j-1,k}}{\Delta y} - J_z^{i,j,k} \tag{4.3.a}$$

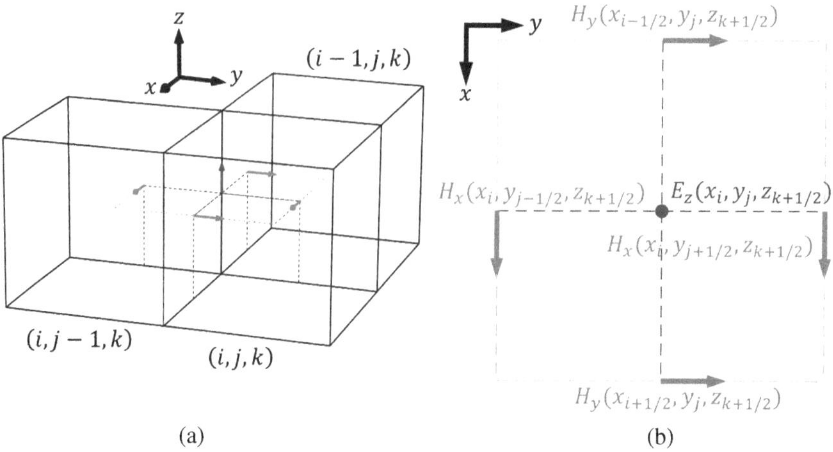

Figure 4.1 The curl operation of E_z in Yee grid. (a) The perspective view. (b) The top view.

Then, the remaining equations in (4.1.a) can be written as

$$\varepsilon_x^{i,j,k}\frac{\partial E_x^{i,j,k}}{\partial t} = \frac{H_z^{i,j,k} - H_z^{i,j-1,k}}{\Delta y} - \frac{H_y^{i,j,k} - H_y^{i,j,k-1}}{\Delta z} - J_x^{i,j,k} \tag{4.3.b}$$

$$\varepsilon_y^{i,j,k}\frac{\partial E_y^{i,j,k}}{\partial t} = \frac{H_x^{i,j,k} - H_x^{i,j,k-1}}{\Delta z} - \frac{H_z^{i,j,k} - H_z^{i-1,j,k}}{\Delta x} - J_y^{i,j,k} \tag{4.3.c}$$

$$\mu_x^{i,j,k}\frac{\partial H_x^{i,j,k}}{\partial t} = -\frac{E_z^{i,j+1,k} - E_z^{i,j,k}}{\Delta y} + \frac{E_y^{i,j,k+1} - E_y^{i,j,k}}{\Delta z} - M_x^{i,j,k} \tag{4.3.d}$$

$$\mu_y^{i,j,k}\frac{\partial H_y^{i,j,k}}{\partial t} = -\frac{E_x^{i,j,k+1} - E_x^{i,j,k}}{\Delta z} + \frac{E_z^{i+1,j,k} - E_z^{i,j,k}}{\Delta x} - M_y^{i,j,k} \tag{4.3.e}$$

$$\mu_z^{i,j,k}\frac{\partial H_z^{i,j,k}}{\partial t} = -\frac{E_y^{i+1,j,k} - E_y^{i,j,k}}{\Delta x} + \frac{E_x^{i,j+1,k} - E_x^{i,j,k}}{\Delta y} - M_z^{i,j,k} \tag{4.3.f}$$

It is not difficult to note that (3.7) and (4.3.a) are both Maxwell's equations treated with central difference approximation of the first-order spatial derivatives, and the only difference is the term $j\omega$ in (3.7) which is replaced by $\frac{\partial}{\partial t}$ in (4.3.a). Different from the FDFD method, (4.3.a) in FDTD are still semi-discrete equations (with only spatial discrete, while without time discrete) and are also need to use numerical methods to deal with the time partial derivative term $\frac{\partial}{\partial t}$. A classical time integration scheme in FDTD, called the leap-frog method, uses the central difference to approximate the time derivative.

$$\left.\frac{\partial E(t)}{\partial t}\right|_{t=(n+1/2)\Delta t} = \frac{E((n+1)\Delta t) - E(n\Delta t)}{\Delta t} \tag{4.4.a}$$

$$\left.\frac{\partial H(t)}{\partial t}\right|_{t=n\Delta t} = \frac{H((n+1/2)\Delta t) - H((n-1/2)\Delta t)}{\Delta t} \tag{4.4.b}$$

Electric and magnetic fields are staggered by half a time interval that E-fields sampling on the integer time interval $n\Delta t$ and H-fields on the half interval $\left(n+\frac{1}{2}\right)\Delta t$. For the sake of brevity, in the rest of this book, the time index n is labeled in the lower right corner of the unknowns, such as $E_{z,n}^{i,j,k}$ and $H_{x,n+1/2}^{i,j,k}$. Using central difference to replace the time derivative in (4.3.a) leads to

$$\varepsilon_x^{i,j,k}\frac{E_{x,n+1}^{i,j,k} - E_{x,n}^{i,j,k}}{\Delta t} = \frac{H_{z,n+1/2}^{i,j,k} - H_{z,n+1/2}^{i,j-1,k}}{\Delta y} - \frac{H_{y,n+1/2}^{i,j,k} - H_{y,n+1/2}^{i,j,k-1}}{\Delta z} - J_{x,n+1/2}^{i,j,k} \tag{4.5.a}$$

$$\varepsilon_y^{i,j,k}\frac{E_{y,n+1}^{i,j,k} - E_{y,n}^{i,j,k}}{\Delta t} = \frac{H_{x,n+1/2}^{i,j,k} - H_{x,n+1/2}^{i,j,k-1}}{\Delta z} - \frac{H_{z,n+1/2}^{i,j,k} - H_{z,n+1/2}^{i-1,j,k}}{\Delta x} - J_{y,n+1/2}^{i,j,k} \tag{4.5.b}$$

$$\varepsilon_z^{i,j,k}\frac{E_{z,n+1}^{i,j,k} - E_{z,n}^{i,j,k}}{\Delta t} = \frac{H_{y,n+1/2}^{i,j,k} - H_{y,n+1/2}^{i-1,j,k}}{\Delta x} - \frac{H_{x,n+1/2}^{i,j,k} - H_{x,n+1/2}^{i,j-1,k}}{\Delta y} - J_{z,n+1/2}^{i,j,k} \tag{4.5.c}$$

$$\mu_x^{i,j,k}\frac{H_{x,n+1/2}^{i,j,k} - H_{x,n-1/2}^{i,j,k}}{\Delta t} = -\frac{E_{z,n}^{i,j+1,k} - E_{z,n}^{i,j,k}}{\Delta y} + \frac{E_{y,n}^{i,j,k+1} - E_{y,n}^{i,j,k}}{\Delta z} - M_{x,n}^{i,j,k} \tag{4.5.d}$$

$$\mu_y^{i,j,k}\frac{H_{y,n+1/2}^{i,j,k} - H_{y,n-1/2}^{i,j,k}}{\Delta t} = -\frac{E_{x,n}^{i,j,k+1} - E_{x,n}^{i,j,k}}{\Delta z} + \frac{E_{z,n}^{i+1,j,k} - E_{z,n}^{i,j,k}}{\Delta x} - M_{y,n}^{i,j,k} \tag{4.5.e}$$

$$\mu_z^{i,j,k}\frac{H_{z,n+1/2}^{i,j,k} - H_{z,n-1/2}^{i,j,k}}{\Delta t} = -\frac{E_{y,n}^{i+1,j,k} - E_{y,n}^{i,j,k}}{\Delta x} + \frac{E_{x,n}^{i,j+1,k} - E_{x,n}^{i,j,k}}{\Delta y} - M_{z,n}^{i,j,k} \tag{4.5.f}$$

One of the key characteristics of the time-domain method is the time-stepping process. Based on the time discrete equation, the program can use the known fields of the previous time steps to obtain the unknown fields at the current time step. Equation (4.5.a) can be written as explicit updating equations.

$$\begin{aligned} E_{x,n+1}^{i,j,k} = E_{x,n}^{i,j,k} &+ \frac{\Delta t}{\varepsilon_x^{i,j,k}\Delta y}\left(H_{z,n+1/2}^{i,j,k} - H_{z,n+1/2}^{i,j-1,k}\right) \\ &- \frac{\Delta t}{\varepsilon_x^{i,j,k}\Delta z}\left(H_{y,n+1/2}^{i,j,k} - H_{y,n+1/2}^{i,j,k-1}\right) - \frac{\Delta t}{\varepsilon_x^{i,j,k}} J_{x,n+1/2}^{i,j,k} \end{aligned} \tag{4.6.a}$$

$$\begin{aligned} E_{y,n+1}^{i,j,k} = E_{y,n}^{i,j,k} &+ \frac{\Delta t}{\varepsilon_y^{i,j,k}\Delta z}\left(H_{x,n+1/2}^{i,j,k} - H_{x,n+1/2}^{i,j,k-1}\right) \\ &- \frac{\Delta t}{\varepsilon_y^{i,j,k}\Delta x}\left(H_{z,n+1/2}^{i,j,k} - H_{z,n+1/2}^{i-1,j,k}\right) - \frac{\Delta t}{\varepsilon_y^{i,j,k}} J_{y,n+1/2}^{i,j,k} \end{aligned} \tag{4.6.b}$$

$$E_{z,n+1}^{i,j,k} = E_{z,n}^{i,j,k} + \frac{\Delta t}{\varepsilon_z^{i,j,k}\Delta x}\left(H_{y,n+1/2}^{i,j,k} - H_{y,n+1/2}^{i-1,j,k}\right)$$

$$- \frac{\Delta t}{\varepsilon_z^{i,j,k} \Delta y} \left(H_{x,n+1/2}^{i,j,k} - H_{x,n+1/2}^{i,j-1,k} \right) - \frac{\Delta t}{\varepsilon_z^{i,j,k}} J_{z,n+1/2}^{i,j,k} \tag{4.6.c}$$

$$H_{x,n+1/2}^{i,j,k} = H_{x,n-1/2}^{i,j,k} - \frac{\Delta t}{\mu_x^{i,j,k} \Delta y} \left(E_{z,n}^{i,j+1,k} - E_{z,n}^{i,j,k} \right) + \frac{\Delta t}{\mu_x^{i,j,k} \Delta z} \left(E_{y,n}^{i,j,k+1} - E_{y,n}^{i,j,k} \right) - \frac{\Delta t}{\mu_x^{i,j,k}} M_{x,n}^{i,j,k} \tag{4.6.d}$$

$$H_{y,n+1/2}^{i,j,k} = H_{y,n-1/2}^{i,j,k} - \frac{\Delta t}{\mu_y^{i,j,k} \Delta z} \left(E_{x,n}^{i,j,k+1} - E_{x,n}^{i,j,k} \right) + \frac{\Delta t}{\mu_y^{i,j,k} \Delta x} \left(E_{z,n}^{i+1,j,k} - E_{z,n}^{i,j,k} \right) - \frac{\Delta t}{\mu_y^{i,j,k}} M_{y,n}^{i,j,k} \tag{4.6.e}$$

$$H_{z,n+1/2}^{i,j,k} = H_{z,n-1/2}^{i,j,k} - \frac{\Delta t}{\mu_z^{i,j,k} \Delta x} \left(E_{y,n}^{i+1,j,k} - E_{y,n}^{i,j,k} \right) + \frac{\Delta t}{\mu_z^{i,j,k} \Delta y} \left(E_{x,n}^{i,j+1,k} - E_{x,n}^{i,j,k} \right) - \frac{\Delta t}{\mu_z^{i,j,k}} M_{z,n}^{i,j,k} \tag{4.6.f}$$

4.1.1.2 Truncation Boundary

All the numerical algorithms based on the PDEs need boundaries to truncate the computational domain. This section introduces three simple boundaries, perfect electrical conductors (PECs), absorption boundary conditions (ABCs), and PML and illustrates how to set them up in the FDTD framework. The PECs and ABCs have been introduced in Chapter 3, and the primary focus is on introducing how to impose them in finite-difference method.

By definition, the electric field on a PEC surface satisfies

$$\hat{\mathbf{n}} \times \mathbf{E} = 0 \tag{4.7}$$

Since (4.7) tells the tangential components of the electric fields are null on the PEC boundary, it is straightforward to allocate the PEC boundary on the faces of the Yee grids, as shown in Figure 4.2. Because the electric field unknowns are placed parallel on the Yee grid's faces, when the face presented in Figure 4.2 is set as PEC, unknowns on the PEC will be forced to zero throughout the entire simulation as

$$E_x^{i,j,k} = E_x^{i,j+1,k} = E_y^{i,j,k} = E_y^{i+1,j,k} = 0 \tag{4.8}$$

which is identified as the Dirichlet boundary condition. Considering a plane wave illuminating vertically on the PEC boundary, because the PEC boundary conditions constrain the tangential electric field on the surface to be zero,

$$\hat{\mathbf{z}} \times (\mathbf{E}_i - \mathbf{E}_r)\big|_{z=0} = 0 \tag{4.9}$$

where $\mathbf{E}_i$ and $\mathbf{E}_r$ denote the incident electric field and the reflected field, respectively. The PEC boundary conditions (4.9) form a reflected plane wave whose amplitude is identical to the incident plane wave, as shown in Figure 4.3(a).

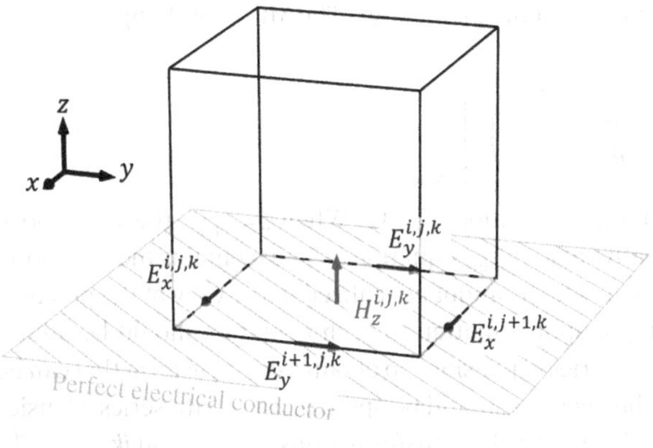

Figure 4.2 The grid cell face on a PEC surface.

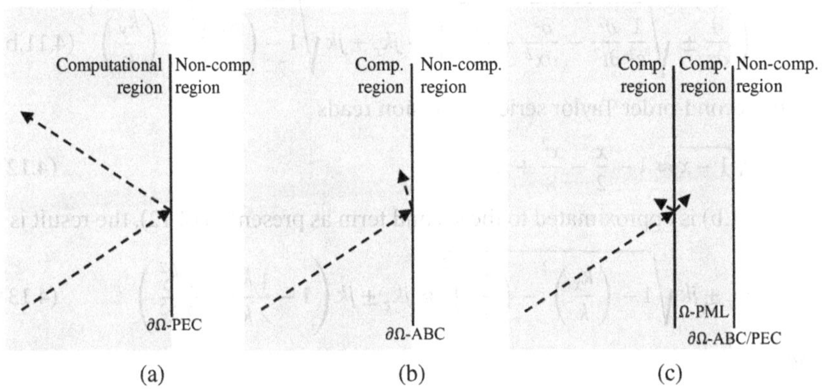

Figure 4.3 The plane wave illuminating on (a) PEC, (b) ABC, and (c) PML.

In order to simulate an infinite region in a finite computational domain Ω, which is also the most common scenario of metasurface simulation, a necessary ability of the truncated boundary is to absorb all the waves (or energy) propagated to it, which is called the absorbing boundary conditions (ABC), as shown in Figure 4.3(b). Here a local[1] absorb boundary condition is introduced, which is a second-order approximation to Engquist Majda ABC [8], and it is also called second-order Mur-ABC.

For any plane wave solution,

$$f(x, y, z, t) = f_0 \exp[j(\omega t - k_x x - k_y y - k_z z)] \tag{4.10}$$

1 It only requires the local fields in the vicinity of the truncation boundary, in contrast to the global ABC, which requires fields of all truncated boundaries bringing quite expensive and complex implementation [9].

where f_0 denotes the amplitude of the plane wave. The Mur-ABC lying at $z = z_{max}$ or z_{min} satisfies

$$\left(\frac{\partial}{\partial z} \pm \sqrt{\frac{1}{c^2}\frac{\partial^2}{\partial t^2} - \frac{\partial^2}{\partial y^2} - \frac{\partial^2}{\partial x^2}} \right) f \Bigg|_{z=z_{max},\, z_{min}} = 0 \tag{4.11.a}$$

where the sign $\pm$ depends on the location of ABC. When $z = z_{max}$, the sign is positive, while the sign is negative at $z = z_{min}$. Equation (4.11.a) contains the derivation operation in the radical sign, which cannot be realized in the view of the practical calculation. The meaning of approximation to such operators should be understood as the result of Taylor series expansion. To obtain a concrete form that can be numerically calculated, the operator should be approximated with series. Considering the plane wave solution (4.10), the transformations $j\omega \to \frac{\partial}{\partial t}$ and $jk_{x/y} \to \frac{\partial}{\partial x/y}$ can be used to rewrite the operator in (4.11.a) as

$$\left(\frac{\partial}{\partial z} \pm \sqrt{\frac{1}{c^2}\frac{\partial^2}{\partial t^2} - \frac{\partial^2}{\partial x^2} - \frac{\partial^2}{\partial y^2}} \right) \to jk_z \pm jk\sqrt{1 - \left(\frac{k_x}{k}\right)^2 - \left(\frac{k_y}{k}\right)^2} \tag{4.11.b}$$

The second-order Taylor series expansion reads

$$\sqrt{1-x} \approx 1 - \frac{x}{2} - \frac{x^2}{8} + \dots, \tag{4.12}$$

If (4.11.b) is approximated to the second term as present in (4.12), the result is

$$jk_z \pm jk\sqrt{1 - \left(\frac{k_x}{k}\right)^2 - \left(\frac{k_y}{k}\right)^2} \approx jk_z \pm jk\left(1 - \frac{1}{2}\frac{k_x^2}{k^2} - \frac{1}{2}\frac{k_y^2}{k^2}\right) \tag{4.13}$$

and

$$\left[jk_z \pm jk\left(1 - \frac{1}{2}\frac{k_x^2}{k^2} - \frac{1}{2}\frac{k_y^2}{k^2}\right)\right] f \Bigg|_{z=0} = 0 \tag{4.14}$$

Applying the transformation from time domain to frequency domain, the equivalent time-domain formulation of (4.14) is obtained.

$$\left(\frac{1}{c}\frac{\partial^2}{\partial z \partial t} \pm \frac{1}{c^2}\frac{\partial^2}{\partial t^2} \mp \frac{1}{2}\frac{\partial^2}{\partial x^2} \mp \frac{1}{2}\frac{\partial^2}{\partial y^2} \right) f \Bigg|_{z=0} = 0 \tag{4.15}$$

This is the second-order approximate ABC employed to avoid reflected waves and to simulate an infinite region. The implementation of ABC (4.15) in the Yee grid is very similar to the implementation of PEC.

Consider Figure 4.4 as an example. The region $z > k_A \Delta z$ is the computational domain and the surface is set as ABC. In this case, (4.14) becomes

$$\left(\frac{\partial^2}{\partial z \partial t} + \frac{1}{c^2}\frac{\partial^2}{\partial t^2} - \frac{1}{2}\frac{\partial^2}{\partial x^2} - \frac{1}{2}\frac{\partial^2}{\partial y^2} \right) E_x \Bigg|_{z=k_A \Delta z} = 0 \tag{4.16.a}$$

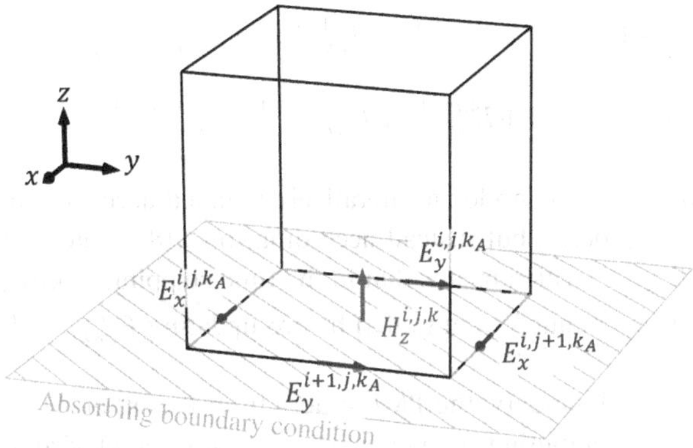

Figure 4.4 The grid cell face on an ABC surface.

$$\left(\frac{\partial^2}{\partial z \partial t} + \frac{1}{c^2}\frac{\partial^2}{\partial t^2} - \frac{1}{2}\frac{\partial^2}{\partial x^2} - \frac{1}{2}\frac{\partial^2}{\partial y^2}\right) E_y \bigg|_{z=k_A \Delta z} = 0 \tag{4.16.b}$$

Discretize the second equation of (4.16.a) at $\left(i + \frac{1}{2}, j, k_A - \frac{1}{2}\right)$ and $n\Delta t$. Each difference can be approximated as

$$\frac{\partial^2 E_x}{\partial z \partial t} = \frac{1}{2\Delta t \Delta z} \times \left(E_{x,n+1}^{i,j,k_A} - E_{x,n+1}^{i,j,k_A-1} - E_{x,n}^{i,j,k_A} + E_{x,n}^{i,j,k_A-1}\right) \tag{4.17.a}$$

$$\frac{\partial^2 E_x}{\partial t^2} = \frac{1}{2\Delta t^2}\left(E_{x,n+1}^{i,j,k_A} - 2E_{x,n}^{i,j,k_A} + E_{x,n-1}^{i,j,k_A} + E_{x,n+1}^{i,j,k_A-1} - 2E_{x,n}^{i,j,k_A-1} + E_{x,n-1}^{i,j,k_A-1}\right) \tag{4.17.b}$$

$$\frac{\partial^2 E_x}{\partial y^2} = \frac{1}{2\Delta y^2}\left(E_{x,n}^{i,j+1,k_A} - 2E_{x,n}^{i,j,k_A} + E_{x,n}^{i,j-1,k_A} + E_{x,n}^{i,j+1,k_A-1} - 2E_{x,n}^{i,j,k_A-1} + E_{x,n}^{i,j-1,k_A-1}\right) \tag{4.17.c}$$

$$\frac{\partial^2 E_x}{\partial x^2} = \frac{1}{2\Delta x^2}\left(E_{x,n}^{i+1,j,k_A} - 2E_{x,n}^{i,j,k_A} + E_{x,n}^{i-1,j,k_A} + E_{x,n}^{i+1,j,k_A-1} - 2E_{x,n}^{i,j,k_A-1} + E_{x,n}^{i-1,j,k_A-1}\right) \tag{4.17.d}$$

The explicit equation of updating E_x on the ABC boundary can be obtained after substituting (4.17.a) into (4.16.a)

$$E_{x,n+1}^{i,j,k_A} = -E_{x,n-1}^{i,j,k_A-1} + \frac{c\Delta t - \Delta z}{c\Delta t + \Delta z}\left(E_{x,n+1}^{i,j,k_A-1} + E_{x,n-1}^{i,j,k_A}\right) - 2\frac{c\Delta t - \Delta z}{\Delta z}\left(E_{x,n}^{i,j,k_A-1} + E_{x,n}^{i,j,k_A}\right)$$

$$+\frac{(c\Delta t)^2}{2\Delta x(c\Delta t+\Delta x)}\left(E_{x,n}^{i-1,j,k_A}+E_{x,n}^{i+1,j,k_A}+E_{x,n}^{i-1,j,k_A-1}+E_{x,n}^{i+1,j,k_A-1}\right)$$
$$+\frac{(c\Delta t)^2}{2\Delta y(c\Delta t+\Delta y)}\left(E_{x,n}^{i,j-1,k_A}+E_{x,n}^{i,j+1,k_A}+E_{x,n}^{i,j-1,k_A-1}+E_{x,n}^{i,j+1,k_A-1}\right)\quad(4.18)$$

As an unknown on ABC, E_x^{i,j,k_A} is no longer iteratively updated according to (4.6.a) in the time-stepping process, but instead according to (4.18). Note that updating $E_{x,n+1}^{i,j,k_A}$ requires $E_{x,n+1}^{i,j,k_A-1}$, which means that in the time-stepping process, the value of $E_{x,n+1}^{i,j,k_A-1}$ must be obtained from (4.6.a) before updating $E_{x,n+1}^{i,j,k_A}$ with (4.18).

Also acting as an absorbing boundary, the PML differs from the other ABCs in that PML terminates the computing domain with a thick layer of virtual absorbing medium, similar to an acoustically absorptive environment. As shown in Figure 4.3(c), PML provides a lossy medium layer instead of a forced boundary condition. PML requires more computational resources than other ABCs, but it can absorb both propagating and evanescent waves, enabling placement of the termination boundary adjacent to the primary solution domain. There are many kinds of PML [7, 10, 11], and the complex frequency shifted (CFS)-PML [12, 13] will be introduced here. Note that it employs the ADE method to deal with the dispersion models in the frequency domain. Based on the coordinate stretched Maxwell's equations, CFS-PML has better absorption of evanescent waves and low-frequency waves. The coordinate stretched operator ∇_s is defined as

$$\nabla_s=\hat{\mathbf{x}}\frac{1}{s_x}\frac{\partial}{\partial x}+\hat{\mathbf{y}}\frac{1}{s_y}\frac{\partial}{\partial y}+\hat{\mathbf{z}}\frac{1}{s_z}\frac{\partial}{\partial z}\quad(4.19)$$

where s_x, s_y, and s_z are coordinate stretched factors, and they are functions of corresponding coordinates as

$$s_w=\kappa_w(w)+\frac{\sigma_w(w)}{a_w(w)+j\omega\varepsilon_0},\quad w=x,\ y,\ z\quad(4.20)$$

where κ, σ, and a are variables in the PML, but equal to 1, 0, and 0 in the regular non-PML domain. Their expressions in the PML region are

$$\sigma_w(w)=\sigma_{\max}\frac{|w-w_0|^m}{d^m}\quad(4.21.a)$$

$$\kappa_w(w)=1+(\kappa_{\max}-1)\frac{|w-w_0|^m}{d^m}\quad(4.21.b)$$

$$a_w(w)=a_{\max}\frac{|w-w_0|^{m_a}}{d^{m_a}}\quad(4.21.c)$$

where d and w_0 denote the thickness of PML and the coordinate of the interface between PML and the computational domain. The empirical values of the parameters are [14].

$$m = 1, 2, 3, \text{ or } 4 \tag{4.22.a}$$

$$m_a = 1 \text{ or } 2 \tag{4.22.b}$$

$$\sigma_{\max} = \frac{m+1}{\sqrt{\varepsilon_r}150\pi\delta} \tag{4.22.c}$$

$$\kappa_{\max} = 1 \sim 15 \tag{4.22.d}$$

$$a_{\max} = 0 \sim 2 \tag{4.22.e}$$

where δ is the size of the Yee cell. The factor s_w is a function of angular frequency, and the ADE method is required to handle (4.20) in time domain.

As shown in Figure 4.5, if the top layer is a regular computational domain with ε_1 and μ_1, which satisfies the Maxwell's equations, the coordinate stretched factors are equal to 1. The bottom layer is the PML with ε_2 and μ_2, which satisfies the coordinate stretched Maxwell's equations, and the coordinate stretched factor are $s_{2,x}$, $s_{2,y}$, and $s_{2,z}$. When the interface between the two layers is perpendicular to the z-axis, the nonreflective conditions are $\varepsilon_1 = \varepsilon_2$, $\mu_1 = \mu_2$, and $s_{1,x} = s_{2,x} = 1$. There is no restriction on the normal factor $s_{2,z}$. The coordinate stretched wave vector $\mathbf{k}_s$ is

$$\mathbf{k}_s = \hat{\mathbf{x}}\frac{k_x}{s_x} + \hat{\mathbf{y}}\frac{k_y}{s_y} + \hat{\mathbf{z}}\frac{k_z}{s_z} \tag{4.23}$$

and the transmitted electric field in the PML will be

$$\mathbf{E}_t = \mathbf{E}e^{-j\mathbf{k}_s\cdot\mathbf{r}} = \mathbf{E}e^{-\mathrm{Im}\{s_{2,z}\}z}e^{-j(k_{2,x}x+k_{2,y}y+\mathrm{Re}\{s_{2,z}\}z)} \tag{4.24}$$

Take the z component of the coordinate stretched Faraday's laws as an example for the implementation of CFS-PML as

$$-j\omega\mu H_z = \frac{1}{s_x}\frac{\partial E_y}{\partial x} - \frac{1}{s_y}\frac{\partial E_x}{\partial y} \tag{4.25}$$

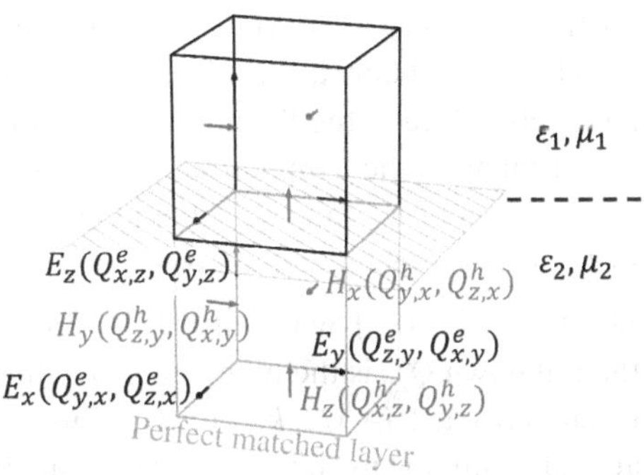

Figure 4.5 The interface between the regular computational domain and PML.

Here, an auxiliary unknown is introduced as

$$Q^e_{x,y} = -\frac{\sigma_x}{\kappa_x}\frac{1}{\kappa_x(j\omega\varepsilon_0 + a_x) + \sigma_x}\frac{\partial E_y}{\partial x} \tag{4.26}$$

Transforming (4.26) into the time domain yields the ADE

$$\kappa_x\varepsilon_0\frac{\partial Q^e_{x,y}}{\partial t} + (\kappa_x a_x + \sigma_x)Q^e_{x,y} = -\frac{\sigma_x}{\kappa_x}\frac{\partial E_y}{\partial x} \tag{4.27}$$

and another auxiliary unknown $Q^e_{y,x}$ is introduced similarly, the corresponding ADE of which is

$$\kappa_y\varepsilon_0\frac{\partial Q^e_{y,x}}{\partial t} + (\kappa_y a_y + \sigma_y)Q^e_{y,x} = -\frac{\sigma_y}{\kappa_y}\frac{\partial E_x}{\partial y} \tag{4.28}$$

Meanwhile, the terms on the right side of Eq. (4.25) can be divided as

$$\frac{1}{s_x}\frac{\partial E_y}{\partial x} = \left(\frac{1}{\kappa_x} - \frac{\sigma_x}{\kappa_x}\frac{1}{\kappa_x(j\omega\varepsilon_0 + a_x) + \sigma_x}\right)\frac{\partial E_y}{\partial x} \tag{4.29.a}$$

$$\frac{1}{s_y}\frac{\partial E_x}{\partial y} = \left(\frac{1}{\kappa_y} - \frac{\sigma_y}{\kappa_y}\frac{1}{\kappa_y(j\omega\varepsilon_0 + a_y) + \sigma_y}\right)\frac{\partial E_x}{\partial y} \tag{4.29.b}$$

Substituting (4.29.a) into (4.25) and transforming it into the time domain leads to

$$-\mu\frac{\partial H_z}{\partial t} = \frac{1}{\kappa_x}\frac{\partial E_y}{\partial x} - \frac{1}{\kappa_y}\frac{\partial E_x}{\partial y} + Q^e_{x,y} - Q^e_{y,x} \tag{4.30}$$

It can be found that the differential equation with complex numbers s_x and s_y in (4.25) becomes the differential equation (4.30) with only real numbers, at the expense of introducing the ADEs in (4.27) and (4.28). This method will also be used later when dealing with dispersive susceptibilities.

As for other PML regions, the auxiliary unknowns will be $Q^e_{x,y}$, $Q^e_{y,x}$, $Q^e_{z,x}$, $Q^e_{x,z}$, $Q^e_{y,z}$, and $Q^e_{z,y}$ based on E_z,E_y, and E_x, and $Q^h_{x,y}$, $Q^h_{y,x}$, $Q^h_{z,x}$, $Q^h_{x,z}$, $Q^h_{y,z}$, and $Q^h_{z,y}$ based on H_z, H_y, and H_x. The ADEs have the same form with (4.27), (4.28), and (4.30), which can be obtained according to the coordinate substitution. Next, these new ADEs are solved in the Yee grid with leapfrog time integration scheme.

Starting with the temporal sampling, as present in (4.27), the unknown $Q^e_{x,y}$ is the function of E_y, thus $Q^e_{x,y}$ should be sampled at the same integral time steps as E_y. This sampling method is also used for other auxiliary unknowns, like Q^h sampled at $\left(n + \frac{1}{2}\right)$ steps with H. For the spatial sampling, as present in (4.30), the unknown $Q^e_{x,y}$ is the function of H_z, thus $Q^e_{x,y}$ should be sampled at the same location $(i+1/2, j+1/2, k)$ with H_z. This sampling method is also used for other auxiliary unknowns, like $Q^e_{y,x}$ locating at $(i+1/2, j+1/2, k)$. For simplicity, the same simplified spatial index is used to record Q^e and Q^h in the memory array, and the explicit updating equations for (4.27) and (4.28) are

$$Q^{e,i,j,k}_{x,y,n+1} = b_x^{i+1/2} Q^{e,i,j,k}_{x,y,n} - \frac{c_x^{i+1/2}}{\Delta x}\left(E^{i+1,j,k}_{y,n+1} - E^{i,j,k}_{y,n+1}\right) \tag{4.31}$$

$$Q^{e,i,j,k}_{y,x,n+1} = b_y^{j+1/2} Q^{e,i,j,k}_{y,x,n} - \frac{c_y^{j+1/2}}{\Delta y}\left(E^{i,j+1,k}_{x,n+1} - E^{i,j,k}_{x,n+1}\right) \tag{4.32}$$

for (4.30) is

$$\begin{aligned} H^{i,j,k}_{z,n+1/2} &= H^{i,j,k}_{z,n-1/2} + Q^{e,i,j,k}_{y,x,n} - Q^{e,i,j,k}_{x,y,n} \\ &+ \frac{\Delta t}{\mu \kappa_y^{j+1/2} \Delta y}\left(E^{i,j+1,k}_{x,n} - E^{i,j,k}_{x,n}\right) + \frac{\Delta t}{\mu \kappa_x^{i+1/2} \Delta x}\left(E^{i+1,j,k}_{y,n} - E^{i,j,k}_{y,n}\right) \end{aligned} \tag{4.33}$$

where

$$b_w^v = \exp\left(-\frac{\Delta t}{\kappa_w(v\Delta w)\varepsilon_0}\right) \tag{4.34.a}$$

$$c_w^v = \frac{\sigma_w(v\Delta w)}{\kappa_w(v\Delta w)} \frac{1}{\kappa_w(v\Delta w) a_w(v\Delta w) + \sigma_w(v\Delta w)} \left(1 - b_w^v\right) \tag{4.34.b}$$

Updating (4.31), (4.32), and (4.33) are the implement of explicit CFS-PML in FDTD, while other updating equations for remaining unknowns can be obtained by coordinate replacement or referring to [14].

4.1.1.3 Source Excitation

Different from the frequency-domain methods using the steady-state or time-harmonic signal as the source excitation, i.e., s = constant, or $s(\omega)$, the time-domain methods generally use impulse or sinusoidal steady-state signal as the source excitation that the source signature should be a time-varying function $s = s(t)$ without any complex number. Meanwhile, the source may have multiple types, such as lumped port source, wave port source, discrete voltage or current, incident plane wave, and so on. In this section, the implementation of the incident plane wave excitation will be introduced, which is the most used scenario in metasurface simulations.

For the source signatures, the Gaussian pulse and the normalized first-order differentiated Blackman-Harris window (BHW1d) are common choices. Readers interested in other signatures, such as the modulated pulse and sinusoidal steady-state source, may refer to [14]. Perhaps, the most prevalent for time-domain methods is the Gaussian pulse

$$s(t) = g(t, t_w) = \exp\left[-\frac{(t - 4t_w)^2}{t_w^2}\right] \tag{4.35}$$

where t_w is its half width. The BHW1d is expressed as [15]

$$s(t) = \text{BHW1d}(t, T) = \begin{cases} -1.47 \sum_{n=1}^{3} a_n n \sin\left(\frac{2\pi nt}{T}\right), & 0 < t < T \\ 0, & \text{otherwise} \end{cases} \tag{4.36}$$

where T represents the effective duration of the BHW1d and

$$a = [-0.488 \quad 0.145 \quad -0.01022222] \tag{4.37}$$

Figure 4.6 shows the time-domain waveform of BHW1d and Gaussian pulse, both with characteristic frequency of $f_c = 10$ GHz. The half-width of the Gaussian pulse and the effective duration of the BHW1d have the following relations:

$$t_w = \frac{1}{\pi f_c} \tag{4.38.a}$$

$$T = \frac{1.55}{f_c} \tag{4.38.b}$$

The frequency-domain energy distribution of the source signal may be more meaningful than the time-domain waveform. Figure 4.7 shows the Fourier transform of these two signals, where the Gaussian pulse has the maximum energy at 0 GHz and the energy drops off to less than 1% (−20 dB) at twice of the characteristic frequency (20 GHz), while the BHW1d has the maximum energy at about 7.38 GHz and less than 1% at 0 GHz and about 21.62 GHz. The frequency-domain distribution illustrates that the Gaussian pulse is suitable for applications that require a direct current (DC) signal component, while BHW1d is suitable for applications that should have zero energy at DC, and BHW1d signal is more concentrated around the characteristic frequency. Both the wavelengths corresponding to

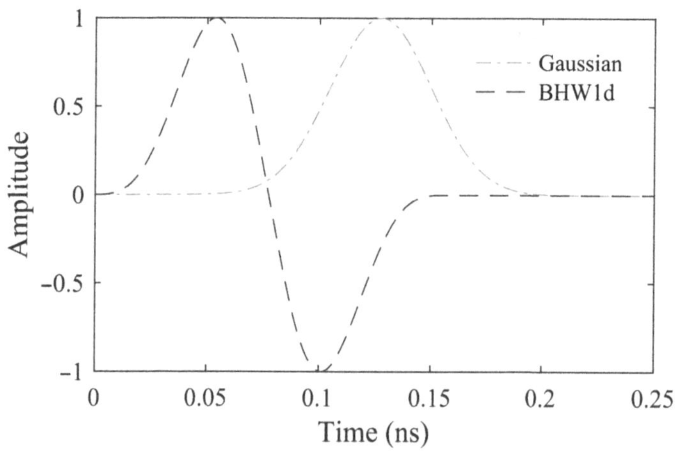

Figure 4.6 The time-domain waveform of BHW1d and Gaussian pulse ($f_c = 10$ GHz).

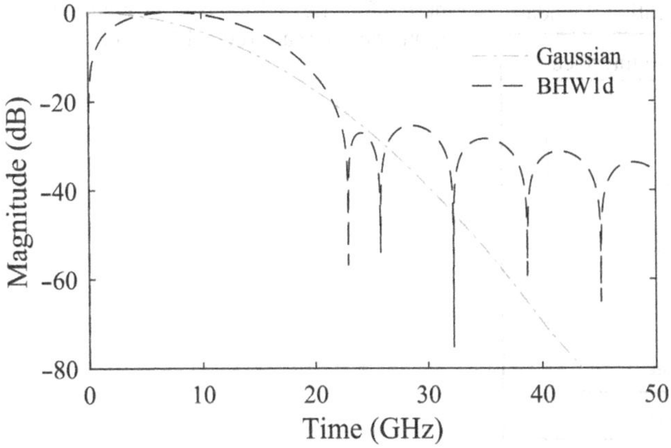

Figure 4.7 The Fourier transform of the BHW1d and Gaussian pulse ($f_c = 10$ GHz).

the maximum frequency (−20 dB) of these two signals are sufficient to determine the spatial FDTD discretization.

Considering that the simulation of metasurfaces often requires ratio results, such as the reflection coefficient and transmission coefficient, which need to be normalized according to the used excitation source. It is important select the appropriate excitation pulse and set the appropriate excitation characteristic frequency, so that the effective frequency range of the source can fully cover the interested frequency range.

After the source signature, the total-field/scattered-field (TF/SF) technique to impress the incident plane wave source is introduced. In most kinds of metasurface problems, the excitation sources are not directly loaded on the metasurface, but are loaded on the computational domain in the form of field sources, such as the plane wave sources and the near-field sources. The TF/SF technique is employed in this situation.

The total electromagnetic field comprises the superposition of the predefined incident field and the induced scattered field produced by the device under test (DUT), which is mainly the metasurface in this book. By artificially defining the TF/SF interface, as shown in Figure 4.8, and setting the corresponding equivalent surface electric and magnetic current on the interface according to Huygens's equivalent principle [16], the incident field can only exist in the total-field region. As shown in Figure 4.9, when there is no DUT, the total-field region has only the preset incident field and the scattered field region is zero.

As illustrated in (4.6.a), the updating of unknowns require the adjacent fields, when the local unknown is in the same region with the adjacent unknowns, the updating equations do not require any changes. When the local unknown is in a different region with the adjacent unknowns, the equivalent current needs

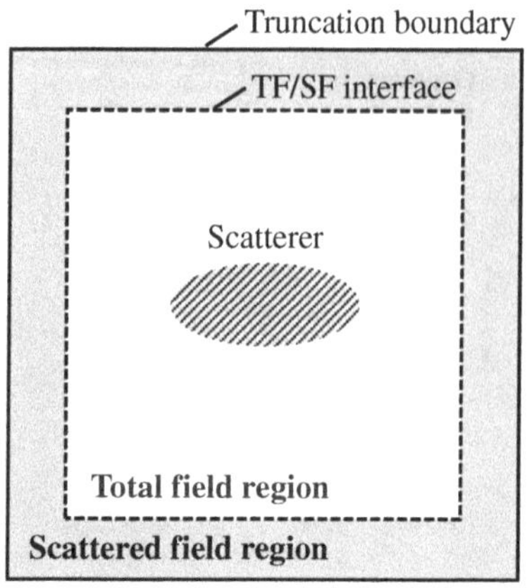

Figure 4.8 The division of the total field region and the scattered field region.

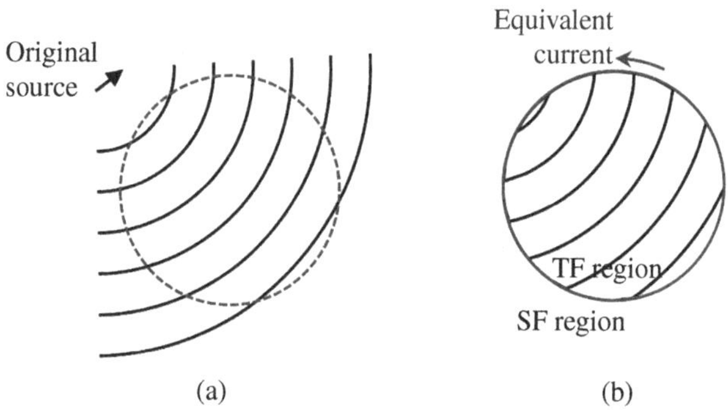

Figure 4.9 Equivalent currents produce preset fields.

to be added or subtracted based on the region where the local unknown are located. If the local unknown is in the total-field region and one of its adjacent unknowns is in the scattered-field region, he desired incident field should be added when updating the local total-field unknowns. On the contrary, if the local unknown is in the scattered-field region and one of its adjacent unknowns is in the total-field region, the incident field should be subtracted when updating the local scattered-field known. Taking Figure 4.10 as an example, the grid with the solid blue lines in (a) represents the cell in the total-field region, and the grid with dash-dot red lines represents the cell in the scattered-field region, these can be regarded as the upper-left area in Figure 4.8. Thus, the unknowns $E_z^{i,j,k}$, $H_x^{i,j,k}$, and $H_y^{i,j,k}$ are total-field unknowns, and $H_x^{i,j-1,k}$ and $H_y^{i-1,j,k}$ are scattered-field ones. When updating the total-field $E_z^{i,j,k}$, the unknowns $H_x^{i,j-1,k}$ and $H_y^{i-1,j,k}$ in the scattered-field region should subtract the incident magnetic fields as

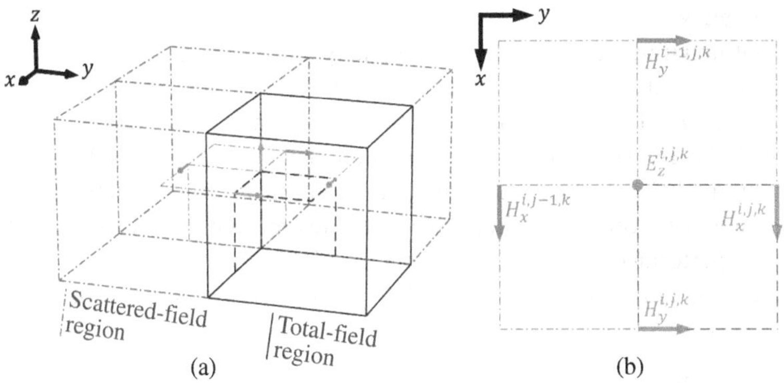

Figure 4.10 The interface of total-field region and scattered-field region: (a) perspective view and (b) top view. The unknowns $E_z^{i,j,k}$, $H_x^{i,j,k}$, and $H_y^{i,j,k}$ are in the total-field region and the unknowns $H_x^{i,j-1,k}$ and $H_y^{i-1,j,k}$ are in the scattered-field region.

$$
\begin{aligned}
E_{z,n+1}^{i,j,k} &= E_{z,n}^{i,j,k} + \frac{\Delta t}{\varepsilon_z^{i,j,k}\Delta x}\left(H_{y,n+1/2}^{i,j,k} - H_{y,n+1/2}^{i-1,j,k}\right) \\
&\quad - \frac{\Delta t}{\varepsilon_z^{i,j,k}\Delta y}\left(H_{x,n+1/2}^{i,j,k} - H_{x,n+1/2}^{i,j-1,k}\right) \\
&\quad - \left[-\frac{\Delta t}{\varepsilon_z^{i,j,k}\Delta x}H_{y,\,inc}^{(i-1,j,k,n+1/2)} + \frac{\Delta t}{\varepsilon_z^{i,j,k}\Delta y}H_{x,\,inc}^{(i,j-1,k,n+1/2)}\right]
\end{aligned}
\tag{4.39}
$$

where the subscript *inc* denotes the known source excitation whose value depends on the location of the unknown and the time. When updating the scattered-field unknown $H_x^{i,j,k}$ or $H_y^{i,j,k}$, $E_z^{i,j,k}$ in the total-field region should add the incident electric fields as

$$
\begin{aligned}
H_{x,n+1/2}^{i,j,k} &= H_{x,n-1/2}^{i,j,k} - \frac{\Delta t}{\mu_x^{i,j,k}\Delta y}\left(E_{z,n}^{i,j+1,k} - E_{z,n}^{i,j,k}\right) + \frac{\Delta t}{\mu_x^{i,j,k}\Delta z}\left(E_{y,n}^{i,j,k+1} - E_{y,n}^{i,j,k}\right) \\
&\quad + \left(\frac{\Delta t}{\mu_x^{i,j,k}\Delta y}E_{z,n,inc}^{i,j,k}\right)
\end{aligned}
\tag{4.40.a}
$$

$$
\begin{aligned}
H_{y,n+1/2}^{i,j,k} &= H_{y,n-1/2}^{i,j,k} - \frac{\Delta t}{\mu_y^{i,j,k}\Delta z}\left(E_{x,n}^{i,j,k+1} - E_{x,n}^{i,j,k}\right) + \frac{\Delta t}{\mu_y^{i,j,k}\Delta x}\left(E_{z,n}^{i+1,j,k} - E_{z,n}^{i,j,k}\right) \\
&\quad + \left(-\frac{\Delta t}{\mu_y^{i,j,k}\Delta x}E_{z,\,inc,\,n}^{i,j,k}\right)
\end{aligned}
\tag{4.40.b}
$$

Equations (4.39) and (4.40) are the TF/SF technique used in this book; the other TF/SF equations at other interface (top, bottom, right, and front) can be derived in the same way.

4.1.1.4 Stability

The explicit updating Eqs. (4.6.a) bring simplicity but also face the challenge of stability, which leads to the need of small time intervals dictated by the Courant–Friedrichs–Lewy (CFL) condition. This condition guarantees the stability of the algorithm. If the time step is too large, the simulation may become unstable, resulting in an infinitely increasing field strength, which is inconsistent with the actual situation. In order to confirm the upper bound of the time interval, a stability analysis must be performed to meet the CFL condition [17].

The updating equations in (4.6.a) can be expressed as two systems of linear equations [14]

$$\mathbf{h}_{n+1/2} = \mathbf{h}_{n-1/2} - \frac{\Delta t}{\mu}\mathbf{C}_e\mathbf{e}_n \tag{4.41.a}$$

$$\mathbf{e}_{n+1} = \mathbf{e}_n + \frac{\Delta t}{\varepsilon}\mathbf{C}_h\mathbf{h}_{n+1/2} \tag{4.41.b}$$

where $\mathbf{h}$ is the vector composed of all the magnetic field unknowns and $\mathbf{e}$ is the vector composed of all the electric field unknowns, while $\mathbf{C}_e$ and $\mathbf{C}_h$ are the sparse matrices representing the discrete curl of the electric and magnetic field unknowns. Combining the linear equations in (4.41.a) leads to

$$\begin{bmatrix}\mathbf{h}_{n+1/2}\\ \mathbf{e}_{n+1}\end{bmatrix} = \begin{bmatrix}\mathbf{I} & -\Delta t\mu^{-1}\mathbf{C}_e\\ \Delta t\varepsilon^{-1}\mathbf{C}_h & \mathbf{I}-(c\Delta t)^2\mathbf{G}\end{bmatrix}\begin{bmatrix}\mathbf{h}_{n-1/2}\\ \mathbf{e}_n\end{bmatrix} \tag{4.42}$$

where $\mathbf{I}$ denotes the identity matrix and $\mathbf{G} = \mathbf{C}_h\mathbf{C}_e$ denotes the discrete curl-curl operator, while $c = \sqrt{\varepsilon\mu}^{-1}$ denotes the speed of light in this material. Defining

$$\mathbf{u}_n = \begin{bmatrix}\mathbf{h}_{n-1/2}\\ \mathbf{e}_n\end{bmatrix}, \tag{4.43}$$

Equation (4.43) can be expressed in a compact form as

$$\mathbf{u}_{n+1} = \mathbf{C}\mathbf{u}_n \tag{4.44}$$

where

$$\mathbf{C} = \begin{bmatrix}\mathbf{I} & -\Delta t\mu^{-1}\mathbf{C}_e\\ \Delta t\varepsilon^{-1}\mathbf{C}_h & \mathbf{I}-(c\Delta t)^2\mathbf{G}\end{bmatrix} \tag{4.45}$$

If the initial value of the field unknown vector is $\mathbf{u}_0$, the vector at n-th time step can be expressed as

$$\mathbf{u}^n = \mathbf{C}^n\mathbf{u}^0 \tag{4.46}$$

To ensure the stability of the passive linear system, an upper bound constant C, should exist as

$$\lim_{n\to\infty}\frac{\|\mathbf{u}^n\|}{\|\mathbf{u}^0\|} \le C \tag{4.47}$$

A necessary condition for (4.47) is that the amplitude of all the eigenvalues of **C** must be not larger than 1, i.e.,

$$|\lambda_c| \leq 1 \tag{4.48}$$

where λ_c denotes the eigenvalues of **C**. Since the curl–curl matrix **G** is positive-definite and symmetric, it is diagonalizable with the projection matrix **P** as

$$\mathbf{P}^{-1}\mathbf{G}\mathbf{P} = \mathbf{D}_g \tag{4.49}$$

where $\mathbf{D}_g$ is a diagonal matrix consisting of λ_g (the eigenvalues of **G**). Now, deduce the constraints on Δx, Δy, and Δz for Δt [18]. The eigenvalue equation of **C** is

$$\mathbf{C}\mathbf{v} = \lambda_c \mathbf{v} \tag{4.50}$$

where **v** denotes the corresponding eigenvector of λ_c. Subtracting **v** from both sides of (4.50) leads to

$$(\mathbf{C} - \mathbf{I})\mathbf{v} = \begin{bmatrix} 0 & -\Delta t\mu^{-1}\mathbf{C}_e \\ \Delta t\varepsilon^{-1}\mathbf{C}_h & -(c\Delta t)^2\mathbf{G} \end{bmatrix} \begin{bmatrix} \mathbf{v}_1 \\ \mathbf{v}_2 \end{bmatrix} = (\lambda_c - 1) \begin{bmatrix} \mathbf{v}_1 \\ \mathbf{v}_2 \end{bmatrix} \tag{4.51}$$

Multiplying the first row of (4.51) by $\Delta t\varepsilon^{-1}\mathbf{C}_h$ and substituting it into the second row leads to

$$\left[(\lambda_c - 1)^2 + \lambda_c(c\Delta t)^2\mathbf{G}\right]\mathbf{v}_2 = 0 \tag{4.52}$$

Considering (4.49), the following formula is obtained:

$$\left[(\lambda_c - 1)^2\mathbf{I} + \lambda_c c^2\Delta t^2\mathbf{D}_g\right](\mathbf{P}^{-1}\mathbf{v}_2) = 0 \tag{4.53}$$

A necessary condition for (4.53) to be zero is that

$$(\lambda_c - 1)^2 + \lambda_c c^2\Delta t^2\lambda_g = 0 \tag{4.54}$$

Solving for λ_c from (4.54) yields

$$\lambda_c = 1 - \frac{c^2\Delta t^2\lambda_g}{2} \pm \sqrt{\left(\frac{c^2\Delta t^2\lambda_g}{2}\right)^2 - c^2\Delta t^2\lambda_g} \tag{4.55}$$

Since **G** is a positive definite matrix, λ_g must be positive real. Then, the stability condition under (4.48) is expressed as

$$\frac{c^2\Delta t^2\lambda_g}{2} \leq 2 \tag{4.56}$$

It can be shown that for a uniformly spaced rectangular grid, the eigenvalue of the curl–curl operator λ_g has the property [2] of

$$\lambda_g \leq 4\left(\frac{1}{\Delta x^2} + \frac{1}{\Delta y^2} + \frac{1}{\Delta z^2}\right) \tag{4.57}$$

Substituting (4.57) into (4.56) leads to

$$\Delta t \leq \frac{1}{c\sqrt{\frac{1}{\Delta x^2}+\frac{1}{\Delta y^2}+\frac{1}{\Delta z^2}}} \tag{4.58}$$

This is the CFL stability limit of the 3-D FDTD based on Yee grids, which restrict the employment of the time interval and the length of the mesh grid. When Δt, Δx, Δy, and Δz are determined, the CFL number (CFLN) N_{CFL} is defined as

$$N_{CFL} = c_{\max}\Delta t\sqrt{\frac{1}{\Delta x^2}+\frac{1}{\Delta y^2}+\frac{1}{\Delta z^2}} \tag{4.59}$$

4.1.2 Dispersive Susceptibility and Parameter Fitting

Modeling dispersive susceptibilities is important to many metasurface simulations. As introduced in Section 2.1.2.2, when the frequency increases, the susceptibility begins to show strong frequency dependence. In this section, some dispersive models will be introduced to describe the temporal dispersion of surface susceptibilities. The method of creating reduced order dispersive models with values sampled from the susceptibilities is called parameter fitting.

The Lorentz dispersive model has been introduced in Section 2.1.2.2. Here, its expression is repeated for the convenience.

$$\chi(\omega;\omega_0,\psi,\Gamma) = \frac{\psi}{\omega_0^2 + j2\omega\Gamma - \omega^2} \tag{4.60}$$

One can also use the sum of Lorentzian functions as the appropriate model

$$\chi(\omega;\ \omega_{0,q},\ \psi_q,\ \Gamma_q) = \sum_{q=1}^{N_q}\frac{\psi_q^2}{\omega_{0,q}^2 + j2\omega\Gamma_q - \omega^2} \tag{4.61}$$

In addition, there are the well-known Debye and Drude dispersion models, whose frequency correlations are rare in the frequency responses of metasurface. In order to deal with the complex dispersion phenomena of metasurfaces, there are two generalized dispersion models, i.e., the rational functions of polynomials

$$\chi(\omega;\ A_q,\ B_q) = \frac{\sum_{q=0}^{N_q} A_q(j\omega)^q}{\sum_{q=0}^{N_q} B_q(j\omega)^q} \tag{4.62}$$

and the complex–conjugate pole-residue (CCPR)

$$\chi(\omega;c_0,\delta_m,\gamma_m) = c_0 + \sum_{q=1}^{N_q}\frac{\delta_q}{j\omega-\gamma_q} \tag{4.63}$$

where A_q, B_q, and c_0 are all real numbers, and δ_q and γ_q can be either real or complex numbers simultaneously. The integer q denotes the index of the dispersive function terms whose total number is N_q. There is no doubt that the dispersion functions in (4.62) and (4.63) can be converted to each other by arithmetic [19]. The difference is the parameter fitting method to obtain the corresponding parameters when approximating them. Considering a series of discrete susceptibilities

$$\chi_k = \chi(j\omega_k), \quad k = 1, 2, \ldots, N_k \tag{4.64}$$

where N_k denotes the number of sampling data. Determine a rational function that approximates the given susceptibility

$$\widetilde{\chi}(\omega) \approx \chi_k, \quad \forall k = 1, \ldots, N_k \tag{4.65}$$

where the rational function $\widetilde{\chi}(\omega)$ is the dispersive model required to represent the original sampled susceptibility χ_k. The Fourier transform pairs of this rational function should satisfy causality, as introduced in Section 2.1.2.2, so that it can be employed in a time-domain algorithm [20]. The Lorentz model (4.60) satisfies these requirements, and the rational models in (4.62) and (4.63) have some limitations in their parameters.

When using the ratio of polynomial functions in (4.62) for the approximations problem in (4.65), one can choose the Levy algorithms [21] or the Sanathanan–Koerner algorithms [22]. Both these algorithms determine the coefficients A_q and B_q by minimizing a l_2-norm of the difference between the sample data and function values. In terms of susceptibilities, that is, to minimize

$$\Delta^2 = \frac{1}{N_q}\sum_{q=1}^{N_q} |\chi_q - \widetilde{\chi}(\omega)|^2 \tag{4.66}$$

The Levy method linearizes (4.66) by neglecting the denominator $b_n(j\omega)^b$ and minimize

$$\Delta^2 = \frac{1}{N_q}\sum_{q=1}^{N_q} \left| \chi_q \sum_{q=1}^{N_q} B_q(j\omega)^n - \sum_{q=1}^{N_q} A_q(j\omega)^n \right|^2 \tag{4.67}$$

which leads to a system of linear equations and is directly solved in the least squares sense. The coefficients A_q and B_q are regarded as the model parameters of $\widetilde{\chi}(\omega)$. Unfortunately, the Levy method requires the denominator to be approximately constant in the sampling frequency, and the system matrix is a Vandermonde matrix that is ill-conditioned [23]. To overcome this issue, the Sanathanan–Koerner method employs an iterative process to improve the quality of the solution, regarding the Levy algorithm as result of the first iteration ($i = 1$)

and minimizing the following linearization in successive iterations ($i \geq 2$)

$$(\Delta^2)^{(i)} = \frac{1}{N_q} \sum_{q=1}^{N_q} \left| \frac{\chi_q \sum_{q=1}^{N_q} B_q^{(i)} (j\omega)^q - \sum_{q=1}^{N_q} A_q^{(i)} (j\omega)^q}{\sum_{q=1}^{N_q} b_q^{(i-1)} (j\omega)^q} \right|^2 \tag{4.68}$$

As the iterative process converges, the coefficients in the last iteration are regarded as the model parameters, i.e., $A_q = A_q^{(i)}$ and $B_q = B_q^{(i)}$.

When using the CCPR in (4.63) to solve (4.65), one can choose the vector fitting (VF) algorithm [24, 25]. The VF method is an iteration method and needs a set of initial poles $\gamma_q^{(0)}$. The fitting process minimizes the following linearization:

$$(\Delta^2)^{(i)} = \frac{1}{N_q} \sum_{q=1}^{N_q} |\chi_q W^{(i)}(\omega) - \widetilde{\chi}^{(i)}(\omega)|^2 \tag{4.69}$$

where $W(\omega)$ is the function with weight coefficient w_m

$$W^{(i)}(\omega) = 1 + \sum_{q=1}^{N_q} w_q \frac{1}{j\omega - \gamma_q^{(i-1)}} \tag{4.70}$$

and the iterative functions $\widetilde{\chi}^{(i)}(\omega)$ are

$$\widetilde{\chi}^{(i)}(\omega) = c^{(i)} + \sum_{q=1}^{N_q} \frac{\delta_q^{(i)}}{j\omega - \gamma_q^{(i-1)}} \tag{4.71}$$

As the iterative process converges, the coefficients in the last iteration of (4.71) are regarded as the model parameters. An implementation of the VF algorithm based on MATLAB is available in [26].

In addition to the above fitting methods, there are also many commercial software with parameter fitting methods, like the "Rational Models" in MATLAB [27] or the "Dielectric Dispersion Fit" in CST. The fitting algorithm is not important in simulation, as long as the appropriate dispersion model is obtained. The model in (4.61) has been used in FDTD-GSTC with the ADE method [28], and the models in (4.62) and (4.63) have been used in FDTD-GSTC with convolution [29, 30].

4.1.3 Strategies for Integrating GSTC in FDTD

In this section, the strategy to incorporate GSTC in FDTD is introduced. As an interface condition, the GSTC is different from the boundary conditions described in Section 4.1.1.2. These mentioned boundary conditions describe the fields (or their derivatives) on only one side of the boundary. In the updating formulas, the unknowns on the boundary are affected by only one-side adjacent unknowns, while by unknowns from both sides in GSTC. Therefore, the implementation of the GSTC is more complex than the boundary conditions introduced in Section 4.1.1.2. To better explain how to impose such a transition condition, an example of how the IBCs are implemented in 1-D FDTD is first presented.

The expression of IBC has been presented in Section 2.1.3.2. Here, a surface with conductivity ($Z_e = \frac{1}{\sigma_s}, Y_m = 0$) is taken as an example [31]. Substituting (2.32) into (2.33) leads to

$$\hat{\mathbf{n}} \times (\mathbf{H}_2 - \mathbf{H}_1) = \sigma_s \mathbf{E}_s \tag{4.72}$$

where the subscripts 1 and 2 denote the two regions separated by the interface with σ_s, and $\hat{\mathbf{n}}$ is the unit vector originating from region 1 to 2 and normal to the interface.

The condition (4.72) tells that the surface with conductivity will cause a jump of tangential magnetic fields, thus the IBC can be implemented at the magnetic field sampling points in the Yee grid [31]. As shown in Figure 4.11(a), an IBC normal to the y-direction is modeled at the $H_x(H_z)$ sampling points. For simplicity, a plane wave traveling along y-direction is considered, and the IBC can be located at H_x^j as presented in Figure 4.11(b). The original unknown H_x^j is hence divided into two new unknowns $H_x^{j^-}$ and $H_x^{j^+}$ for capacity of describing the discontinuity caused by surface conductivity. Then, a virtual electric field unknown $E_z^{j^0}$ is added to maintain the characteristic spatial stagger of the electric field and magnetic field in the Yee grid. Discretizing Faraday's laws at j^- leads to

$$\frac{\mu}{\Delta t}\left(H_{x,n+1/2}^{j^-} - H_{x,n-1/2}^{j^-}\right) = \frac{2}{\Delta y}\left(E_{z,n}^{j^0} - E_{z,n}^{j}\right) \tag{4.73}$$

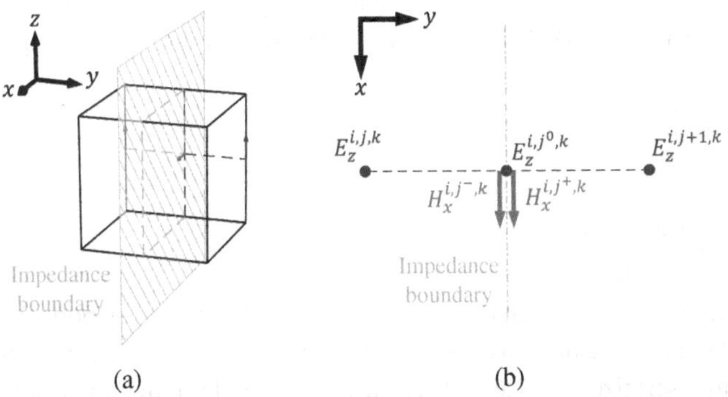

Figure 4.11 Implementation of the y-direction IBC in Yee grid. (a) Perspective view and (b) top view. The hatched surface in (a) and the dot-dashed line in (b) both denote the impedance boundary.

and at j^+ leads to

$$\frac{\mu}{\Delta t}\left(H^{j^+}_{x,n+1/2} - H^{j^+}_{x,n-1/2}\right) = \frac{2}{\Delta y}\left(E^{j+1}_{z,n} - E^{j^0}_{z,n}\right) \tag{4.74}$$

Discretizing (4.72) at j^0 leads to

$$\frac{1}{2}\left(H^{j^+}_{x,n+1/2} + H^{j^+}_{x,n-1/2} - H^{j^-}_{x,n+1/2} - H^{j^-}_{x,n-1/2}\right) = \sigma_s E^{j^0}_{z,n} \tag{4.75}$$

Combining (4.73), (4.74), and (4.75) leads to the updating equations of the new unknowns

$$\begin{aligned} H^{j^-}_{x,n+1/2} &= \frac{\Delta t}{(\mu\Delta y\sigma_s + 2\Delta t)} \\ &\quad \times \left(\frac{2\Delta t}{\mu\Delta y}E^{j+1}_{z,n} - \frac{2(\mu\Delta y\sigma_s + \Delta t)}{\mu\Delta y}E^{j}_{z,n} + H^{j^-}_{x,n-1/2} + 2H^{j^+}_{x,n-1/2}\right) \end{aligned} \tag{4.76.a}$$

$$\begin{aligned} H^{j^+}_{x,n+1/2} &= \frac{\Delta t}{(\mu\Delta y\sigma_s + 2\Delta t)} \\ &\quad \times \left(\frac{2\Delta t}{\mu\Delta y}E^{j}_{z,n} - \frac{2(\mu\Delta y\sigma_s + \Delta t)}{\mu\Delta y}E^{j+1}_{z,n} + H^{j^+}_{x,n-1/2} + 2H^{j^-}_{x,n-1/2}\right) \end{aligned} \tag{4.76.b}$$

As can be seen from (4.76.a), $E^{j^0}_{z,n}$ is not required to update $H^{j^-}_x$ and $H^{j^+}_x$, so it is called the virtual unknown, because it does not really exist in the time-stepping process. Equation (4.76.a) is the implementation of IBC in 1-D FDTD, in which two new magnetic field unknowns $H^{j^-}_x$ and $H^{j^+}_x$ take the place of H^{j}_x. Besides, different from (4.8) and (4.18), the updating of unknowns on the interface boundary needs data from both sides of the interface, which will be one of the properties of GSTC implementation. Next, the GSTC implementation in FDTD will be introduced.

4.1.3.1 Basic GSTC Implementation in FDTD

First, implementing GSTC with only tangential anisotropic surface susceptibilities $\overline{\overline{\chi}}_{ee,\parallel}$ and $\overline{\overline{\chi}}_{mm,\parallel}$ in 1-D FDTD [32] is considered. The time-domain anisotropic GSTC writes

$$\hat{\mathbf{n}} \times \Delta\mathbf{H} = \varepsilon_0 \frac{\partial \overline{\overline{\chi}}_{ee,\parallel}\mathbf{E}_{\parallel}}{\partial t} \tag{4.77.a}$$

$$\hat{\mathbf{n}} \times \Delta\mathbf{E} = -\mu_0 \frac{\partial \overline{\overline{\chi}}_{mm,\parallel}\mathbf{H}_{\parallel}}{\partial t} \tag{4.77.b}$$

where $\parallel$ denotes the tangential components. Since the GSTC describes both the magnetic field and the electric field discontinuities brought by the surface susceptibilities, it is best to discretize (4.77) at neither the electric field nor magnetic field sampling points. Instead, a good choice is the middle of the two sampling points, as shown in Figure 4.12(a). The virtual field points $E^{i,j^g,k}_z$ and $H^{i,j^g,k}_x$ are introduced on both sides of the generalized sheet to describe the field discontinuities and

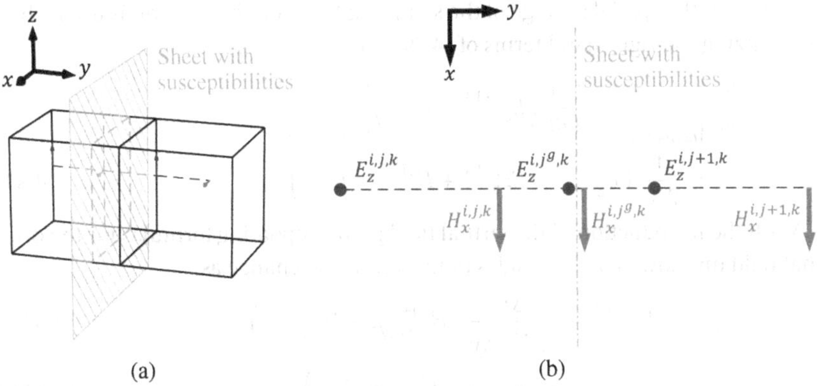

Figure 4.12 Implementation of the y-direction GSTC in 1-D FDTD. (a) Perspective view and (b) top view. The hatched surface in (a) and the dot-dashed line in (b) both denote the generalized sheet with surface susceptibilities.

to ensure the Yee grid's characteristic spatial stagger, as shown in Figure 4.12(b). Hence, condition (4.77.a) can be written as

$$\begin{bmatrix} -\Delta H_x \\ \Delta H_z \end{bmatrix} = \varepsilon_0 \begin{bmatrix} \dfrac{\partial}{\partial t}\left(\chi_{ee}^{zx} E_x + \chi_{ee}^{zz} E_z\right) \\ \dfrac{\partial}{\partial t}\left(\chi_{ee}^{xx} E_x + \chi_{ee}^{xz} E_z\right) \end{bmatrix} \tag{4.78.a}$$

$$\begin{bmatrix} \Delta E_x \\ -\Delta E_z \end{bmatrix} = \mu_0 \begin{bmatrix} \dfrac{\partial}{\partial t}\left(\chi_{mm}^{zx} H_x + \chi_{mm}^{zz} H_z\right) \\ \dfrac{\partial}{\partial t}\left(\chi_{mm}^{xx} H_x + \chi_{mm}^{xz} H_z\right) \end{bmatrix} \tag{4.78.b}$$

Since the 1-D FDTD is with only E_z and H_x components, and the surface susceptibilities are time-invariant, (4.78.a) can be simplified as

$$-\Delta H_x = \varepsilon_0 \chi_{ee}^{zz} \frac{\partial E_z}{\partial t} \tag{4.79.a}$$

$$-\Delta E_z = \mu_0 \chi_{mm}^{xx} \frac{\partial H_x}{\partial t} \tag{4.79.b}$$

Discretizing (4.79.a) at the location of $E_z^{i,j^g,k}$ leads to

$$-\left(H_{x,n+1/2}^{i,j^g,k} - H_{x,n+1/2}^{i,j,k}\right) = \frac{\varepsilon_0 \chi_{ee}^{zz}}{2\Delta t}\left(E_{z,n+1}^{i,j+1,k} - E_{z,n}^{i,j+1,k} + E_{z,n+1}^{i,j,k} - E_{z,n}^{i,j,k}\right) \tag{4.80}$$

Discretizing (4.79.b) at the location of $H_x^{i,j^g,k}$ leads to

$$-\left(E_{z,n}^{i,j+1,k} - E_{z,n+1}^{i,j^g,k}\right) = \frac{\mu_0 \chi_{mm}^{xx}}{2\Delta t}\left(H_{x,n+1/2}^{i,j+1,k} - H_{x,n-1/2}^{i,j+1,k} + H_{x,n+1/2}^{i,j,k} - H_{x,n-1/2}^{i,j,k}\right) \tag{4.81}$$

Note that the spatial average of the sampling fields on the Yee grid is used when discretizing the right-hand terms of (4.79.a) as

$$\begin{aligned}\left.\frac{\partial E_z}{\partial t}\right|_{t=(n+1/2)} &= \left.\frac{1}{2}\frac{\partial}{\partial t}\left(E_{z,}^{i,j+1,k} + E_z^{i,j,k}\right)\right|_{t=(n+1/2)} \\ &= \frac{1}{2\Delta t}\left(E_{z,n+1}^{i,j+1,k} - E_{z,n}^{i,j+1,k} + E_{z,n+1}^{i,j,k} - E_{z,n}^{i,j,k}\right) \end{aligned} \tag{4.82}$$

With the introduction of the virtual field points, updating formulas for the original field unknowns on both sides of the sheet also change as

$$E_{z,n+1}^{i,j+1,k} = E_{z,n}^{i,j+1,k} - \frac{\Delta t}{\varepsilon_z^{i,j+1,k}\Delta y}\left(H_{x,n+1/2}^{i,j+1,k} - H_{x,n+1/2}^{i,j^g,k}\right) \tag{4.83.a}$$

$$H_{x,n+1/2}^{i,j,k} = H_{x,n-1/2}^{i,j,k} - \frac{\Delta t}{\mu_x^{i,j,k}\Delta y}\left(E_{z,n}^{i,j^g,k} - E_{z,n}^{i,j,k}\right) \tag{4.83.b}$$

where $H_{z,n+1/2}^{i,j^g,k}$ and $E_{z,n}^{i,j^g,k}$ replace $H_{z,n+1/2}^{i,j,k}$ and $E_{z,n}^{i,j+1,k}$ in the original updating equations, respectively. Conventional Maxwell's equations can no longer describe the relationship between the unknowns across the generalized sheet, and they are only effective on the same side of the sheet. Substituting (4.80) and (4.81) into (4.83.a) leads to

$$\begin{aligned}\left(1 + \frac{\chi_{ee}^{zz}}{2\varepsilon_{r,z}^{i,j+1,k}\Delta y}\right)E_{z,n+1}^{i,j+1,k} &= \left(1 + \frac{\chi_{ee}^{zz}}{2\varepsilon_{r,z}^{i,j+1,k}\Delta y}\right)E_{z,n}^{i,j+1,k} - \frac{\Delta t}{\varepsilon_z^{i,j+1,k}\Delta y} \\ &\times\left(H_{x,n+1/2}^{i,j+1,k} - H_{x,n+1/2}^{i,j^g,k}\right) - \frac{\chi_{ee}^{zz}}{2\varepsilon_z^{i,j+1,k}\Delta y}\left(E_{z,n+1}^{i,j,k} - E_{z,n}^{i,j,k}\right)\end{aligned} \tag{4.84.a}$$

$$\begin{aligned}\left(1 + \frac{\chi_{mm}^{xx}}{2\mu_{r,x}^{i,j,k}\Delta y}\right)H_{x,n+1/2}^{i,j,k} &= \left(1 + \frac{\chi_{mm}^{xx}}{2\mu_{r,x}^{i,j,k}\Delta y}\right)H_{x,n-1/2}^{i,j,k} - \frac{\Delta t}{\mu_x^{i,j,k}\Delta y} \\ &\times\left(E_{z,n}^{i,j^g,k} - E_{z,n}^{i,j,k}\right) - \frac{\chi_{mm}^{xx}}{2\mu_{r,x}^{i,j,k}\Delta y}\left(H_{x,n+1/2}^{i,j+1,k} - H_{x,n-1/2}^{i,j+1,k}\right)\end{aligned} \tag{4.84.b}$$

where ε_r and μ_r denote the relative permittivity and permeability at the corresponding sampling points. Note that updating $E_{z,n+1}^{i,j+1,k}$ according to (4.84.a) needs $E_{z,n+1}^{i,j,k}$, which means that although these two unknowns are at the same time step, the former should be updated after the latter one, and vice versa for $H_{x,n+1/2}^{i,j,k}$ and $H_{x,n+1/2}^{i,j+1,k}$. Some readers may have doubts about the spatial discretization in (4.83.a). For example, in the discretization of $\frac{\partial H_x}{\partial y}$, the distance from $H_x^{i,j+1,k}$ to $E_z^{i,j+1,k}$ is $\frac{\Delta y}{2}$ and the distance from $H_x^{i,j^g,k}$ to $E_z^{i,j+1,k}$ is $\frac{\Delta y}{4}$. The choice of using Δy or $\frac{3}{4}\Delta y$ in spatial approximation is discussed in detail in [33], and the conclusion is that the approximation of using Δy has negligible influence in accuracy. Furthermore, there is an advantage of using the length of the original Yee grid is that it is more convenient to add metasurfaces in a standard FDTD code, without the need to add additional subgrids to the original Yee grids. The updating equation in (4.84.a) is the implementation of tangential anisotropic GSTC in 1-D FDTD. Note that the new field components $E_z^{i,j^g,k}$ and $H_x^{i,j^g,k}$ are virtual fields, just like $E_z^{j^0}$ in IBC implementation, which do not really exist in the time-stepping loop and do not take up any memory.

There is a common class of dispersive metasurfaces, for a normally incident plane wave, there is a fixed transmittance T and reflectivity R in a certain frequency range, without changing the wavefront or phase. The expressions of this class of metasurfaces in the corresponding frequency band are [34]

$$\overline{\overline{\chi}}_{ee,\parallel}(\omega) = \frac{1}{j\omega}\overline{\overline{K}}_{ee}, \quad \overline{\overline{\chi}}_{mm,\parallel}(\omega) = \frac{1}{j\omega}\overline{\overline{K}}_{mm} \tag{4.85}$$

where $\overline{\overline{K}}$ denotes a tensor with all constant elements, and its value depends on the transmission and reflection coefficients. For FDTD-GSTC, such types of surface susceptibilities have a different updating equation from (4.84.a). Consider the frequency-domain form of (4.79.a)

$$-\Delta H_x = \varepsilon_0 j\omega \chi_{ee}^{zz} E_z \tag{4.86.a}$$

$$-\Delta E_z = \mu_0 j\omega \chi_{mm}^{xx} H_x \tag{4.86.b}$$

Substituting (4.85) into (4.86.a) leads to

$$-\Delta H_x = \varepsilon_0 K_{ee}^{zz} E_z \tag{4.87.a}$$

$$-\Delta E_z = \mu_0 K_{mm}^{xx} H_x \tag{4.87.b}$$

Discretizing (4.87.a) at the location of $E_z^{i,j^g,k}$ leads to

$$-\left(H_{x,n+1/2}^{i,j^g,k} - H_{x,n+1/2}^{i,j,k}\right) = \frac{\varepsilon_0 K_{ee}^{zz}}{4}\left(E_{z,n+1}^{i,j+1,k} + E_{z,n}^{i,j+1,k} + E_{z,n+1}^{i,j,k} + E_{z,n}^{i,j,k}\right) \tag{4.88}$$

Discretizing (4.87.b) at the location of $H_x^{i,j^g,k}$ leads to

$$-\left(E_{z,n}^{i,j+1,k} - E_{z,n}^{i,j^g,k}\right) = \frac{\mu_0 K_{mm}^{xx}}{4}\left(H_{x,n+1/2}^{i,j+1,k} + H_{x,n-1/2}^{i,j+1,k} + H_{x,n+1/2}^{i,j,k} + H_{x,n-1/2}^{i,j,k}\right) \tag{4.89}$$

Substituting (4.88) and (4.89) into (4.83.a), the updating equations of surface susceptibilities with the form $\chi(\omega) = K/j\omega$ is obtained

$$\left(1 + \frac{K_{ee}^{zz}\Delta t}{4\varepsilon_{r,z}^{i,j+1,k}\Delta y}\right)E_{z,n+1}^{i,j+1,k} = \left(1 - \frac{K_{ee}^{zz}\Delta t}{4\varepsilon_{r,z}^{i,j+1,k}\Delta y}\right)E_{z,n}^{i,j+1,k} - \frac{\Delta t}{\varepsilon_z^{i,j+1,k}\Delta y}$$
$$\times\left(H_{x,n+1/2}^{i,j+1,k} - H_{x,n+1/2}^{i,j^g,k}\right) - \frac{K_{ee}^{zz}\Delta t}{4\varepsilon_{r,z}^{i,j+1,k}\Delta y}\left(E_{z,n+1}^{i,j,k} + E_{z,n}^{i,j,k}\right) \tag{4.90.a}$$

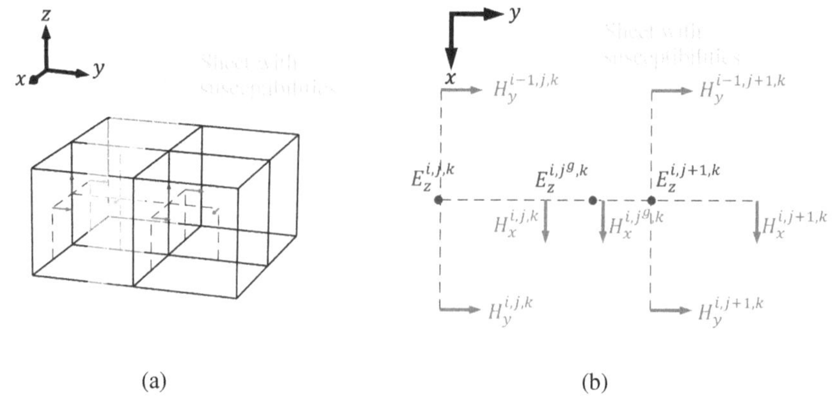

Figure 4.13 Implementation of the y-direction GSTC in 2-D FDTD. (a) Perspective view and (b) top view. The hatched surface in (a) and the dot-dashed line in (b) both denote the generalized sheet with surface susceptibilities.

$$\left(1+\frac{K_{mm}^{xx}\Delta t}{4\mu_{r,x}^{i,j,k}\Delta y}\right)H_{x,n+1/2}^{i,j,k}=\left(1+\frac{K_{mm}^{xx}\Delta t}{4\mu_{r,x}^{i,j,k}\Delta y}\right)H_{x,n-1/2}^{i,j,k}-\frac{\Delta t}{\mu_x^{i,j,k}\Delta y}$$
$$\times\left(E_{z,n}^{i,j^g,k}-E_{z,n}^{i,j,k}\right)-\frac{K_{mm}^{xx}\Delta t}{4\mu_{r,x}^{i,j,k}\Delta y}\left(H_{x,n+1/2}^{i,j+1,k}+H_{x,n-1/2}^{i,j+1,k}\right) \tag{4.90.b}$$

In addition to (4.85), there are other forms of dispersion, such as $\chi(\omega)=\frac{1}{-\omega^2+j\gamma\omega+\omega_0^2}$ and $\chi(\omega)=\frac{\sum_{a=0}^{N_a}p_a(j\omega)^a}{\sum_{b=0}^{N_b}q_b(j\omega)^b}$. The surface susceptibilities of these forms of dispersion require the introduction of ADE and auxiliary unknowns, and this will be introduced in Section 4.1.3.3. Now, the updating equations to 2-D scenarios will be extended. There are TE and TM modes in the 2-D problem. Without loss of generality, consider the TE mode problem ($E_x = E_y = H_z = 0$) here. As shown in Figure 4.13(a), same as the 1-D FDTD-GSTC, the virtual field points $E_z^{i,j^g,k}$ and $H_x^{i,j^g,k}$ are introduced on both sides of the sheet to describe the field discontinuities and to ensure the Yee grid's characteristic spatial stagger, as shown in Figure 4.13(b).

There is no additional virtual field point $H_y^{i,j^g,k}$ because the surface susceptibilities only bring tangential discontinuities. For the basic tangential anisotropic GSTC in (4.79.a), the discretization is the same as (4.80) and (4.81) for 2-D situation. The discretizations of the Ampere's and Faraday's laws are

$$E_{z,n+1}^{i,j+1,k}=E_{z,n}^{i,j+1,k}-\frac{\Delta t}{\varepsilon_z^{i,j+1,k}\Delta y}\left(H_{x,n+1/2}^{i,j+1,k}-H_{x,n+1/2}^{i,j^g,k}\right)$$
$$+\frac{\Delta t}{\varepsilon_z^{i,j+1,k}\Delta x}\left(H_{y,n+1/2}^{i,j+1,k}-H_{y,n+1/2}^{i-1,j+1,k}\right) \tag{4.91.a}$$

$$H_{x,n+1/2}^{i,j,k} = H_{x,n-1/2}^{i,j,k} - \frac{\Delta t}{\mu_x^{i,j,k}\Delta y}\left(E_{z,n}^{i,j^g,k} - E_{z,n}^{i,j,k}\right) \tag{4.91.b}$$

$$H_{y,n+1/2}^{i,j,k} = H_{y,n-1/2}^{i,j,k} + \frac{\Delta t}{\mu_y^{i,j,k}\Delta x}\left(E_{z,n}^{i,j+1,k} - E_{z,n}^{i,j,k}\right) \tag{4.91.c}$$

Where in the third equations, updating H_y is the same as the original equations in (4.6.a), because the unknowns $E_z^{i,j,k}$ and $E_z^{i+1,j,k}$ used to update $H_y^{i,j,k}$ belong to the same side of the generalized sheet with surface susceptibilities. Substituting (4.80) into (4.91.a) and (4.81) into (4.91.b) leads to

$$\begin{aligned}
&\left(1 + \frac{\chi_{ee}^{zz}}{2\varepsilon_{r,z}^{i,j+1,k}\Delta y}\right) E_{z,n+1}^{i,j+1,k} \\
&= \left(1 + \frac{\chi_{ee}^{zz}}{2\varepsilon_{r,z}^{i,j+1,k}\Delta y}\right) E_{z,n}^{i,j+1,k} - \frac{\Delta t}{\varepsilon_z^{i,j+1,k}\Delta y}\left(H_{x,n+1/2}^{i,j+1,k} - H_{x,n+1/2}^{i,j^g,k}\right) \\
&+ \frac{\Delta t}{\varepsilon_z^{i,j+1,k}\Delta x}\left(H_{y,n+1/2}^{i,j+1,k} - H_{y,n+1/2}^{i-1,j+1,k}\right) - \frac{\chi_{ee}^{zz}}{2\varepsilon_z^{i,j+1,k}\Delta y}\left(E_{z,n+1}^{i,j,k} - E_{z,n}^{i,j,k}\right)
\end{aligned} \tag{4.92.a}$$

$$\begin{aligned}
&\left(1 + \frac{\chi_{mm}^{xx}}{2\mu_{r,x}^{i,j,k}\Delta y}\right) H_{x,n+1/2}^{i,j,k} = \left(1 + \frac{\chi_{mm}^{xx}}{2\mu_{r,x}^{i,j,k}\Delta y}\right) H_{x,n-1/2}^{i,j,k} \\
&- \frac{\Delta t}{\mu_x^{i,j,k}\Delta y}\left(E_{z,n}^{i,j^g,k} - E_{z,n}^{i,j,k}\right) - \frac{\chi_{mm}^{xx}}{2\mu_{r,x}^{i,j,k}\Delta y}\left(H_{x,n+1/2}^{i,j+1,k} - H_{x,n-1/2}^{i,j+1,k}\right)
\end{aligned} \tag{4.92.b}$$

This is the 2-D FDTD-GSTC updating equations when the surface susceptibilities are constant tensors. The corresponding equations for generalized sheets normal to other directions, or for 3-D FDTD-GSTC, or for other coordinate systems [35], can be obtained following the same procedure. While the simplified characteristics, including complex dispersion models, bi-isotropic tensors, and normal polarization, will be introduced in the following sections. FDTD is a time-domain full-wave algorithm with abundant exploitable space. Many scholars have developed corresponding technical plug-ins for FDTD to solve specific problems [14, 36]. For FDTD, GSTC is also a homogenization technology plug-in for metasurface or structures with electrical small thicknesses, and it can certainly be combined with the existing FDTD framework to solve more complex problems. For example, in [37], researchers combined local coordinate transformation [38] with FDTD-GSTC to realize the simulation of curved metasurfaces. In [39], researchers combined PBCs [40] with FDTD-GSTC to simulate an infinite metasurface array. Therefore, the FDTD-GSTC formulas (4.84.a) and (4.92.a) are not generalized for all scenarios. Their novelty and importance lies in how to describe the discontinuities of both electric and magnetic fields in the finite difference scheme, and, as introduced in the next section, how to describe the discontinuities affected by the normal components, which is particularly important for the artificial metasurfaces, but beyond the ability of the previous boundary conditions.

4.1.3.2 FDTD-GSTC with Normal Surface Polarization

In this section, the implementation of the generalized sheet with normal surface susceptibilities in FDTD will be introduced. For simplicity, the surface susceptibilities tensors are assumed to be diagonal with elements χ_{ee}^{xx}, χ_{ee}^{yy}, χ_{ee}^{zz}, χ_{mm}^{xx}, χ_{mm}^{yy}, and χ_{mm}^{zz}. The corresponding time-domain GSTC is

$$\hat{\mathbf{n}} \times \Delta\mathbf{H} = \varepsilon_0 \frac{\partial \overline{\overline{\chi}}_{ee,\parallel}\mathbf{E}_\parallel}{\partial t} - \hat{\mathbf{n}} \times \nabla_\parallel(\chi_{mm,n}H_n) \tag{4.93.a}$$

$$\hat{\mathbf{n}} \times \Delta\mathbf{E} = -\mu_0 \frac{\partial \overline{\overline{\chi}}_{mm,\parallel}\mathbf{H}_\parallel}{\partial t} - \hat{\mathbf{n}} \times \nabla_\parallel(\chi_{ee,n}E_n) \tag{4.93.b}$$

The normal direction is assigned as the z-direction, and the equation in (4.93.a) can be written as

$$\begin{bmatrix} -\Delta H_y \\ \Delta H_x \end{bmatrix} = \begin{bmatrix} \varepsilon_0 \dfrac{\partial}{\partial t}\left(\chi_{ee}^{xx}E_x\right) + \dfrac{\partial}{\partial y}\left(\chi_{mm}^{zz}H_z\right) \\ \varepsilon_0 \dfrac{\partial}{\partial t}\left(\chi_{ee}^{yy}E_y\right) - \dfrac{\partial}{\partial x}\left(\chi_{mm}^{zz}H_z\right) \end{bmatrix} \tag{4.94.a}$$

$$\begin{bmatrix} -\Delta E_y \\ \Delta E_x \end{bmatrix} = \begin{bmatrix} -\mu_0 \dfrac{\partial}{\partial t}\left(\chi_{mm}^{xx}H_x\right) + \dfrac{\partial}{\partial y}\left(\chi_{ee}^{zz}E_z\right) \\ -\mu_0 \dfrac{\partial}{\partial t}\left(\chi_{mm}^{yy}H_y\right) - \dfrac{\partial}{\partial x}\left(\chi_{ee}^{zz}E_z\right) \end{bmatrix} \tag{4.94.b}$$

First, consider the TE mode ($H_x = E_y = E_z = 0$) situation in 2-D FDTD, as shown in Figure 4.14. Equations in (4.94.a) are simplified as

$$-\Delta H_y = \varepsilon_0 \chi_{ee}^{xx} \frac{\partial E_x}{\partial t} + \chi_{mm}^{zz} \frac{\partial H_z}{\partial y} \tag{4.95.a}$$

$$\Delta E_x = -\mu_0 \chi_{mm}^{yy} \frac{\partial H_y}{\partial t} \tag{4.95.b}$$

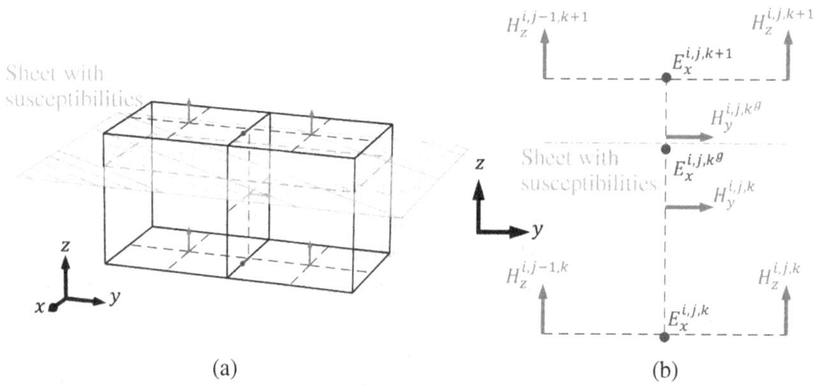

Figure 4.14 Implementation of the z-direction GSTC with normal susceptibilities in 2-D FDTD. (a) Perspective view and (b) front view. The hatched surface in (a) and the dot-dashed line in (b) both denote the generalized sheet with surface susceptibilities.

Discretizing the first equation of (4.95.a) at the location of E_x^{i,j,k^g} leads to

$$-\left(H_{y,n+1/2}^{i,j,k^g}-H_{y,n+1/2}^{i,j,k}\right)=\frac{\varepsilon_0\chi_{ee}^{xx}}{2\Delta t}\left(E_{x,n+1}^{i,j,k+1}-E_{x,n}^{i,j,k+1}+E_{x,n+1}^{i,j,k}-E_{x,n}^{i,j,k}\right)$$
$$+\frac{\chi_{mm}^{zz}}{2\Delta y}\left(H_{z,n+1/2}^{i,j,k+1}-H_{z,n+1/2}^{i,j-1,k+1}+H_{z,n+1/2}^{i,j,k}-H_{z,n+1/2}^{i,j-1,k}\right) \tag{4.96}$$

and discretizing the second equation of (4.95.b) at the location of H_y^{i,j,k^g} leads to

$$\left(E_{x,n}^{i,j,k+1}-E_{x,n}^{i,j,k^g}\right)=-\frac{\mu_0\chi_{mm}^{yy}}{2\Delta t}\left(H_{y,n+1/2}^{i,j,k+1}-H_{y,n-1/2}^{i,j,k+1}+H_{y,n+1/2}^{i,j,k}-H_{y,n-1/2}^{i,j,k}\right) \tag{4.97}$$

The difference between (4.97) and (4.89) are only the normal direction of the generalized sheet; hence, the updating equations of $H_{y,n+1/2}^{i,j,k}$ is similar to (4.84.a), which will not be derived here. The important thing is the updating equation for $E_{x,n}^{i,j,k+1}$, which relates to the component H_z. The discretization of Ampere's law at the location of $E_x^{i,j,k+1}$ is

$$E_{x,n+1}^{i,j,k+1}=E_{x,n}^{i,j,k+1}+\frac{\Delta t}{\varepsilon_x^{i,j,k+1}\Delta y}\left(H_{z,n+1/2}^{i,j,k+1}-H_{z,n+1/2}^{i,j-1,k+1}\right)$$
$$-\frac{\Delta t}{\varepsilon_x^{i,j,k+1}\Delta z}\left(H_{y,n+1/2}^{i,j,k+1}-H_{y,n+1/2}^{i,j,k^g}\right) \tag{4.98}$$

Substituting (4.96) into (4.98) leads to

$$E_{x,n+1}^{i,j,k+1}=E_{x,n}^{i,j,k+1}-\frac{\chi_{ee}^{xx}}{2\varepsilon_{r,x}^{i,j,k+1}\Delta z+\chi_{ee}^{xx}}\left(E_{x,n+1}^{i,j,k}-E_{x,n}^{i,j,k}\right)$$
$$-\frac{2\Delta t}{\varepsilon_0\left(2\varepsilon_{r,x}^{i,j,k+1}\Delta z+\chi_{ee}^{xx}\right)}\left(H_{y,n+1/2}^{i,j,k+1}-H_{y,n+1/2}^{i,j,k}\right)$$
$$+\frac{\Delta t\left(\Delta z-\chi_{mm}^{zz}\right)}{\varepsilon_0\Delta y\left(2\varepsilon_{r,x}^{i,j,k+1}\Delta z+\chi_{ee}^{xx}\right)}\left(H_{z,n+1/2}^{i,j,k+1}-H_{z,n+1/2}^{i,j-1,k+1}\right)$$
$$-\frac{\Delta t\chi_{mm}^{zz}}{\varepsilon_0\Delta y\left(2\varepsilon_{r,x}^{i,j,k+1}\Delta z+\chi_{ee}^{xx}\right)}\left(H_{z,n+1/2}^{i,j,k}-H_{z,n+1/2}^{i,j-1,k}\right) \tag{4.99}$$

This is the updating equation for $E_x^{i,j,k+1}$, and note that the auxiliary field components $E_{x,n}^{i,j,k^g}$ and $H_{y,n+1/2}^{i,j,k^g}$ are virtual fields.

The spatial average in (4.100) is employed to discretize $\partial H_z/\partial y$ to obtain the right-hand side of (4.96)

$$\left.\frac{\partial H_z}{\partial y}\right|_{\substack{y=j\Delta y\\ z=k^g\Delta z}}=\left.\frac{H_z^{i,j,z}-H_z^{i,j-1,z}}{\Delta y}\right|_{z=k^g\Delta z}$$
$$=\frac{1}{\Delta y}\left(\frac{H_z^{i,j,k+1}+H_z^{i,j,k}}{2}-\frac{H_z^{i,j-1,k+1}+H_z^{i,j-1,k}}{2}\right) \tag{4.100}$$

This spatial average can be simplified by considering the continuity of normal fields on both sides of the generalized sheet, i.e.,

$$H_z^{i,j,z}\Big|_{z=k^g\Delta z} = H_z^{i,j,k+1}, \quad H_z^{i,j-1,z}\Big|_{z=k^g\Delta z} = H_z^{i,j-1,k+1} \tag{4.101}$$

With this new approximation, the updating equation in (4.99) is simplified as

$$\begin{aligned} E_{x,n+1}^{i,j,k+1} = E_{x,n}^{i,j,k+1} &- \frac{\chi_{ee}^{xx}}{2\varepsilon_{r,x}^{i,j,k+1}\Delta z + \chi_{ee}^{xx}}\left(E_{x,n+1}^{i,j,k} - E_{x,n}^{i,j,k}\right) \\ &- \frac{2\Delta t}{\varepsilon_0\left(2\varepsilon_{r,x}^{i,j,k+1}\Delta z + \chi_{ee}^{xx}\right)}\left(H_{y,n+1/2}^{i,j,k+1} - H_{y,n+1/2}^{i,j,k}\right) \\ &+ \frac{2\Delta t\left(\Delta z - \chi_{mm}^{zz}\right)}{\varepsilon_0\Delta y\left(2\varepsilon_{r,x}^{i,j,k+1}\Delta z + \chi_{ee}^{xx}\right)}\left(H_{z,n+1/2}^{i,j,k+1} - H_{z,n+1/2}^{i,j-1,k+1}\right) \end{aligned} \tag{4.102}$$

A more concise updating equation can be obtained by choosing one single side field to approximate the tangential surrounding fields. That is, $E_x^{i,j,z}\Big|_{z=k^g\Delta z} = \left(E_x^{i,j,k+1} + E_x^{i,j,k}\right)/2$ can be replaced by $E_x^{i,j,z}\Big|_{z=k^g\Delta z} = E_x^{i,j,k+1}$. Then, (4.102) can be further simplified as

$$\begin{aligned} E_{x,n+1}^{i,j,k+1} = E_{x,n}^{i,j,k+1} &- \frac{\Delta t}{\varepsilon_0\left(\varepsilon_{r,x}^{i,j,k+1}\Delta z + \chi_{ee}^{xx}\right)}\left(H_{y,n+1/2}^{i,j,k+1} - H_{y,n+1/2}^{i,j,k}\right) \\ &+ \frac{\Delta t\left(\Delta z - \chi_{mm}^{zz}\right)}{\varepsilon_0\Delta y\left(\varepsilon_{r,x}^{i,j,k+1}\Delta z + \chi_{ee}^{xx}\right)}\left(H_{z,n+1/2}^{i,j,k+1} - H_{z,n+1/2}^{i,j-1,k+1}\right) \end{aligned} \tag{4.103}$$

The equation in (4.103) has the same expression as the conclusion obtained by using the induced surface current method in [39]. Updating equations in (4.99), (4.102), and (4.103) are all implementations of GSTC with normal surface susceptibilities in 2-D FDTD. Their difference lies merely in the approximation of the surrounding fields. Concluded from numerical cases, the difference between these three updating schemes is negligible.

The derivation of 3-D FDTD-GSTC updating equations is the same as the above 2-D scenario. A generalized sheet with surface susceptibilities χ_{ee}^{xx}, χ_{ee}^{yy}, χ_{ee}^{zz}, χ_{mm}^{xx}, χ_{mm}^{yy}, and χ_{mm}^{zz} normal to the z-direction is set at the lower half part of the cell (i, j, k), as shown in Figure 4.15. The equations in (4.6.a), (4.6.b), (4.6.d), and (4.6.e) need to be rewritten as

$$E_{x,n+1}^{i,j,k} = E_{x,n}^{i,j,k} + C_{eh}^{xz}\left(H_{z,n+1/2}^{i,j,k} - H_{z,n+1/2}^{i,j-1,k}\right) - C_{eh}^{xy}\left(H_{y,n+1/2}^{i,j,k} - H_{y,n+1/2}^{i,j,k-1}\right) \tag{4.104.a}$$

$$E_{y,n+1}^{i,j,k} = E_{y,n}^{i,j,k} + C_{eh}^{yx}\left(H_{x,n+1/2}^{i,j,k} - H_{x,n+1/2}^{i,j,k-1}\right) - C_{eh}^{yz}\left(H_{z,n+1/2}^{i,j,k} - H_{z,n+1/2}^{i-1,j,k}\right) \tag{4.104.b}$$

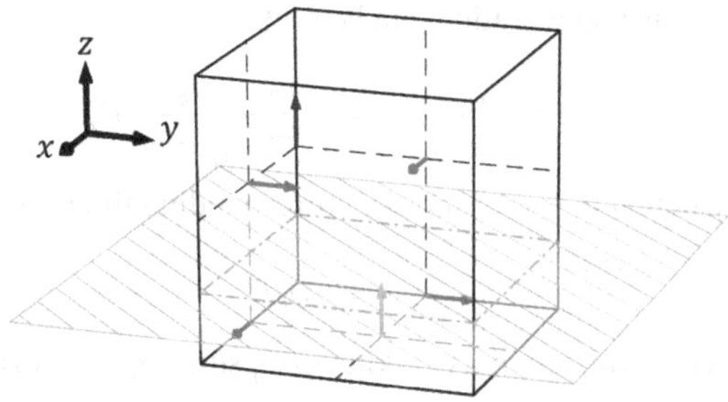

Figure 4.15 Implementation of the z-direction GSTC with normal susceptibilities in 3-D FDTD. The hatched surface denotes the generalized sheet with surface susceptibilities.

$$H^{i,j,k}_{x,n+1/2} = H^{i,j,k}_{x,n-1/2} - C^{xz}_{he}\left(E^{i,j+1,k}_{z,n} - E^{i,j,k}_{z,n}\right) + C^{xy}_{he}\left(E^{i,j,k+1}_{y,n} - E^{i,j,k}_{y,n}\right) \tag{4.104.c}$$

$$H^{i,j,k}_{y,n+1/2} = H^{i,j,k}_{y,n-1/2} - C^{yx}_{he}\left(E^{i,j,k+1}_{x,n} - E^{i,j,k}_{x,n}\right) + C^{yz}_{he}\left(E^{i+1,j,k}_{z,n} - E^{i,j,k}_{z,n}\right) \tag{4.104.d}$$

where

$$C^{xz}_{eh} = \frac{\Delta z \Delta t - \chi^{zz}_{mm}\Delta t}{\varepsilon_x \Delta y \Delta z + \varepsilon_0 \Delta y \chi^{xx}_{ee}}, \quad C^{xy}_{eh} = \frac{\Delta t}{\varepsilon_x \Delta z + \varepsilon_0 \chi^{xx}_{ee}} \tag{4.105.a}$$

$$C^{yz}_{eh} = \frac{\Delta z \Delta t - \chi^{zz}_{mm}\Delta t}{\varepsilon_y \Delta x \Delta z + \varepsilon_0 \Delta x \chi^{yy}_{ee}}, \quad C^{yx}_{eh} = \frac{\Delta t}{\varepsilon_y \Delta z + \varepsilon_0 \chi^{yy}_{ee}} \tag{4.105.b}$$

$$C^{xz}_{he} = \frac{\Delta z \Delta t - \chi^{zz}_{ee}\Delta t}{\mu_x \Delta y \Delta z + \mu_0 \Delta y \chi^{xx}_{mm}}, \quad C^{xy}_{he} = \frac{\Delta t}{\mu_x \Delta z + \mu_0 \chi^{xx}_{mm}} \tag{4.105.c}$$

$$C^{yz}_{he} = \frac{\Delta z \Delta t - \chi^{zz}_{ee}\Delta t}{\mu_y \Delta x \Delta z + \mu_0 \Delta x \chi^{yy}_{mm}}, \quad C^{yx}_{he} = \frac{\Delta t}{\mu_y \Delta z + \mu_0 \chi^{yy}_{mm}} \tag{4.105.d}$$

The surface susceptibilities in (4.105.a) need to be constant in the simulation frequency range, or the variation is small enough to be regarded as constants. If susceptibilities are dispersive in the simulation frequency range, one should employ the ADE, which will be introduced in the next section.

4.1.3.3 Methods for Dispersive Susceptibilities

Section 4.1.3.1 introduces the basic GSTC implementation in FDTD where the surface susceptibilities are constant. In this section, the method of handling dispersive susceptibilities in FDTD-GSTC will be introduced. For the dispersive models in (4.61), (4.62), and (4.63), the ADE method for the Lorentz model (4.61) and the convolution method for the CCPR model (4.63) will be introduced. The piecewise linear recursive convolution (PLRC) method can avoid the quadratic growth of complexity with problem size caused by convolution operation [29].

For the susceptibilities approximated by Lorentz dispersive model, it would be more compatible with the leapfrog time integration when $\Gamma = 0$ [31]; hence, the surface polarization can be written as

$$P_a = \varepsilon_0 \chi_{ee}^{aa}(\omega) E_a = \varepsilon_0 \sum_{k=1}^{K} \frac{\psi_k}{\omega_{0,k}^2 - \omega^2} E_a \tag{4.106}$$

where $a = x, y, z$, and the corresponding time-domain version of (4.106) is

$$\frac{\partial^2 P_{k,a}}{\partial t^2} + \omega_{0,k}^2 P_{k,a} = \varepsilon_0 \psi_k E_a \tag{4.107}$$

where the auxiliary unknown P_a sums up the surface polarizations

$$P_a = \sum_{k=1}^{K} P_{k,a} \tag{4.108}$$

For the kth unknown, the equation in (4.107) can be discretized by the central difference approximation

$$\frac{1}{\Delta t^2}\left(P_{k,a}^{n+1} - 2P_{k,a}^{n} + P_{k,a}^{n-1}\right) + \omega_{0,k}^2 P_{k,a}^{n} = \varepsilon_0 \psi_k E_a^n \tag{4.109}$$

Hence, the updating equation of $P_{k,a}$ is

$$P_{k,a}^{n+1} = \left[2 - \omega_{0,k}^2 \Delta t^2\right] P_{k,a}^{n} - P_{k,a}^{n-1} + \Delta t^2 \varepsilon_0 \psi_k E_a^n \tag{4.110}$$

The surface magnetization can be handled in the same way, and the updating equations of the kth auxiliary surface magnetization unknown is

$$M_{k,a}^{n+1/2} = \left[2 - \omega_{0,k}^2 \Delta t^2\right] M_{k,a}^{n-1/2} - M_{k,a}^{n-3/2} + \Delta t^2 \psi_k H_a^{n-1/2} \tag{4.111}$$

Note that the new unknowns P_a and M_a are updated by themselves and corresponding field unknowns; hence, their sampling locations are the same as their corresponding field unknowns. Rewrite (4.94.a) in the form that includes surface polarization and magnetization.

$$\begin{bmatrix} -\Delta H_y \\ \Delta H_x \end{bmatrix} = \begin{bmatrix} \dfrac{\partial}{\partial t}(P_x) + \dfrac{\partial}{\partial y}(M_z) \\ \dfrac{\partial}{\partial t}(P_y) - \dfrac{\partial}{\partial x}(M_z) \end{bmatrix} \tag{4.112.a}$$

$$\begin{bmatrix} -\Delta E_y \\ \Delta E_x \end{bmatrix} = \begin{bmatrix} -\mu_0 \dfrac{\partial}{\partial t}(M_x) + \dfrac{1}{\varepsilon_0}\dfrac{\partial}{\partial y}(P_z) \\ -\mu_0 \dfrac{\partial}{\partial t}(M_y) - \dfrac{1}{\varepsilon_0}\dfrac{\partial}{\partial x}(P_z) \end{bmatrix} \tag{4.112.b}$$

Using the virtual fields method, one can discretize the first row of (4.112.a) around the sheet

$$-\left(H_{y,n+1/2}^{i,j,k} - H_{y,n+1/2}^{i,j,kg}\right) = \frac{1}{\Delta t}\left(P_{x,n+1}^{i,j,k} - P_{x,n}^{i,j,k}\right) + \frac{1}{\Delta y}\left(M_{z,n+1/2}^{i,j,k} - M_{z,n+1/2}^{i,j-1,k}\right) \tag{4.113}$$

whose schematic is the same as Figure 4.15. Discretizing the second row of (4.112.b) at the location of $E_x^{i,j,k}$ leads to

$$E_{x,n+1}^{i,j,k} = E_{x,n}^{i,j,k} + \frac{\Delta t}{\varepsilon_x^{i,j,k+1} \Delta y}\left(H_{z,n+1/2}^{i,j,k} - H_{z,n+1/2}^{i,j-1,k}\right)$$

$$- \frac{\Delta t}{\varepsilon_x^{i,j,k+1} \Delta z} \left(H_{y,n+1/2}^{i,j,k^g} - H_{y,n+1/2}^{i,j,k-1} \right) \tag{4.114}$$

Substituting (4.113) into (4.114) leads to

$$\begin{aligned} E_{x,n+1}^{i,j,k} &= E_{x,n}^{i,j,k} + \frac{\Delta t}{\varepsilon_x^{i,j,k+1} \Delta y} \left(H_{z,n+1/2}^{i,j,k} - H_{z,n+1/2}^{i,j-1,k} \right) \\ &- \frac{\Delta t}{\varepsilon_x^{i,j,k+1} \Delta z} \left(H_{y,n+1/2}^{i,j,k} - H_{y,n+1/2}^{i,j,k-1} \right) \\ &- \frac{1}{\varepsilon_x^{i,j,k+1} \Delta z} \left(P_{x,n+1}^{i,j,k} - P_{x,n}^{i,j,k} \right) - \frac{\Delta t}{\varepsilon_x^{i,j,k+1} \Delta y \Delta z} \left(M_{z,n+1/2}^{i,j,k} - M_{z,n+1/2}^{i,j-1,k} \right) \end{aligned} \tag{4.115}$$

Following the same procedure, the updating equations for other field components can be obtained

$$\begin{aligned} E_{y,n+1}^{i,j,k} &= E_{y,n}^{i,j,k} + \frac{\Delta t}{\varepsilon_y^{i,j,k} \Delta z} \left(H_{x,n+1/2}^{i,j,k} - H_{x,n+1/2}^{i,j,k-1} \right) \\ &- \frac{\Delta t}{\varepsilon_y^{i,j,k} \Delta x} \left(H_{z,n+1/2}^{i,j,k} - H_{z,n+1/2}^{i-1,j,k} \right) \\ &- \frac{1}{\varepsilon_y^{i,j,k+1} \Delta z} \left(P_{y,n+1}^{i,j,k} - P_{y,n}^{i,j,k} \right) + \frac{\Delta t}{\varepsilon_y^{i,j,k+1} \Delta x \Delta z} \left(M_{z,n+1/2}^{i,j,k} - M_{z,n+1/2}^{i,j-1,k} \right) \end{aligned} \tag{4.116.a}$$

$$\begin{aligned} H_{x,n+1/2}^{i,j,k} &= H_{x,n-1/2}^{i,j,k} - \frac{\Delta t}{\mu_x^{i,j,k} \Delta y} \left(E_{z,n}^{i,j+1,k} - E_{z,n}^{i,j,k} \right) + \frac{\Delta t}{\mu_x^{i,j,k} \Delta z} \left(E_{y,n}^{i,j,k+1} - E_{y,n}^{i,j,k} \right) \\ &- \frac{1}{\mu_{x,r}^{i,j,k} \Delta z} \left(M_{x,n+1/2}^{i,j,k} - M_{x,n-1/2}^{i,j,k} \right) + \frac{\Delta t}{\mu_x^{i,j,k} \Delta x \Delta z} \left(P_{z,n}^{i,j,k} - P_{z,n}^{i,j-1,k} \right) \end{aligned} \tag{4.116.b}$$

$$\begin{aligned} H_{y,n+1/2}^{i,j,k} &= H_{y,n-1/2}^{i,j,k} - \frac{\Delta t}{\mu_y^{i,j,k} \Delta z} \left(E_{x,n}^{i,j,k+1} - E_{x,n}^{i,j,k} \right) + \frac{\Delta t}{\mu_y^{i,j,k} \Delta x} \left(E_{z,n}^{i+1,j,k} - E_{z,n}^{i,j,k} \right) \\ &- \frac{1}{\mu_{ry}^{i,j,k} \Delta z} \left(M_{y,n+1/2}^{i,j,k} - M_{y,n-1/2}^{i,j,k} \right) - \frac{\Delta t}{\mu_y^{i,j,k} \Delta x \Delta z} \left(P_{z,n}^{i,j,k} - P_{z,n}^{i-1,j,k} \right) \end{aligned} \tag{4.116.c}$$

The updating equations (4.110), (4.111), (4.115), and (4.116.a) are the implementation of the z-direction generalized sheet with dispersive surface susceptibilities in FDTD. The ADEs in (4.110) and (4.111) bring in the additional unknowns that are updated with electric and magnetic field unknowns in the time-stepping process.

Next, the susceptibilities approximated by the CCPR model are introduced. Considering the discontinuity of the x-component of the electric field across a generalized sheet in the frequency domain

$$\Delta E_x = -\mu_0 j\omega \chi_{mm}^{yy} H_y \tag{4.117}$$

where the normal susceptibility is neglected for simplicity. It is important to realize that parameter fitting is to fit a frequency-dependent function and does not necessarily approximate directly to surface susceptibilities, which means that $j\omega\chi_{mm}^{yy}(\omega)$ can be approximated directly

$$j\omega\chi_{mm}^{yy}(\omega) \approx c_0^{mmyy} + \sum_{q=1}^{N_q}\left(\frac{\delta_q^{mmyy}}{j\omega - \gamma_q^{mmyy}}\right) \tag{4.118}$$

where δ_q and γ_q are parameters of the CCPR model and the imaginary parts of the sum are zero. Equation (4.117) can be approximated by the CCPR as

$$\frac{1}{\mu_0}\Delta E_x = -c_0^{mmyy}H_y - \sum_{q=1}^{N_q}\left(\frac{\delta_q^{mmyy}}{j\omega - \gamma_q^{mmyy}}\right)H_y \tag{4.119}$$

Owing to the convolution theorem, the equation in (4.119) can be converted to the temporal representation via inverse Fourier transform

$$\frac{1}{\mu_0}\Delta E_x(t) = -H_y(t)c_0^{mmyy} - H_y(t) * \sum_{q=1}^{N_q}\delta_q^{mmyy}e^{-\gamma_q^{mmyy}t} \tag{4.120}$$

where * denotes the convolution operation defined as

$$x(t) = h(t) * g(t) = \int_{-\infty}^{t} h(t-\tau)g(\tau)d\tau \tag{4.121}$$

For the magnetic unknowns at integer and a half time steps, their convolution with any continuous function $\zeta(t)$ equals to the unknowns at integer time steps

$$g(t) * H(t)|_{t=n\Delta t} = \sum_{n'=0}^{n}\frac{1}{2}(H_{n+1/2-n'} + H_{n-1/2-n'})\int_{n'\Delta t}^{(n'+1)\Delta t} g(\tau)d\tau \tag{4.122}$$

According to (4.122), the integration of the time-domain CCPR expressions over a temporal step is need. If δ_q and γ_q are all real numbers,

$$\begin{aligned}\psi_{mm}^{yy}\left(n'; \delta_q^{mmyy}, \gamma_q^{mmyy}; \Delta t\right) &= \int_{n'\Delta t}^{(n'+1)\Delta t} \delta_q^{mmyy}e^{-\gamma_q^{mmyy}\tau}d\tau \\ &= \frac{\delta_q^{mmyy}}{\gamma_q^{mmyy}}e^{-\gamma_q^{mmyy}n'\Delta t}\left(1 - e^{-\gamma_q^{mmyy}\Delta t}\right)\end{aligned} \tag{4.123.a}$$

and if δ_q and γ_q are complex–conjugate pair,

$$\begin{aligned}\phi_q(n'; \delta_q, \gamma_q; \Delta t) &= \int_{n'\Delta t}^{(n'+1)\Delta t} \frac{2e^{-\mathrm{Re}\{\gamma_q\}t}}{|\delta_q|}\sin\left(\mathrm{Im}\{\gamma_q\}t - \arctan\frac{\mathrm{Re}\{\delta_q\}}{\mathrm{Im}\{\delta_q\}}\right)d\tau \\ &= \frac{\varphi_q(n'\Delta t + \Delta t) - \varphi_q(n'\Delta t)}{|\delta_q|}\end{aligned} \tag{4.123.b}$$

where

$$\varphi_q(t) = 2e^{-\mathrm{Re}\{\gamma_q\}t}\mathrm{Re}\{\gamma_q\}\sin\left(\mathrm{Im}\{\gamma_q\}t + \arctan\frac{\mathrm{Re}\{\delta_q\}}{\mathrm{Im}\{\delta_q\}}\right) + 2e^{-\mathrm{Re}\{\gamma_q\}t}\mathrm{Im}\{\gamma_q\}\cos\left(\mathrm{Im}\{\gamma_q\}t + \arctan\frac{\mathrm{Re}\{\delta_q\}}{\mathrm{Im}\{\delta_q\}}\right) \tag{4.124}$$

The integrations ψ_{mm}^{yy} is independent of unknown fields; therefore, they can be confirmed once the CCPR parameters and time interval are determined. Substituting (4.122) and (4.123.a) into (4.120) leads to

$$E_{x,n}^{i,j,k} = E_{x,n}^{i,j,k_g} - \frac{\mu_0}{2}\left(c_0^{mmyy} + \sum_q^{N_q}\phi_{q,0}^{mmyy} + \sum_q^{N_q}\varphi_{q,0}^{mmyy}\right)\left(H_{y,n+1/2}^{i,j,k} + H_{y,n-1/2}^{i,j,k}\right) - \frac{\mu_0}{2}\sum_{n'=1}^{n}\left[\left(H_{y,n+1/2-n'}^{i,j,k} + H_{y,n-1/2-n'}^{i,j,k}\right)\left(\sum_q^{N_q}\phi_{q,n'}^{mmyy} + \sum_q^{N_q}\varphi_{q,n'}^{mmyy}\right)\right] \tag{4.125}$$

Following the virtual field elimination process, the updating equation can be obtained

$$\begin{aligned}\left(1+\Phi_{mm}^{yy}\right)H_{y,n+1/2}^{i,j,k} &= \left(1-\Phi_{mm}^{yy}\right)H_{y,n-1/2}^{i,j,k} \\ &- \frac{\Delta t}{\mu_y^{i,j,k}\Delta z}\left(E_{x,n}^{i,j,k+1} - E_{x,n}^{i,j,k}\right) + \frac{\Delta t}{\mu_y^{i,j,k}\Delta x}\left(E_{z,n}^{i+1,j,k} - E_{z,n}^{i,j,k}\right) \\ &- \frac{\Delta t}{2\mu_{r,y}^{i,j,k}\Delta z}\sum_{n'=1}^{n}\left[\left(H_{y,n+1/2-n'}^{i,j,k} + H_{y,n-1/2-n'}^{i,j,k}\right)\left(\sum_q^{N_q}\phi_{q,n'}^{mmyy} + \sum_q^{N_q}\varphi_{q,n'}^{mmyy}\right)\right]\end{aligned} \tag{4.126}$$

where

$$\Phi_{mm}^{yy} = \frac{\Delta t}{2\mu_{r,y}^{i,j,k}\Delta z}\left(c_0^{mmyy} + \sum_q^{N_q}\phi_{q,0}^{mmyy} + \sum_q^{N_q}\varphi_{q,0}^{mmyy}\right) \tag{4.127}$$

The updating equations for E_x, E_y, and H_x can be obtained in the same way. The last term of the right-hand side of the equation in (4.126) is a sum operation, which means that the time complexity of (4.126) is $O(n^2)$ with respect to the time step n, which is the disadvantage of the combination of convolution and CCPR mode. In this situation, the recursive convolution (RC) method can be employed to achieve a linearly increasing complexity.

Taking the updating Eq. (4.115) as the example, ignoring the M_z components and abbreviating the spatial difference of the magnetic field curl as $\nabla_x \times H_{n+1/2}^{i,j,k}$ lead to

$$
\begin{aligned}
E_{x,n+1}^{i,j,k} &= E_{x,n}^{i,j,k} + \frac{\Delta t}{\varepsilon_0}\nabla_x \times H_{n+1/2}^{i,j,k} \\
&\quad - \frac{\Delta t}{4\Delta z}\left(c_0^{eexx} + \sum_{n'=0}^{n}\sum_{q=1}^{N_q}\frac{\delta_q^{eexx}}{\gamma_q^{eexx}}\left[e^{\gamma_q^{eexx}(n'+1)\Delta t} - e^{\gamma_q^{eexx}n'\Delta t}\right]\right) \\
&\quad \cdot \left(E_{x,n+1-n'}^{i,j,k} + E_{x,n-n'}^{i,j,k} + E_{x,n+1-n'}^{i,j,k+1} + E_{x,n-n'}^{i,j,k+1}\right)
\end{aligned} \tag{4.128}
$$

Define that

$$
G_{q,n}^{eexx} = \frac{dt}{4dz}\sum_{n'=1}^{n}\frac{\delta_q^{eexx}}{\gamma_q^{eexx}}e^{\gamma_q^{eexx}n'dt}\left(e^{\gamma_q^{eexx}dt} - 1\right)E^{n-n'} \tag{4.129}
$$

and there exists a recursive relationship

$$
\begin{aligned}
G_{q,n+1}^{eexx} &= \frac{dt}{4dz}\sum_{n'=1}^{n+1}\frac{\delta_q^{eexx}}{\gamma_q^{eexx}}e^{\gamma_q^{eexx}n'dt}\left(e^{\gamma_q^{eexx}dt} - 1\right)E_{x,n+1-n'} \\
&= \frac{dt}{4dz}\frac{\delta_q^{eexx}}{\gamma_q^{eexx}}\left(e^{2\gamma_q^{eexx}dt} - e^{\gamma_q^{eexx}dt}\right)E_n + e^{\gamma_q^{eexx}dt}G_{q,n}^{eexx}
\end{aligned} \tag{4.130}
$$

Define the parameters

$$
u_{ee}^{xx} = \frac{dt}{4dz}\left(c_0^{eexx} + \sum_{q=1}^{N_q}\frac{\delta_q^{eexx}}{\gamma_q^{eexx}}\left[e^{\gamma_q^{eexx}\Delta t} - 1\right]\right) \tag{4.131.a}
$$

$$
v_{ee,q}^{xx} = \frac{dt}{4dz}\frac{\delta_q^{eexx}}{\gamma_q^{eexx}}\left(e^{2\gamma_q^{eexx}dt} - e^{\gamma_q^{eexx}dt}\right) \tag{4.131.b}
$$

$$
w_{ee,q}^{xx} = e^{\gamma_q^{eexx}dt} \tag{4.131.c}
$$

Thus, Eq. (4.128) can be rewritten as

$$
\begin{aligned}
E_{x,n+1}^{i,j,k} &= E_{x,n}^{i,j,k} + \frac{dt}{\varepsilon_0}\nabla_x \times H_{n+1/2}^{i,j,k} \\
&\quad - \frac{2u_{ee}^{xx}}{1+u_{ee}^{xx}}E_{x,n}^{i,j,k} - \frac{u_{ee}^{xx}}{1+u_{ee}^{xx}}\frac{dt}{\varepsilon_0}\nabla_x \times H_{n+1/2}^{i,j,k} - \frac{u_{ee}^{xx}}{1+u_{ee}^{xx}}\left(E_{x,n}^{i,j,k+1}\right) \\
&\quad + \frac{1}{1+u_{ee}^{xx}}\left(\sum_{q}^{N_q}\frac{v_{ee,q}^{xx}}{w_{ee,q}^{xx}}\right)\left(E_{x,n+1}^{i,j,k+1} + E_{x,n}^{i,j,k+1}\right) + \frac{-u_{ee}^{xx}}{1+u_{ee}^{xx}}\left(E_{x,n+1}^{i,j,k+1}\right) \\
&\quad - \frac{1}{1+u_{ee}^{xx}}\sum_{q}^{N_q}\left[\frac{w_{ee,q}^{xx}+1}{w_{ee,q}^{xx}}\left(G_{k,z,n+1}^{i,j,k} + G_{k,z,n+1}^{i,j,k+1}\right)\right]
\end{aligned} \tag{4.132}
$$

As presented in (4.131.a), in the time sequence step, apart from the conventional E_n and $H_{n+1/2}$, it only needs to record additional G_n and also avoids the summation of time as in (4.126).

4.1.4 Simulation Examples

This section presents two numerical examples. Categorized according to the susceptibility determination outlined in Section 2.3.4, the first example is the *susceptibility synthesis* aiming to validate the accuracy and robustness of the GSTC-FDTD algorithm. The second example is the *susceptibility extraction* applications that demonstrate the computational efficiency advantage of GSTC-FDTD method over conventional FDTD.

4.1.4.1 A Wideband Refraction Metasurface

The first example is to solve the response of a wideband refraction metasurface using GSTC-FDTD. This wideband refraction metasurface is hypothetical, and it exists mainly in the form of space- and frequency-dependent surface susceptibilities. This is a suitable case for testing the accuracy and the robustness of the algorithms incorporating GSTC, which is also employed as the numerical example in the previous sections (FDFD-GSTC in 3.2.3.1 and SEM-GSTC in 3.3.3.1).

This metasurface deflects an incident plane wave ($\mathbf{E}_i$, $\mathbf{H}_i$, $\mathbf{k}_i$) into a transmitted plane wave ($\mathbf{E}_t$, $\mathbf{H}_t$, $\mathbf{k}_t$), as shown in Figure 4.16, with the incident and transmitted angles θ_i and θ_t, respectively. The susceptibilities are the same as those in (3.35), where the parameters are $\theta_i = \pi/3$, $\theta_t = \pi/6$, and $T = 0.5$, and there is no reflected wave. The effective frequency range is 5–10 GHz.

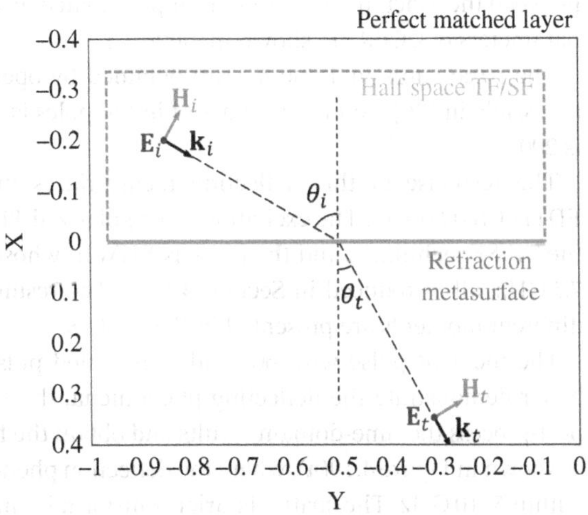

Figure 4.16 The sketch of the simulation of the wideband deflecting metasurface in 2-D FDTD.

The susceptibilities are the functions of y-coordinate and angular frequency ω as

$$\chi_{ee}^{zz}(\omega, y) = \frac{2c_0}{j\omega} \frac{\frac{1}{2}e^{-\frac{j\omega\sqrt{3}y}{2c_0}} - \frac{\sqrt{3}}{4}e^{-\frac{j\omega y}{2c_0}}}{e^{-\frac{j\omega\sqrt{3}y}{2c_0}} + \frac{1}{2}e^{-\frac{j\omega y}{2c_0}}} \tag{4.133.a}$$

$$\chi_{mm}^{yy}(\omega, y) = \frac{2c_0}{j\omega} \frac{e^{-\frac{j\omega\sqrt{3}y}{2c_0}} - \frac{1}{2}e^{-\frac{j\omega y}{2c_0}}}{\frac{1}{2}e^{-\frac{j\omega\sqrt{3}y}{2c_0}} + \frac{\sqrt{3}}{4}e^{-\frac{j\omega y}{2c_0}}} \tag{4.133.b}$$

Mapping this dispersive function of susceptibilities to y-coordinate is straightforward in the FDTD framework with $y = j\Delta y$. While making it compatible with the time-stepping process needs an additional process, i.e., the parameters fitting process introduced in Section 4.1.2. Here, the sum of CCPR dispersive models in (4.63) is used to approximate the product of χ and $j\omega$

$$\tilde{f}(\omega; c_0, \delta_q, \gamma_q) = c_0 + \sum_{q=1}^{N_q} \frac{\delta_q}{j\omega - \gamma_q} \approx j\omega\chi(\omega, y) \tag{4.134}$$

To present this approximating process in detail, the fitting results of $j\omega\chi_{ee}^{zz}$ are presented and the exact analytic values can be obtained from (4.133.a) in Figure 4.17.

Figure 4.17(a) shows the exact and approximated amplitude (Abs) and phase (Angle) of $j\omega\chi_{ee}^{zz}$, and these approximations are actually done independently for each individual y coordinate. In this case, the metasurface is set at -0.96 m $\leq y \leq -0.08$ m. So, there are 441 functions of $j\omega\chi_{ee}^{zz}(\omega, y)$ that need to be fitted. Figure 4.17(b) presents the comparison of the amplitude of $j\omega\chi_{ee}^{zz}$ at $y = -0.8$ m between the exact function and the approximated dispersive model, and the fitted parameters of CCPR are shown in Table 4.1.

The parameters in Table 4.1 are obtained by open-source vector fitting code accessible in [26], and the number of initial poles is 20 and the iteration number is 200.

The response of this deflecting metasurfaces model is solved in the 2-D FDTD-GSTC solver. The excitation is a z-polarized TE mode plane wave added by the TF/SF technique, and the pulse is BHW1d whose characteristic frequency is 7.5 GHz, all introduced in Section 4.1.1.3. The results of E-field distributions in different moments are presented in Figure 4.18.

The incident pulse with 60° and transmitted pulse with 30° are obvious. To better demonstrate the deflecting phenomena, the Fourier is used transform to post-process the time-domain results and obtain the frequency-domain response, as shown in Figure 4.19(a), where the deflection phenomenon is simulated clearly within 5–10 GHz. The spatial Fourier transform is employed to obtain the weight

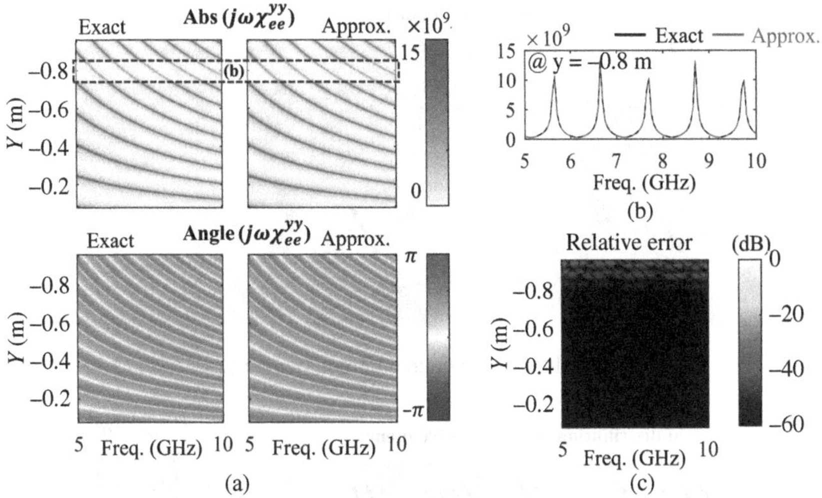

Figure 4.17 The vector fitting result of dispersive tangential electric surface susceptibility $j\omega\chi_{ee}^{zz}$. (a) The exact and approximated amplitude (Abs) and phase (Angle) of susceptibility in different frequency and y coordinates. (b) The comparison between the exact and approximated susceptibility at $y = -0.8$ m within 5–10 GHz. (c) The relative error of susceptibility in different frequency and y coordinates.

Table 4.1 The fitted parameters of CCPR for $j\omega\chi_{ee}^{zz}$ at $y = -0.8$ m within 5–10 GHz corresponding to Figure 4.17(b).

c_0 = 1.1188e9			
Re{δ_q}	**Im{δ_q}**	**Re{γ_q}**	**Im{γ_q}**
4.5604e18	0	−3.7848e9	0
−5.7584e20	0	−4.7780e11	0
1.2735e18	−1.8069e17	−1.0926e9	2.1845e10
8.3678e17	1.1610e15	−7.0804e8	2.8956e10
8.3895e17	−1.9967e12	−7.0982e8	3.5389e10
8.3895e17	6.9079e11	−7.0982e8	4.1823e10
8.3895e17	−4.6792e11	−7.0982e8	4.8257e10
8.3895e17	−9.7671e10	−7.0982e8	5.4692e10
8.3895e17	5.3904e12	−7.0982e8	6.1126e10
9.3985e17	6.2829e16	−8.0284e8	6.7674e10
4.4732e18	1.2526e18	−3.9142e9	8.2780e10

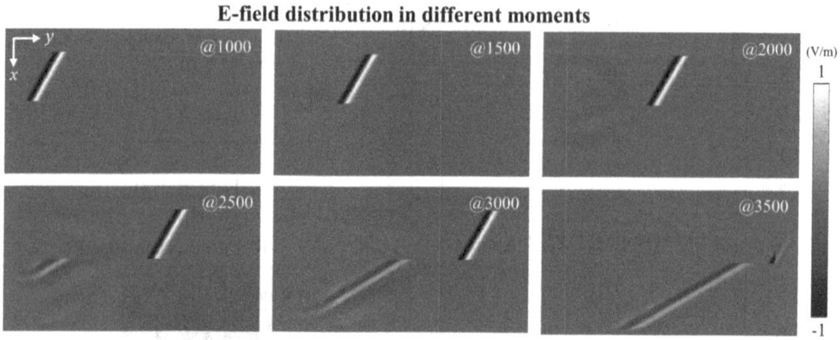

Figure 4.18 The electric field distribution in different moments.

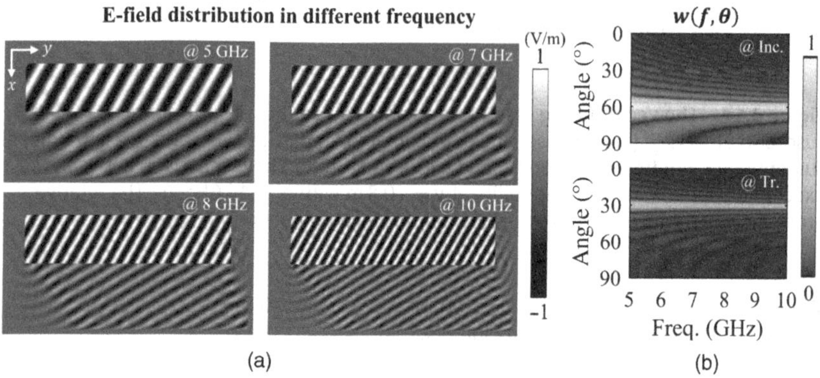

Figure 4.19 (a) The electric field distribution in different frequencies, obtained by the Fourier transform of the temporal results. (b) The weight of electric fields in different frequencies and propagating angles.

of angular spatial frequency in spectral domain of both the incident region and transmitted region, as shown in Figure 4.19(b).

$$w(f,\theta) = \frac{\left|\int_{-0.9}^{-0.1} E_z(x=\pm 0.06, y)e^{j2\pi f \sin(\theta)y/c_0}dy\right|}{\left|\int_{-0.9}^{-0.1} e^{j2\pi f y/c_0}dy\right|} \tag{4.135}$$

The weight coefficients (4.135) can reveal the proportion of waves propagating at different angles in the integrated region. Figure 4.19(b) shows that the incident waves are mainly propagating at 60°, and there are no significant reflected components, while the transmitted waves are mainly propagating at 30° and the amplitude is half of the incident wave. This example is mainly implemented according to (4.125), which demonstrates the capability of the convolution FDTD-GSTC in the simulation of complicated dispersive susceptibilities.

4.1.4.2 Array of Split-Ring Resonators

In 2000, D. R. Smith demonstrated a composite medium, based on a periodic array of interspaced conducting nonmagnetic split-ring resonators (SRR) and continuous wires, which exhibit a frequency range in the microwave regime with simultaneously negative values of effective permeability and permittivity. This work started the research upsurge of metamaterial and metasurface for the past 20 years. The transmitted power of an array of SRRs is needed to determine the negative permeability within a frequency range. More research details can be found in [41]. Here, the stop-band of finitely large SRR arrays is detected using FDTD-GSTC, and CST simulation of the actual structure will be the reference.

Figures 4.20 and 4.21 show the simulation results, and Table 4.2 shows the computing resource used. It can be seen that the error of GSTC-equivalence in the predicted stop-band is about 0.02 GHz (3%), but the peak memory and time consumption are much less than that from CST simulation of the actual SRR structure. Because the minimum dimension size of SRR ring is 0.4 mm, which is about 0.009 wavelength (corresponding to 7 GHz), while the size of the entire array in the x-direction is 3.17 wavelengths. This is a very typical multiscale problem, which is the main reason for the intense resource consumption when using commercial software to simulate metasurfaces.

The equivalent susceptibilities of the SRR are approximated by the CCPR in (4.63) within 2–8 GHz as

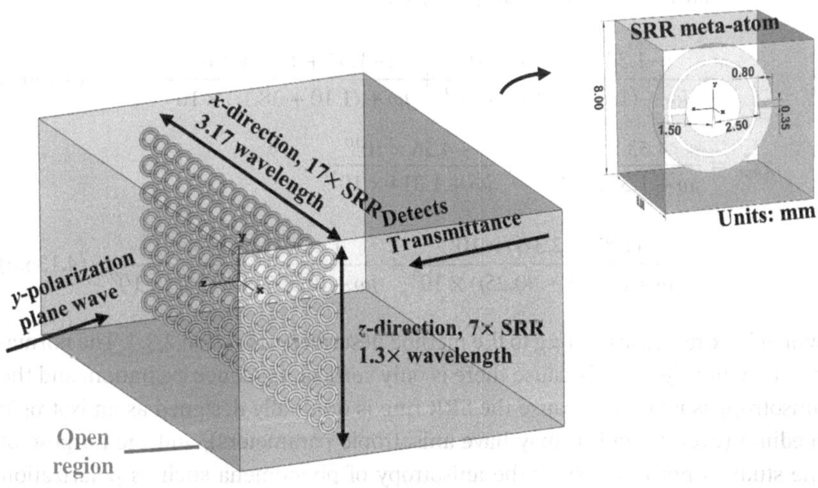

Figure 4.20 The sketch of simulation of the finite SRR array and the dimension of SRR units.

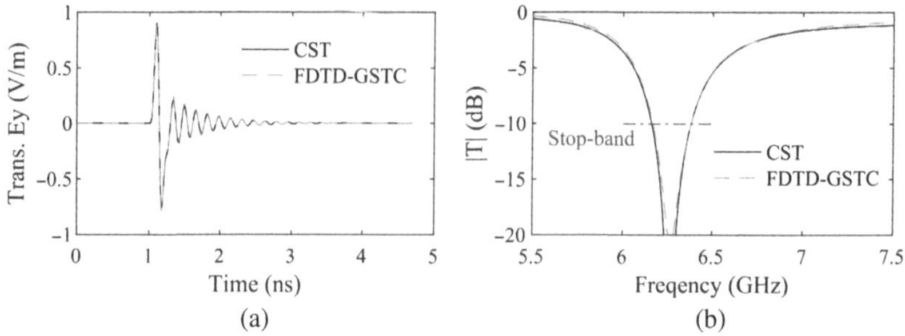

Figure 4.21 (a) The time-domain and (b) frequency-domain results of the y-component of the transmitted electric field of the finite SRR array obtained from CST and FDTD-GSTC method.

Table 4.2 Calculated stop-band and computing resources consumption of Example 4.1.4.3.

	Stop-Band (GHz)	Memory (MB)	Time (s)
FDTD-GSTC	6.167–6.379	43	108
CST	6.170–6.381	5707	2236 (=37 min)

$$\chi_{ee}^{yy} = \frac{3.14 \times 10^{8}}{j\omega + 4.62 \times 10^{11}} + \frac{2.84 \times 10^{9}}{j\omega + 1.26 \times 10^{12}} + \frac{(-1.34 - 1.54j) \times 10^{7}}{j\omega + (1.10 - 38.1j) \times 10^{9}} + \frac{(-1.34 + 1.54j) \times 10^{7}}{j\omega + (1.10 + 38.1j) \times 10^{9}} \tag{4.136.a}$$

$$\chi_{mm}^{xx} = \frac{4.53 \times 10^{10}}{j\omega + 1.311 \times 10^{11}} + \frac{-4.56 \times 10^{10}}{j\omega + 1.314 \times 10^{11}} + \frac{(1.47 - 2.34j) \times 10^{7}}{j\omega + (0.555 - 40.25j) \times 10^{9}} + \frac{(1.47 + 2.34j) \times 10^{7}}{j\omega + (0.555 + 40.25j) \times 10^{9}} \tag{4.136.b}$$

which is extracted according to the method described in Section 2.3.4. The normal component is ignored because there is only vertical incidence excitation, and the anisotropy is ignored because the SRR ring is originally designed as an isotropic medium (even though it may have anisotropic parameters), and the purpose of the study is not to consider the anisotropy of phenomena such as polarization rotation. So, this simplification does not affect the result of the equivalent simulation.

4.2 DGTD Method with GSTCs

4.2.1 Introduction of the DGTD Method

The FDTD-GSTC method effectively addresses most requirements for fast time-domain simulation of metasurfaces. However, for metasurfaces composed of elements arranged on a curved surface, as shown in Figure 4.22(a), or for conformal metasurfaces [42, 43], an algorithm supporting unstructured mesh becomes essential. To meet this need, the DGTD method is introduced, which integrates GSTC into it. As the final algorithm introduced in this book, DGTD emerged later in the field of computational electromagnetic (CEM) compared to previously introduced algorithms. For readers familiar with DGTD, it is often described as a combination of both the finite element time-domain (FETD) method and the finite volume

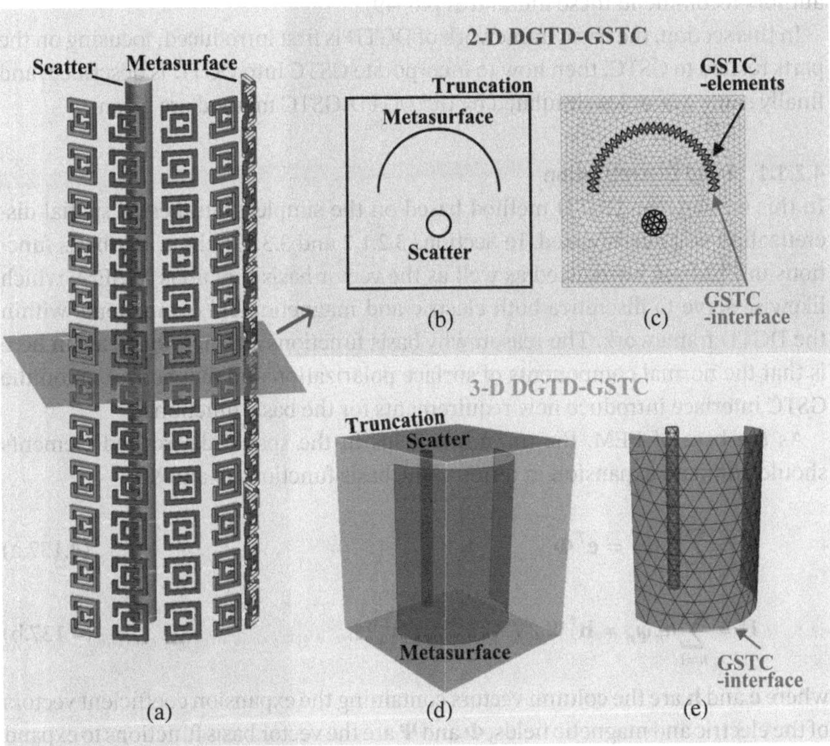

Figure 4.22 The schematic of 2-D and 3-D versions of DGTD-GSTC. (a) The real 3-D metasurface. (b) The corresponding 2-D model, in which the metasurface is modeled as a 1-D curve. (c) The mesh of 2-D DGTD-GSTC. (d) The corresponding 3-D model, in which the metasurface is modeled as a 2-D surface. (e) The mesh of 3-D DGTD-GSTC.

time-domain (FVTD) method, since DGTD inherits the basis function expansion concept from FETD to achieve high-order spatial approximation, while adopting the discontinuous interface treatment from FVTD to avoid computationally expensive large matrix inversions. Compared to FDTD, DGTD provides geometric flexibility while preserving the benefits of spatial explicitness, inherent parallelizability, and computational cost scaling linearly with element count. Although more complex to implement, DGTD provides enhanced capabilities for electromagnetic simulations.

The discontinuous Galerkin (DG) method was first proposed by W. H. Reed in 1973 and was used to solve the neutron transport equation [44]. Around 2000, the DGTD method began to be applied to computational electromagnetism. J. S. Hesthaven applied tetrahedral mesh and node scalar basis function to solve the Maxwell's equation, which may be the first paper of DGTD in computational electromagnetism [45]. As for the basic concept and latest progress of DGTD, the authors recommend these literatures [46–49].

In this section, the basic framework of DGTD is first introduced, focusing on the parts related to GSTC, then how to incorporate GSTC into DGTD is described, and finally some examples simulated by the DGTD-GSTC method are given.

4.2.1.1 Field Discretization

In this section, the DGTD method based on the simplest tetrahedral spatial discretization will be discussed. In Sections 3.2.1.2 and 3.3.1.2, the scalar basis functions in FEM are introduced as well as the vector basis functions in SEM, which likewise serve to discretize both electric and magnetic field components within the DGTD framework. The reason why basis functions are mentioned again here is that the normal components of surface polarization and magnetization on the GSTC interface introduce new requirements for the basis functions.

As the base of FEM, the unknown fields in the spatial discretized elements should admit an expansion in terms of the basis functions $\mathbf{\Phi}$ and $\mathbf{\Psi}$

$$\mathbf{E} = \sum_{n=1}^{N_\Phi} e_n \mathbf{\Phi}_n = \mathbf{e}^T \mathbf{\Phi} \tag{4.137.a}$$

$$\mathbf{H} = \sum_{n=1}^{N_\Psi} h_n \mathbf{\Psi}_n = \mathbf{h}^T \mathbf{\Psi} \tag{4.137.b}$$

where $\mathbf{e}$ and $\mathbf{h}$ are the column vectors containing the expansion coefficient vectors of the electric and magnetic fields, $\mathbf{\Phi}$ and $\mathbf{\Psi}$ are the vector basis functions to expand the electric and magnetic fields, and N_Φ and N_Ψ denote the numbers of DoFs. Like FEM-GSTC, the surface polarization and magnetization are placed on the boundary of elements, or in other words, the generalized sheet with the surface susceptibilities is treated as a geometry with one dimension lower than the computational space, as shown in Figure 4.22(b,c), where the metasurface is treated as a 1-D edge in 2-D DGTD-GSTC, and as a 2-D surface in 3-D DGTD-GSTC, as shown in Figure 4.22(d,e).

As for the tangential and normal surface polarizations, they should also be expressed by the basis functions, respectively,

$$\mathbf{P}_{\|} = \sum_{n=1}^{N_\gamma} p_{\|,n}\boldsymbol{\gamma}_n = \mathbf{p}_{\|}^T\boldsymbol{\gamma} \tag{4.138.a}$$

$$\mathbf{M}_{\|} = \sum_{n=1}^{N_\gamma} m_{\|,n}\boldsymbol{\gamma}_n = \mathbf{m}_{\|}^T\boldsymbol{\gamma} \tag{4.138.b}$$

$$P_n = \sum_{n=1}^{N_\xi} p_{\perp,n}\zeta_n = \mathbf{p}_{\perp}^T\boldsymbol{\zeta} \tag{4.138.c}$$

$$M_n = \sum_{n=1}^{N_\xi} m_{\perp,n}\zeta_n = \mathbf{m}_{\perp}^T\boldsymbol{\zeta} \tag{4.138.d}$$

where $\boldsymbol{\gamma}$ denotes the set of vector basis functions to expand the tangential surface polarization, $\mathbf{p}_{\|}$ and $\mathbf{m}_{\|}$ are the column vectors containing the expansion coefficients, $\boldsymbol{\zeta}$ denotes the set of scalar basis functions to expand the normal surface polarization, $\mathbf{p}_{\perp}$ and $\mathbf{m}_{\perp}$ are the column vectors containing the expansion coefficients, and N_γ and N_ξ denote the DoFs for the basis functions $\boldsymbol{\gamma}$ and $\boldsymbol{\zeta}$, respectively. The vector basis functions based on tetrahedral and triangular, and the scalar basis functions based on triangles and lines in this book can be of interpolatory type [50] or hierarchical type [51]. The subscript $\perp$ employed in this section denotes the normal direction.

4.2.1.2 Discontinuous Galerkin Method and the Numerical Flux

Let Γ denotes the bounded computational domain whose boundary is $\partial\Gamma$. The domain Γ is divided into M non-overlapping and conformal tetrahedral (triangle in 2-D DGTD method) elements. The i-th element is denoted by γ_i, and the boundaries are denoted by $\partial\gamma_i$. The constitutive parameters are uniform within each element (ε_i, μ_i, σ_i). Among all the boundaries, there are N_a GSTC-related boundaries (GSTC-boundaries), and the surface susceptibilities are also uniform in g-th GSTC-boundary ($\overline{\overline{\chi}}_{ee,g}$, $\overline{\overline{\chi}}_{em,g}$, $\overline{\overline{\chi}}_{mm,g}$, and $\overline{\overline{\chi}}_{me,g}$). Considering the vector Maxwell's curl equation in isotropic linear media,

$$\varepsilon\frac{\partial \mathbf{E}}{\partial t} = \nabla\times\mathbf{H} - \mathbf{J}_e \tag{4.139.a}$$

$$\mu\frac{\partial \mathbf{H}}{\partial t} = -\nabla\times\mathbf{E} - \mathbf{J}_h \tag{4.139.b}$$

Using the Galerkin's method to test (4.139.a) in each element leads to

$$\int_{\gamma} \boldsymbol{\Phi} \cdot \left(\varepsilon \frac{\partial \mathbf{E}}{\partial t} - \nabla \times \mathbf{H} + \mathbf{J}_e \right) = 0 \tag{4.140.a}$$

$$\int_{\gamma} \boldsymbol{\Psi} \cdot \left(\mu \frac{\partial \mathbf{H}}{\partial t} + \nabla \times \mathbf{E} + \mathbf{J}_h \right) = 0 \tag{4.140.b}$$

Applying the vector identity and the Gauss divergence law to the curl terms leads to

$$\int_{\gamma} \boldsymbol{\Phi} \cdot \varepsilon \frac{\partial \mathbf{E}}{\partial t} dV - \int_{\gamma} \nabla \times \boldsymbol{\Phi} \cdot \mathbf{H} dV + \int_{\gamma} \boldsymbol{\Phi} \cdot \mathbf{J}_e dV = \int_{\partial\gamma} \boldsymbol{\Phi} \cdot (\hat{\mathbf{n}} \times \mathbf{H}) dS \tag{4.141.a}$$

$$\int_{\gamma} \boldsymbol{\Psi} \cdot \mu \frac{\partial \mathbf{H}}{\partial t} dV + \int_{\gamma} \nabla \times \boldsymbol{\Psi} \cdot \mathbf{E} dV + \int_{\gamma} \boldsymbol{\Psi} \cdot \mathbf{J}_h dV = -\int_{\partial\gamma} \boldsymbol{\Psi} \cdot (\hat{\mathbf{n}} \times \mathbf{E}) dS \tag{4.141.b}$$

Equation (4.141.a) is called the weak form of (4.139.a). Enforcing the tangential field continuity condition between adjacent elements

$$\hat{\mathbf{n}}^i \times \mathbf{E}^i \Big|_{\partial\gamma_i} = \hat{\mathbf{n}}^i \times \mathbf{E}^j \Big|_{\partial\gamma_i} \tag{4.142.a}$$

$$\hat{\mathbf{n}}^i \times \mathbf{H}^i \Big|_{\partial\gamma_i} = \hat{\mathbf{n}}^i \times \mathbf{H}^j \Big|_{\partial\gamma_i} \tag{4.142.b}$$

where the superscript j denotes the adjacent element of the local i-th element across the interface $\partial\gamma$. One of the core theories of DGTD, inherited from the FVTD, is the numerical flux that relaxes the tangential field continuity condition. Using the numerical flux $\hat{\mathbf{n}}^i \times \mathbf{E}^* \Big|_{\partial\gamma_i}$ and $\hat{\mathbf{n}}^i \times \mathbf{H}^* \Big|_{\partial\gamma_i}$ to participate in the integration of the right-hand side of (4.141.a) leads to

$$\int_{\gamma} \boldsymbol{\Phi} \cdot \varepsilon \frac{\partial \mathbf{E}}{\partial t} dV - \int_{\gamma} \nabla \times \boldsymbol{\Phi} \cdot \mathbf{H} dV + \int_{\gamma} \boldsymbol{\Phi} \cdot \mathbf{J}_e dV = \int_{\partial\gamma} \boldsymbol{\Phi} \cdot (\hat{\mathbf{n}} \times \mathbf{H}^*) dS \tag{4.143.a}$$

$$\int_{\gamma} \boldsymbol{\Psi} \cdot \mu \frac{\partial \mathbf{H}}{\partial t} dV + \int_{\gamma} \nabla \times \boldsymbol{\Psi} \cdot \mathbf{E} dV + \int_{\gamma} \boldsymbol{\Psi} \cdot \mathbf{J}_m dV = -\int_{\partial\gamma} \boldsymbol{\Psi} \cdot (\hat{\mathbf{n}} \times \mathbf{E}^*) dS \tag{4.143.b}$$

The numerical flux depends on the field values across the element interface of $\partial\gamma_m$

$$\hat{\mathbf{n}}^i \times \mathbf{E}^* \Big|_{\partial\gamma_i} = \hat{\mathbf{n}}^i \times f_E(\mathbf{E}^i, \mathbf{H}^i, \mathbf{E}^j, \mathbf{H}^j) \tag{4.144.a}$$

$$\hat{\mathbf{n}}^i \times \mathbf{H}^* \Big|_{\partial\gamma_i} = \hat{\mathbf{n}}^i \times f_H(\mathbf{E}^i, \mathbf{H}^i, \mathbf{E}^j, \mathbf{H}^j) \tag{4.144.b}$$

Here, the widely employed numerical flux in (4.145.a), i.e., the upwind flux [46], and its derivation are presented. Other numerical fluxes, such as the central flux and the partially penalized flux, can be found in [47].

$$Y^{ij}\hat{\mathbf{n}}^i \times \mathbf{E}^* = \hat{\mathbf{n}}^i \times (Y^i\mathbf{E}^i + Y^j\mathbf{E}^j) - \hat{\mathbf{n}}^i \times [\hat{\mathbf{n}}^i \times (\mathbf{H}^i - \mathbf{H}^j)] \tag{4.145.a}$$

$$Z^{ij}\hat{\mathbf{n}}^i \times \mathbf{H}^* = \hat{\mathbf{n}}^i \times (Z^i\mathbf{H}^i + Z^j\mathbf{H}^j) + \hat{\mathbf{n}}^i \times [\hat{\mathbf{n}}^i \times (\mathbf{E}^i - \mathbf{E}^j)] \tag{4.145.b}$$

where the coefficients are

$$Z^i = \sqrt{\frac{\mu^i}{\varepsilon^i}}, Z^j = \sqrt{\frac{\mu^j}{\varepsilon^j}}, Y^i = \frac{1}{Z^i}, Y^j = \frac{1}{Z^j}, Z^{ij} = Z^i + Z^j, Y^{ij} = Y^i + Y^j \tag{4.146}$$

The upwind flux is first used in FVTD [52] and its derivation is based on the Rankine–Hugoniot jump condition. Rewrite (4.139.a) in the following form:

$$\frac{\partial}{\partial t}\mathbf{q} + \mathbf{A}_x\frac{\partial}{\partial x}\mathbf{q} + \mathbf{A}_y\frac{\partial}{\partial y}\mathbf{q} + \mathbf{A}_z\frac{\partial}{\partial z}\mathbf{q} = 0 \tag{4.147}$$

where

$$\mathbf{q} = \left[E_x\ E_y\ E_z\ H_x\ H_y\ H_z\right]^T \tag{4.148}$$

and

$$\mathbf{A}_x = \begin{bmatrix} & & & & & 0 \\ & & & & 0 & -\varepsilon^{-1} \\ & & & 0 & \varepsilon^{-1} & \\ & & 0 & 0 & & \\ & 0 & \mu^{-1} & & & \\ 0 & -\mu^{-1} & & & & \end{bmatrix}, \quad \mathbf{A}_y = \begin{bmatrix} & & & & & \varepsilon^{-1} \\ & & & & 0 & \\ & & & -\varepsilon^{-1} & & \\ & & -\mu^{-1} & & & \\ & 0 & & & & \\ \mu^{-1} & & & & & \end{bmatrix},$$

$$\mathbf{A}_z = \begin{bmatrix} & & & & -\varepsilon^{-1} & 0 \\ & & & \varepsilon^{-1} & 0 & \\ & & 0 & 0 & & \\ & \mu^{-1} & 0 & & & \\ -\mu^{-1} & 0 & & & & \\ 0 & & & & & \end{bmatrix} \tag{4.149}$$

Equation (4.147) is a classical 3-D hyperbolic problem, and the discontinuity of field at the points of element interfaces (4.144.a) is a fundamental topic in finite volume method called the Riemann problem. For the linear hyperbolic systems, the Riemann problem can be solved in terms of the eigenvalues and eigenvectors of the coefficient matrix [53].

Consider an arbitrary volume Ω and the surface bounding, this volume is $\partial\Omega$. To determine the relationship between the data q inside and outside the surface $\partial\Omega$, the function $\mathbf{f}(q) = \left[f_x(q)\ f_y(q)\ f_z(q)\right]$ is defined as the flux vector, where $f_x(q)$, $f_y(q)$, and $f_z(q)$ represent the flux of q in the x-, y-, and z-direction across $\partial\Omega$.

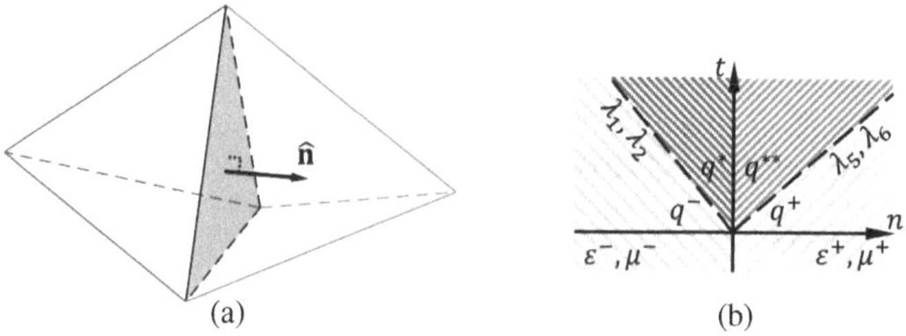

Figure 4.23 (a) The normal unit vector at the interface of two elements. (b) The construction of the solution of the Riemann problem at the interface.

Denote $\hat{\mathbf{n}} = n_x\hat{\mathbf{x}} + n_y\hat{\mathbf{y}} + n_z\hat{\mathbf{z}}$ as the outward unit normal vector to $\partial\Omega$ at a point $\mathbf{r}$ on $\partial\Omega$. Then, the flux at $\mathbf{r}$ along the direction $\hat{\mathbf{n}}$ is

$$\hat{\mathbf{n}} \cdot \mathbf{f}(q) = n_x f_x(q) + n_y f_y(q) + n_z f_z(q) \tag{4.150}$$

In Figure 4.23(a), substituting (4.150) into (4.147) leads to

$$\frac{\partial}{\partial t}\mathbf{q} + \mathbf{A}\frac{\partial}{\partial n}\mathbf{q} = 0 \tag{4.151}$$

where $\mathbf{A}$ is a linear combination of matrices $\mathbf{A}_x$, $\mathbf{A}_y$, and $\mathbf{A}_z$ that has been restricted to a direction specified by $\hat{\mathbf{n}}$

$$\mathbf{A} = n_x\mathbf{A}_x + n_y\mathbf{A}_y + n_z\mathbf{A}_z$$
$$= \begin{bmatrix} & & & & -n_z\varepsilon^{-1} & n_y\varepsilon^{-1} \\ & & & n_z\varepsilon^{-1} & 0 & -n_x\varepsilon^{-1} \\ & & 0 & -n_y\varepsilon^{-1} & n_x\varepsilon^{-1} & \\ & n_z\mu^{-1} & -n_y\mu^{-1} & 0 & & \\ -n_z\mu^{-1} & 0 & n_x\mu^{-1} & & & \\ n_y\mu^{-1} & -n_x\mu^{-1} & & & & \end{bmatrix} \tag{4.152}$$

Since the matrix $\mathbf{A}$ is diagonalizable with real eigenvalues $\lambda_1 = \lambda_2 = -c$, $\lambda_3 = \lambda_4 = 0$, $\lambda_5 = \lambda_6 = c$, and $c = \sqrt{\varepsilon\mu}^{-1}$, any vector $\mathbf{q}$ can be represented uniquely as a superposition of the eigenmodes, whose eigenvalues specify the propagation speeds of $\mathbf{q}$ propagating in the medium. It is important that (4.151) represents a linear system with piecewise-constant coefficients that the matrix A can be different in two elements sharing the same interface. Now, according to the Rankine–Hugoniot condition, for the jump between $\mathbf{q}^-$ and $\mathbf{q}^+$, there will be two intermediate vectors $\mathbf{q}^*$ and $\mathbf{q}^{**}$. The plot of system (4.151) is presented in Figure 4.23(b) for simplicity.

The relationships between the field vectors described by the eigenvalues are

(1) Across λ_1, λ_2 in the region with ε^-, μ^-

$$\lambda_1(\mathbf{q}^- - \mathbf{q}^*) - [\mathbf{A}^-\mathbf{q}^- - \mathbf{A}^*\mathbf{q}^*] = 0 \tag{4.153}$$

(2) Across λ_3, λ_4 between the two regions

$$\lambda_3(\mathbf{q}^* - \mathbf{q}^{**}) - [\mathbf{A}^-\mathbf{q}^* - \mathbf{A}^+\mathbf{q}^{**}] = 0 \tag{4.154}$$

(3) Across λ_5, λ_6 in the region with ε^+, μ^+

$$\lambda_5(\mathbf{q}^{**} - \mathbf{q}^+) - [\mathbf{A}^+\mathbf{q}^{**} - \mathbf{A}^+\mathbf{q}^+] = 0 \tag{4.155}$$

Besides, there are

$$\mathbf{Aq} = \begin{bmatrix} \varepsilon^{-1}(n_yH_z - n_zH_y) \\ \varepsilon^{-1}(n_zH_x - n_xH_z) \\ \varepsilon^{-1}(n_xH_y - n_yH_x) \\ \mu^{-1}(n_zE_y - n_yE_z) \\ \mu^{-1}(n_xE_z - n_zE_x) \\ \mu^{-1}(n_yE_x - n_xE_y) \end{bmatrix} = \begin{bmatrix} \varepsilon^{-1}\hat{\mathbf{n}} \times \mathbf{H} \\ -\mu^{-1}\hat{\mathbf{n}} \times \mathbf{E} \end{bmatrix} \tag{4.156.a}$$

and

$$\mathbf{AAq} = \frac{-1}{\varepsilon\mu}\begin{bmatrix} n_y(n_xE_y - n_yE_x) - n_z(n_zE_x - n_xE_z) \\ n_z(n_yE_z - n_zE_y) - n_x(n_xE_y - n_yE_x) \\ n_x(n_zE_x - n_xE_z) - n_y(n_yE_z - n_zE_y) \\ n_y(n_xH_y - n_yH_x) - n_z(n_zH_x - n_xH_z) \\ n_z(n_yH_z - n_zH_y) - n_x(n_xH_y - n_yH_x) \\ n_x(n_zH_x - n_xH_z) - n_y(n_yH_z - n_zH_y) \end{bmatrix} = \frac{-1}{\varepsilon\mu}\begin{bmatrix} \hat{\mathbf{n}} \times (\hat{\mathbf{n}} \times \mathbf{H}) \\ \hat{\mathbf{n}} \times (\hat{\mathbf{n}} \times \mathbf{E}) \end{bmatrix}. \tag{4.156.b}$$

The result of (4.153)–(4.155) is the upwind flux (4.145.a). This derivation process will play an important role in the following section that (4.154) can handle additional jump introduced by the boundary conditions.

Substituting (4.145.a) into (4.143.a) leads to the semi-discrete equations for the i-th element as

$$\begin{aligned} \int_{\gamma^i} \mathbf{\Phi}^i \cdot \varepsilon^i \frac{\partial \mathbf{E}^i}{\partial t} dV = {} & \int_{\gamma^i} \mathbf{\Phi}^i \cdot \nabla \times \mathbf{H}^i dV + \frac{Z^j}{Z^{ij}} \int_{\partial\gamma^i} \mathbf{\Phi}^i \cdot \hat{\mathbf{n}}^i \times (\mathbf{H}^j - \mathbf{H}^i) dS \\ & + \frac{1}{Z^{ij}} \int_{\partial\gamma^i} \mathbf{\Phi}^i \cdot \hat{\mathbf{n}}^i \times [\hat{\mathbf{n}}^i \times (\mathbf{E}^i - \mathbf{E}^j)] dS - \int_{\gamma^i} \mathbf{\Phi}^i \cdot \mathbf{J}_e^i dV \end{aligned} \tag{4.157.a}$$

$$\begin{aligned} \int_{\gamma^i} \mathbf{\Psi}^i \cdot \mu^i \frac{\partial \mathbf{H}^i}{\partial t} dV = {} & -\int_{\gamma^i} \mathbf{\Psi}^i \cdot \nabla \times \mathbf{E}^i dV + \frac{Y^j}{Y^{ij}} \int_{\partial\gamma^i} \mathbf{\Psi}^i \cdot \hat{\mathbf{n}}^i \times (\mathbf{E}^i - \mathbf{E}^j) dS \\ & + \frac{1}{Y^{ij}} \int_{\partial\gamma^i} \mathbf{\Psi}^i \cdot \hat{\mathbf{n}}^i \times [\hat{\mathbf{n}}^i \times (\mathbf{H}^i - \mathbf{H}^j)] dS - \int_{\gamma^i} \mathbf{\Psi}^i \cdot \mathbf{J}_h^i dV \end{aligned} \tag{4.157.b}$$

Expanding the unknown fields with basis function leads to

$$\mathbf{M}_e^i \frac{\partial}{\partial t}\mathbf{e}^i = \mathbf{S}_{eh}^{ii}\mathbf{h}^i + \mathbf{L}_{eh}^{ii}\mathbf{h}^i + \sum \mathbf{L}_{eh}^{ij}\mathbf{h}^j + \mathbf{L}_{ee}^{ii}\mathbf{e}^i + \sum \mathbf{L}_{ee}^{ij}\mathbf{e}^j + \mathbf{j}_e^i \tag{4.158.a}$$

$$\mathbf{M}_h^i \frac{\partial}{\partial t}\mathbf{h}^i = \mathbf{S}_{he}^{ii}\mathbf{e}^i + \mathbf{L}_{he}^{ii}\mathbf{e}^i + \sum \mathbf{L}_{he}^{ij}\mathbf{e}^j + \mathbf{L}_{hh}^{ii}\mathbf{h}^i + \sum \mathbf{L}_{hh}^{ij}\mathbf{h}^j + \mathbf{j}_h^i \tag{4.158.b}$$

Note that the superscript j refers to any adjacent elements. Besides, there are four adjacent elements for a tetrahedron element, so the summation in (4.158.a) means that all the fields in the four adjacent elements should be counted. The elements of matrices in (4.158.a) are

$$\mathbf{M}_{e,n'n}^i = \int_{\gamma^i} \mathbf{\Phi}_{n'}^i \cdot \varepsilon^i \mathbf{\Phi}_n^i dV \quad \mathbf{M}_{h,n'n}^i = \int_{\gamma^i} \mathbf{\Psi}_{n'}^i \cdot \mu^i \mathbf{\Psi}_n^i dV$$

$$\mathbf{S}_{eh,n'n}^{ii} = \int_{\gamma^i} \mathbf{\Phi}_{n'}^i \cdot \nabla \times \mathbf{\Psi}_n^i dV \quad \mathbf{S}_{he,n'n}^{ii} = -\int_{\gamma^i} \mathbf{\Psi}_{n'}^i \cdot \nabla \times \mathbf{\Phi}_n^i dV$$

$$\mathbf{L}_{ee,n'n}^{ii} = \sum_j \frac{1}{Z^{ij}} \int_{\partial\gamma^{ij}} \mathbf{\Phi}_{n'}^i \cdot \hat{\mathbf{n}}^i \times \left(\hat{\mathbf{n}}^i \times \mathbf{\Phi}_n^i\right) dS \quad \mathbf{L}_{hh,n'n}^{ii} = \sum_j \frac{Y^j}{Y^{ij}} \int_{\partial\gamma^{ij}} \mathbf{\Psi}_{n'}^i \cdot \hat{\mathbf{n}}^i \times \mathbf{\Phi}_n^i dS$$

$$\mathbf{L}_{ee,n'n}^{ij} = -\frac{1}{Z^{ij}} \int_{\partial\gamma^{ij}} \mathbf{\Phi}_{n'}^i \cdot \hat{\mathbf{n}}^i \times \left(\hat{\mathbf{n}}^i \times \mathbf{\Phi}_n^j\right) dS$$

$$\mathbf{L}_{hh,n'n}^{ij} = -\frac{1}{Y^{ij}} \int_{\partial\gamma^{ij}} \mathbf{\Psi}_{n'}^i \cdot \hat{\mathbf{n}}^i \times \left(\hat{\mathbf{n}}^i \times \mathbf{\Psi}_n^j\right) dS$$

$$\mathbf{L}_{eh,n'n}^{ii} = -\sum_j \frac{Z^j}{Z^{ij}} \int_{\partial\gamma^{ij}} \mathbf{\Phi}_{n'}^i \cdot \hat{\mathbf{n}}^i \times \mathbf{\Psi}_n^i dS \quad \mathbf{L}_{he,n'n}^{ii} = \sum_j \frac{1}{Y^{ij}} \int_{\partial\gamma^{ij}} \mathbf{\Psi}_{n'}^i \cdot \hat{\mathbf{n}}^i \times \left(\hat{\mathbf{n}}^i \times \mathbf{\Psi}_n^i\right) dS$$

$$\mathbf{L}_{eh,n'n}^{ij} = \frac{Z^j}{Z^{ij}} \int_{\partial\gamma^{ij}} \mathbf{\Phi}_{n'}^i \cdot \hat{\mathbf{n}}^i \times \mathbf{\Psi}_n^j dS \quad \mathbf{L}_{he,n'n}^{ij} = -\frac{Y^j}{Y^{ij}} \int_{\partial\gamma^{ij}} \mathbf{\Psi}_{n'}^i \cdot \hat{\mathbf{n}}^i \times \mathbf{\Phi}_n^j dS$$

$$\mathbf{j}_{e,n'}^i = -\int_{\gamma^i} \mathbf{\Phi}_{n'}^i \cdot \mathbf{J}_e^i dV \quad \mathbf{j}_h^i = -\int_{\gamma^i} \mathbf{\Psi}_{n'}^i \cdot \mathbf{J}_h^i dV \tag{4.159}$$

Now, there are only one step away from the solution, the time-stepping process. In the next section, two temporal integration methods will be introduced. One is the leap-frog method, which is very similar to the one in FDTD and the other is the Runge–Kutta (RK) method.

4.2.1.3 Time Integration Scheme

There are two major families of time integration schemes, the implicit and explicit methods. From the perspective of computer program implementation, the difference between explicit and implicit methods is that in each time step, the unknowns are updated simultaneously (solving a large linear system of equations) or can be updated separately (solving multiple small linear systems of equations). The leap-frog method in FDTD introduced above is a typical explicit method, which can be updated point by point for the unknowns in (4.6.a). The advantage of the implicit method is that it can be unconditionally stable [54] and its time interval is not constrained by the spatial discretization that leads to a fewer time-stepping iterations, while the time interval in the explicit method is limited by the size of spatial discretization, as described in Section 4.1.1.4. Considering that avoiding solving a large matrix is one of the advantages of DGTD, two explicit stepping methods are mainly introduced, i.e., the leap-frog method and the explicit RK method.

Using the central difference scheme in (4.4.a) to replace the time derivative in (4.158.a) leads to

$$\mathbf{e}^i_{n+1} = \mathbf{e}^i_n + \Delta t\left(\mathbf{M}^i_e\right)^{-1} \times \left(\mathbf{S}^{ii}_{eh}\mathbf{h}^i_{n+1/2} + \mathbf{L}^{ii}_{eh}\mathbf{h}^i_{n-1/2} + \sum_j \mathbf{L}^{ij}_{eh}\mathbf{h}^j_{n+1/2} + \mathbf{L}^{ii}_{ee}\mathbf{e}^i_n + \sum_j \mathbf{L}^{ij}_{ee}\mathbf{e}^j_n + \mathbf{j}^i_{e,n}\right) \tag{4.159.a}$$

$$\mathbf{h}^i_{n+1/2} = \mathbf{h}^i_{n-1/2} + \Delta t\left(\mathbf{M}^i_h\right)^{-1} \times \left(\mathbf{S}^{ii}_{he}\mathbf{e}^i_n + \mathbf{L}^{ii}_{he}\mathbf{e}^i_n + \sum_j \mathbf{L}^{ij}_{he}\mathbf{e}^j_n + \mathbf{L}^{ii}_{hh}\mathbf{h}^i_{n-1/2} + \sum_j \mathbf{L}^{ij}_{hh}\mathbf{h}^j_{n-1/2} + \mathbf{j}^i_{h,n-1/2}\right) \tag{4.159.b}$$

where the dissipative terms $\mathbf{L}^{ii}_{ee}\mathbf{e}^i_n$, $\mathbf{L}^{ij}_{ee}\mathbf{e}^j_n$, $\mathbf{L}^{ii}_{hh}\mathbf{h}^i_{n-1/2}$, and $\mathbf{L}^{ij}_{hh}\mathbf{h}^j_{n-1/2}$ and the impress source terms $\mathbf{J}^i_{e,n}$ and $\mathbf{J}^i_{h,n-1/2}$ are all approximated by the backward approximations as

$$\mathbf{e}|_{t=(n+1/2)\Delta t} \approx \mathbf{e}_n,\ \mathbf{h}|_{t=n\Delta t} \approx \mathbf{h}_{n-1/2},\ \mathbf{j}_e|_{t=(n+1/2)\Delta t} \approx \mathbf{j}_{e,n},\ \mathbf{j}_h|_{t=n\Delta t} \approx \mathbf{j}_{h,n-1/2} \tag{4.160}$$

Another time integration scheme is the RK scheme [55], whose main advantages compared to the leap-frog method are (1) adaptive order matching between temporal integration and spatial approximation is straightforward to implement and (2) it is more robust when additional unknowns in the ADE methods are needed to handle the dispersive susceptibilities. For a N-dimensional system of differential equations

$$\mathbf{y}' = \mathbf{f}(t, \mathbf{y}) \tag{4.161}$$

where

$$\mathbf{y}' = \begin{bmatrix} y'_1 \\ y'_2 \\ \vdots \\ y'_N \end{bmatrix}, \mathbf{f} = \begin{bmatrix} f_1(t, y_1, \cdots, y_N) \\ f_2(t, y_1, \cdots, y_N) \\ \vdots \\ f_N(t, y_1, \cdots, y_N) \end{bmatrix} \tag{4.162}$$

and $y'_n = \frac{\partial y_n}{\partial t}$ denotes the partial difference with respect to the independent variable t. To solve the equation in (4.161) with the initial condition $\mathbf{y}(t_0) = \mathbf{y}_0$, the RK method can be written as

$$\mathbf{y}_{n+1} = \mathbf{y}_n + h\sum_{k=1}^{s} b_k \mathbf{Y}_k \tag{4.163.a}$$

$$\mathbf{Y}_i = \mathbf{f}\left(t_n + c_i h, \mathbf{y}_n + h\sum_{j=1}^{i-1} a_{i,j}\mathbf{Y}_j\right) \tag{4.163.b}$$

where the coefficients a, b, and c are fixed real number, and their values are determined by the order of accuracy. The parameter h denotes the step size of the independent variable, and it equals to the time interval as $h = \Delta t$ in DGTD. The coefficients of classical fourth-order RK are

$$\begin{array}{c|cccc} 0 & & & & \\ c_2 & a_{2,1} & & & \\ c_3 & a_{3,1} & a_{3,2} & & \\ c_4 & a_{4,1} & a_{4,2} & a_{4,3} & \\ \hline & b_1 & b_2 & b_3 & b_4 \end{array} \rightarrow \begin{array}{c|cccc} 0 & & & & \\ 1/2 & 1/2 & & & \\ 1/2 & 0 & 1/2 & & \\ 1 & 0 & 0 & 1 & \\ \hline & 1/6 & 1/3 & 1/3 & 1/6 \end{array} \tag{4.164}$$

Substituting the coefficients in (4.164) into (4.163.a) leads to the explicit RK updating equations

$$\mathbf{y}_{n+1} = \mathbf{y}_n + \frac{\Delta t}{6}(\mathbf{Y}_1 + 2\mathbf{Y}_2 + 2\mathbf{Y}_3 + \mathbf{Y}_4) \tag{4.164.b}$$

$$\mathbf{Y}_1 = \mathbf{f}(t_n, \mathbf{y}_n) \tag{4.164.c}$$

$$\mathbf{Y}_2 = \mathbf{f}\left(t_n + \frac{1}{2}\Delta t, \quad \mathbf{y}_n + \frac{1}{2}\Delta t\mathbf{Y}_1\right) \tag{4.164.d}$$

$$\mathbf{Y}_3 = \mathbf{f}\left(t_n + \frac{1}{2}\Delta t, \quad \mathbf{y}_n + \frac{1}{2}\Delta t\mathbf{Y}_2\right) \tag{4.164.e}$$

$$\mathbf{Y}_4 = \mathbf{f}(t_n + \Delta t, \quad \mathbf{y}_n + \Delta t\mathbf{Y}_3) \tag{4.164.f}$$

Using the RK method, one should abandon the idea that the half-step staggered temporal sampling method of electric and magnetic fields in the previous leap-frog method but merge them into a vector of unknowns at the same sampling time steps instead. Rewrite (4.158.a) as a system of PDEs as

$$\mathbf{M}_q^i \frac{\partial}{\partial t}\mathbf{q}^i = \mathbf{S}_q^i \mathbf{q}^i + \sum_j \mathbf{L}_{qq}^{ij}\mathbf{q}^j + \mathbf{j}^i \tag{4.165}$$

where

$$\mathbf{q}^i = \left[\mathbf{e}^i \ \mathbf{h}^i\right]^T, \quad \mathbf{j}^i = \left[\mathbf{j}_e^i \ \mathbf{j}_h^i\right]^T$$

$$\mathbf{M}_q^i = \begin{bmatrix} \mathbf{M}_e^i & \\ & \mathbf{M}_h^i \end{bmatrix}, \quad \mathbf{S}_q^i = \begin{bmatrix} \mathbf{L}_{ee}^{ii} & \mathbf{S}_{eh}^{ii} + \mathbf{L}_{eh}^{ii} \\ \mathbf{S}_{he}^{ii} + \mathbf{L}_{he}^{ii} & \mathbf{L}_{hh}^{ii} \end{bmatrix}, \quad \mathbf{L}_{qq}^{ij} = \begin{bmatrix} \mathbf{L}_{ee}^{ij} & \mathbf{L}_{eh}^{ij} \\ \mathbf{L}_{he}^{ij} & \mathbf{L}_{hh}^{ij} \end{bmatrix} \tag{4.166}$$

and the updating equations are

$$\mathbf{q}_{n+1}^i = \mathbf{q}_n^i + \frac{\Delta t}{6}\left(\mathbf{Q}_1^i + 2\mathbf{Q}_2^i + 2\mathbf{Q}_3^i + \mathbf{Q}_4^i\right) \tag{4.167.a}$$

$$\mathbf{M}_q^i \mathbf{Q}_1^i = \mathbf{S}_q^i \mathbf{q}_n^i + \sum_j \mathbf{L}_{qq}^{ij} \mathbf{q}_n^j + \mathbf{j}_n^i \tag{4.167.b}$$

$$\mathbf{M}_q^i \mathbf{Q}_2^i = \mathbf{S}_q^i \left(\mathbf{q}_n^i + \frac{1}{2}\Delta t \mathbf{Q}_1^i\right) + \sum_j \mathbf{L}_{qq}^{ij} \left(\mathbf{q}_n^j + \frac{1}{2}\Delta t \mathbf{Q}_1^j\right) + \mathbf{j}_{n+1/2}^i \tag{4.167.c}$$

$$\mathbf{M}_q^i \mathbf{Q}_3^i = \mathbf{S}_q^i \left(\mathbf{q}_n^i + \frac{1}{2}\Delta t \mathbf{Q}_2^i\right) + \sum_j \mathbf{L}_{qq}^{ij} \left(\mathbf{q}_n^j + \frac{1}{2}\Delta t \mathbf{Q}_2^j\right) + \mathbf{j}_{n+1/2}^i \tag{4.167.d}$$

$$\mathbf{M}_q^i \mathbf{Q}_4^i = \mathbf{S}_q^i \left(\mathbf{q}_n^i + \Delta t \mathbf{Q}_3^i\right) + \sum_j \mathbf{L}_{qq}^{ij} \left(\mathbf{q}_n^j + \Delta t \mathbf{Q}_3^j\right) + \mathbf{j}_{n+1}^i \tag{4.167.e}$$

Note that updating intermediate stage $\mathbf{Q}_{k+1}^i$ requires the previous stage of its own, $\mathbf{Q}_k^i$, and all the adjacent stages $\mathbf{Q}_k^{j_1}$, $\mathbf{Q}_k^{j_2}$, $\mathbf{Q}_k^{j_3}$, and $\mathbf{Q}_k^{j_4}$. Hence, the element-wise loops should be within the stage-loops in the time-stepping program.

The fourth-order RK introduced above actually requires four times memory at each time step n to record $\mathbf{Q}_{1,2,3,4}$. If one is in pursuit of stage-independent memory, the low-storage RK method is a good choice [56].

The time step size of both the leap-frog and the RK methods are bounded by the spatial discretization. Since the tetrahedral mesh is unstructured, the stability condition is not as simple as that for the FDTD method. The stability condition of the leap-frog DGTD in the isotropic medium is given as [57]

$$c_i \Delta t_i \left[\frac{4}{3}\sqrt{\frac{5}{2}} + \frac{8}{3}\max\left(\sqrt{\frac{\mu_i}{\mu_j}}, \sqrt{\frac{\varepsilon_i}{\varepsilon_j}}\right)\right] < \frac{4V_i}{P_i} \tag{4.168}$$

where c_i, ε_i, and μ_i are the light speed, permittivity, and permeability of the i-th element, and ε_j and μ_j denote the permittivity and permeability of its adjacent element. The coefficients V_i and P_i denote the volume and the surface area of the i-th element. The default time step is generated by the minimal value of Δt_i according to (4.168), and $\Delta t = \frac{4}{5}\min(\Delta t_i)$ is usually chosen in practice.

As for the RK DGTD method, the stability condition of the fourth-order RK method is presented in [55]. Synthesize all the unknowns into a vector $\mathbf{q} = [\mathbf{q}^1, \mathbf{q}^2, \cdots, \mathbf{q}^N]^T$, and synthesize all the differential equations (4.165) into a large system of PDEs corresponding to the entire computational domain

$$\mathbf{M}\frac{\partial}{\partial t}\mathbf{q} = \mathbf{S}\mathbf{q} + \mathbf{j} \tag{4.169}$$

Finally, calculate the maximal absolute value of the eigenvalues of the system matrix as

$$\lambda_{\max} = \max(|\text{eig}(\mathbf{M}^{-1}\mathbf{S})|) \tag{4.170}$$

As long as this maximal eigenvalue and the time step size meet the following stability condition in (4.173.a), the RK DGTD method will be stable.

$$f(z) = \left|1 + z + \frac{z^2}{2} + \frac{z^3}{6} + \frac{z^4}{24}\right| < 1, z = \Delta t \lambda_{\max} \tag{4.171}$$

Considering that the calculation of (4.170) can be complex, a trick is to use the half of the value obtained from (4.168) directly. This method has the fourth order of convergence, but if the order of the basis function is lower than 4, this advantage will not be fully exerted [55]. Another advantage of the RK method is more attractive, i.e., the robustness. In contrast, the leap-frog method requires careful design of the iteration sequence of field variables and other auxiliary unknowns, which can be complicated for the dispersive susceptibilities.

Till here, the basis theory of DGTD is introduced, including the spatial basis functions, the numerical flux, and the time integration scheme. The other contents, such as the truncation boundary conditions and the source excitation, involve the same computational electromagnetic principle with the FDTD method. For example, the PML is based on the virtual absorbing medium theory whose implementation employs the ADE method, and the TF/SF is based on the Huygens's equivalence principle whose implementation employs the equivalent surface current method. These will not be introduced in this book, and the implementation method can be found in [47].

4.2.2 Strategies for Integrating GSTCs in DGTD

This chapter introduces the coupling method of DGTD and GSTC. The implements of isotropic (with and without normal components) and dispersive anisotropic susceptibilities are introduced in this section.

4.2.2.1 Basic GSTC Implementation in DGTD

The original research combining DGTD and GSTC also employed the virtual field theory as introduced in Section 4.1.3 [58, 59]. In recent work, it is more in line with the framework of DG, which is strictly derived from the theory of numerical flux [60–62]. Both derivations are effective, and here the derivation based on the concept of numerical flux is introduced.

As introduced in Section 4.1.1.2, the volume inside the elements will not bring in any jump between the intermediate vectors $\mathbf{q}^*$ and $\mathbf{q}^{**}$ on the interface according to (4.154). However, when there is a zero-thickness sheet with surface susceptibilities

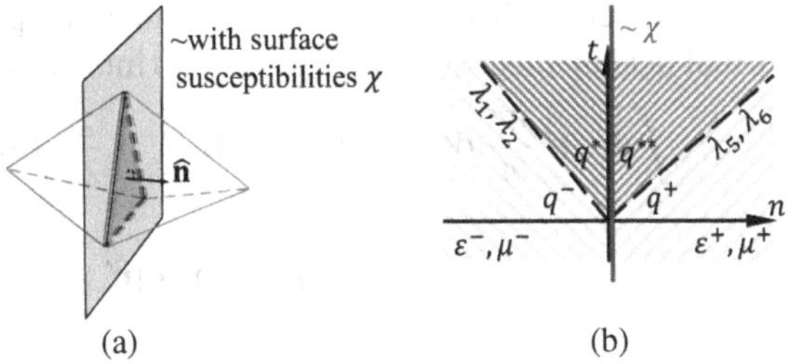

Figure 4.24 (a) The zero-thickness generalized sheet placed at the interface between two elements. (b) The construction of the solution to the Riemann problem on the interface.

between two elements, as shown in Figure 4.4(a), there will be a jump between $\mathbf{q}^*$ and $\mathbf{q}^{**}$, as shown in Figure 4.24(b), and its relationship is directly described by the GSTC without normal susceptibilities as

$$\hat{\mathbf{n}} \times (\mathbf{q}^{**} - \mathbf{q}^*) = \begin{bmatrix} -\mu_0 \dfrac{\partial}{\partial t}\mathbf{M}_\parallel \\ \dfrac{\partial}{\partial t}\mathbf{P}_\parallel \end{bmatrix} \tag{4.172}$$

Combining (4.153), (4.155), (4.155), (4.156), and (4.172) leads to the following equations:

$$Y^{ij}\hat{\mathbf{n}}^i \times \mathbf{E}^* = \hat{\mathbf{n}}^i \times \left(Y^i\mathbf{E}^i + Y^j\mathbf{E}^j - \frac{\partial}{\partial t}\mathbf{P}_\parallel\right) + \hat{\mathbf{n}}^i \times \left[\hat{\mathbf{n}}^i \times \left(\mathbf{H}^j - \mathbf{H}^i - \mu_0 Y^j \frac{\partial}{\partial t}\mathbf{M}_\parallel\right)\right] \tag{4.173.a}$$

$$Z^{ij}\hat{\mathbf{n}}^i \times \mathbf{H}^* = \hat{\mathbf{n}}^i \times \left(Z^i\mathbf{H}^i + Z^j\mathbf{H}^j - \mu_0\frac{\partial}{\partial t}\mathbf{M}_\parallel\right) + \hat{\mathbf{n}}^i \times \left[\hat{\mathbf{n}}^i \times \left(\mathbf{E}^i - \mathbf{E}^j + Z^j \frac{\partial}{\partial t}\mathbf{P}_\parallel\right)\right] \tag{4.173.b}$$

This is the numerical flux for the metasurface with the tangential surface polarization. Considering the simplest constitutive relations

$$\mathbf{P}_\parallel = \frac{1}{2}\varepsilon_0\chi_{ee}^{\tau\tau}\left(\mathbf{E}_\parallel^i + \mathbf{E}_\parallel^j\right), \quad \mathbf{M}_\parallel = \frac{1}{2}\chi_{mm}^{\tau\tau}\left(\mathbf{H}_\parallel^i + \mathbf{H}_\parallel^j\right) \tag{4.174}$$

where the surface susceptibility tensors are simplified as the tangential scalar isotropic susceptibilities $\chi^{\tau\tau}$. Substituting (4.174) into (4.173.a) and using $\hat{\mathbf{n}} \times \mathbf{E}_\parallel = (\hat{\mathbf{n}} \times \mathbf{E})$ lead to

$$\begin{aligned} Y^{ij}\hat{\mathbf{n}} \times \mathbf{E}^* &= \hat{\mathbf{n}}^i \times (Y^i\mathbf{E}^i + Y^j\mathbf{E}^j) + \hat{\mathbf{n}}^i \times [\hat{\mathbf{n}}^i \times (\mathbf{H}^j - \mathbf{H}^i)] \\ &\quad - \frac{\varepsilon_0\chi_{ee}^{\tau\tau}}{2}\frac{\partial}{\partial t}\hat{\mathbf{n}}^i \times (\mathbf{E}^i + \mathbf{E}^j) - \frac{\mu_0\chi_{mm}^{\tau\tau}Y^j}{2}\frac{\partial}{\partial t}\hat{\mathbf{n}}^i \times [\hat{\mathbf{n}}^i \times (\mathbf{H}^i + \mathbf{H}^j)] \end{aligned} \tag{4.175.a}$$

$$\begin{aligned} Z^{ij}\hat{\mathbf{n}} \times \mathbf{H}^* &= \hat{\mathbf{n}}^i \times (Z^i\mathbf{H}^i + Z^j\mathbf{H}^j) + \hat{\mathbf{n}}^i \times [\hat{\mathbf{n}}^i \times (\mathbf{E}^i - \mathbf{E}^j)] \\ &\quad - \frac{\mu_0\chi_{mm}^{\tau\tau}}{2}\frac{\partial}{\partial t}\hat{\mathbf{n}}^i \times (\mathbf{H}^i + \mathbf{H}^j) + \frac{\varepsilon_0\chi_{ee}^{\tau\tau}Z^j}{2}\frac{\partial}{\partial t}\hat{\mathbf{n}}^i \times [\hat{\mathbf{n}}^i \times (\mathbf{E}^i + \mathbf{E}^j)] \end{aligned} \tag{4.175.b}$$

The numerical flux in (4.175.a) is the same as Eq. (4.36) in [58] derived by the virtual field method. Substituting (4.175.a) into (4.143.a) leads to

$$\int_{\gamma^i} \mathbf{\Phi}^i \cdot \varepsilon^i \frac{\partial \mathbf{E}^i}{\partial t} dV = \int_{\gamma^i} \mathbf{\Phi}^i \cdot \nabla \times \mathbf{H}^i dV + \frac{Z^j}{Z^{ij}} \int_{\partial\gamma^i} \mathbf{\Phi}^i \cdot \hat{\mathbf{n}}^i \times (\mathbf{H}^j - \mathbf{H}^i) dS$$
$$+ \frac{1}{Z^{ij}} \int_{\partial\gamma^i} \mathbf{\Phi}^i \cdot \hat{\mathbf{n}}^i \times [\hat{\mathbf{n}}^i \times (\mathbf{E}^i - \mathbf{E}^j)] dS - \int_{\gamma^i} \mathbf{\Phi}^i \cdot \mathbf{J}_e^i dV$$
$$- \frac{\mu_0 \chi_{mm}^{\tau\tau}}{2Z^{ij}} \int_{\partial\gamma^i} \mathbf{\Phi}^i \cdot \frac{\partial}{\partial t} \hat{\mathbf{n}}^i \times (\mathbf{H}^i + \mathbf{H}^j) dS$$
$$+ \frac{\varepsilon_0 \chi_{ee}^{\tau\tau} Z^j}{2Z^{ij}} \int_{\partial\gamma^i} \mathbf{\Phi}^i \cdot \frac{\partial}{\partial t} \hat{\mathbf{n}}^i \times [\hat{\mathbf{n}}^i \times (\mathbf{E}^i + \mathbf{E}^j)] dS \tag{4.176.a}$$

$$\int_{\gamma^i} \mathbf{\Psi}^i \cdot \mu^i \frac{\partial \mathbf{H}^i}{\partial t} dV = -\int_{\gamma^i} \mathbf{\Psi}^i \cdot \nabla \times \mathbf{E}^i dV + \frac{Y^j}{Y^{ij}} \int_{\partial\gamma^i} \mathbf{\Psi}^i \cdot \hat{\mathbf{n}}^i \times (\mathbf{E}^i - \mathbf{E}^j) dS$$
$$+ \frac{1}{Y^{ij}} \int_{\partial\gamma^i} \mathbf{\Psi}^i \cdot \hat{\mathbf{n}}^i \times [\hat{\mathbf{n}}^i \times (\mathbf{H}^i - \mathbf{H}^j)] dS - \int_{\gamma^i} \mathbf{\Psi}^i \cdot \mathbf{J}_h^i dV$$
$$- \frac{\varepsilon_0 \chi_{ee}^{\tau\tau}}{2Z^{ij}} \int_{\partial\gamma^i} \mathbf{\Psi}^i \cdot \frac{\partial}{\partial t} \hat{\mathbf{n}}^i \times (\mathbf{E}^i + \mathbf{E}^j) dS$$
$$- \frac{\mu_0 \chi_{mm}^{\tau\tau} Y^j}{2Z^{ij}} \int_{\partial\gamma^i} \mathbf{\Psi}^i \cdot \frac{\partial}{\partial t} \hat{\mathbf{n}}^i \times [\hat{\mathbf{n}}^i \times (\mathbf{H}^i + \mathbf{H}^j)] dS \tag{4.176.b}$$

and the corresponding semi-discrete equations in the matrix form are

$$\mathbf{M}_e^i \frac{\partial}{\partial t} \mathbf{e}^i - \mathbf{G}_{ee}^{ii} \frac{\partial}{\partial t} \mathbf{e}^i - \mathbf{G}_{ee}^{ij} \frac{\partial}{\partial t} \mathbf{e}^j - \mathbf{G}_{eh}^{ii} \frac{\partial}{\partial t} \mathbf{h}^i - \mathbf{G}_{eh}^{ij} \frac{\partial}{\partial t} \mathbf{h}^j$$
$$= \mathbf{S}_{eh}^{ii} \mathbf{h}^i + \mathbf{L}_{eh}^{ii} \mathbf{h}^i + \sum \mathbf{L}_{eh}^{ij} \mathbf{h}^j + \mathbf{L}_{ee}^{ii} \mathbf{e}^i + \sum \mathbf{L}_{ee}^{ij} \mathbf{e}^j + \mathbf{j}_e^i \tag{4.177.a}$$

$$\mathbf{M}_h^i \frac{\partial}{\partial t} \mathbf{h}^i - \mathbf{G}_{he}^{ii} \frac{\partial}{\partial t} \mathbf{e}^i - \mathbf{G}_{he}^{ij} \frac{\partial}{\partial t} \mathbf{e}^j - \mathbf{G}_{hh}^{ii} \frac{\partial}{\partial t} \mathbf{h}^i - \mathbf{G}_{hh}^{ij} \frac{\partial}{\partial t} \mathbf{h}^j$$
$$= \mathbf{S}_{he}^{ii} \mathbf{e}^i + \mathbf{L}_{he}^{ii} \mathbf{e}^i + \sum \mathbf{L}_{he}^{ij} \mathbf{e}^j + \mathbf{L}_{hh}^{ii} \mathbf{h}^i + \sum \mathbf{L}_{hh}^{ij} \mathbf{h}^j + \mathbf{j}_h^i \tag{4.177.b}$$

where the elements of the new matrices corresponding to the surface susceptibilities are

$$\mathbf{G}_{ee,n'n}^{ii} = \frac{\varepsilon_0 \chi_{ee}^{\tau\tau} Z^j}{2Z^{ij}} \int_{\partial\gamma_g^i} \mathbf{\Phi}_{n'}^i \cdot \hat{\mathbf{n}}^i \times \left(\hat{\mathbf{n}}^i \times \mathbf{\Phi}_n^i \right) dS \tag{4.178.a}$$

$$\mathbf{G}_{ee,n'n}^{ij} = \frac{\varepsilon_0 \chi_{ee}^{\tau\tau} Z^j}{2Z^{ij}} \int_{\partial\gamma_g^i} \mathbf{\Phi}_{n'}^i \cdot \hat{\mathbf{n}}^i \times \left(\hat{\mathbf{n}}^i \times \mathbf{\Phi}_n^j \right) dS \tag{4.178.b}$$

$$\mathbf{G}^{ii}_{eh,n'n} = -\frac{\mu_0\chi^{\tau\tau}_{mm}}{2Z^{ij}}\int_{\partial\gamma^i_g}\mathbf{\Phi}^i_{n'}\cdot\hat{\mathbf{n}}^i\times\mathbf{\Psi}^i_n dS \tag{4.178.c}$$

$$\mathbf{G}^{ij}_{eh,n'n} = -\frac{\mu_0\chi^{\tau\tau}_{mm}}{2Z^{ij}}\int_{\partial\gamma^i_g}\mathbf{\Phi}^i_{n'}\cdot\hat{\mathbf{n}}^i\times\mathbf{\Psi}^j_n dS \tag{4.178.d}$$

$$\mathbf{G}^{ii}_{he,n'n} = -\frac{\varepsilon_0\chi^{\tau\tau}_{ee}}{2Z^{ij}}\int_{\partial\gamma^i_g}\mathbf{\Psi}^i_{n'}\cdot\hat{\mathbf{n}}^i\times\mathbf{\Phi}^i_n dS \tag{4.178.e}$$

$$\mathbf{G}^{ij}_{he,n'n} = -\frac{\varepsilon_0\chi^{\tau\tau}_{ee}}{2Z^{ij}}\mathbf{\Psi}\int_{\partial\gamma^i_g}\mathbf{\Phi}^i_{n'}\cdot\hat{\mathbf{n}}^i\times\mathbf{\Phi}^j_n dS \tag{4.178.f}$$

$$\mathbf{G}^{ii}_{hh,n'n} = -\frac{\mu_0\chi^{\tau\tau}_{mm}Y^j}{2Z^{ij}}\int_{\partial\gamma^i_g}\mathbf{\Psi}^i_{n'}\cdot\hat{\mathbf{n}}^i\times\left(\hat{\mathbf{n}}^i\times\mathbf{\Phi}^i_n\right)dS \tag{4.178.g}$$

$$\mathbf{G}^{ij}_{hh,n'n} = -\frac{\mu_0\chi^{\tau\tau}_{mm}Y^j}{2Z^{ij}}\int_{\partial\gamma^i_g}\mathbf{\Psi}^i_{n'}\cdot\hat{\mathbf{n}}^i\times\left(\hat{\mathbf{n}}^i\times\mathbf{\Phi}^j_n\right)dS \tag{4.178.h}$$

The symbol $\partial\gamma^i_g$ denotes the boundary containing the generalized sheet with surface susceptibilities (GSTC-interface), which means the matrices (4.178.a) only exist on GSTC-interface.

Using the central difference approximation in (4.177.a) will result in local implicity, i.e., the two elements related to the GSTC-interface need to be updated simultaneously. This does not lead to large matrix inversion. The advantage of the local implicity is that it provides better stability, while the disadvantage is that its implementation and time-stepping iteration are complex, and the detailed information can be referred to [61]. To incorporate GSTC as a plug-in to the classical formulations in (4.159), backward differences for the GSTC-interface-related derivative terms are employed at the expense of worse stability and accuracy.

$$\mathbf{G}^{ij}_{ee}\frac{\partial}{\partial t}\mathbf{e}^j\bigg|_{t=\left(n+\frac{1}{2}\right)\Delta t} \approx \mathbf{G}^{ij}_{ee}\frac{\mathbf{e}^j_n-\mathbf{e}^j_{n-1}}{\Delta t} \tag{4.179.a}$$

$$\mathbf{G}^{ij}_{eh}\frac{\partial}{\partial t}\mathbf{h}^j\bigg|_{t=\left(n+\frac{1}{2}\right)\Delta t} \approx \mathbf{G}^{ij}_{eh}\frac{\mathbf{h}^j_{n+1/2}-\mathbf{h}^j_{n-1/2}}{\Delta t} \tag{4.179.b}$$

$$\mathbf{G}^{ij}_{he}\frac{\partial}{\partial t}\mathbf{e}^j\bigg|_{t=n\Delta t} \approx \mathbf{G}^{ij}_{he}\frac{\mathbf{e}^j_n-\mathbf{e}^j_{n-1}}{\Delta t} \tag{4.179.c}$$

$$\mathbf{G}^{ij}_{eh}\frac{\partial}{\partial t}\mathbf{h}^j\bigg|_{t=n\Delta t} \approx \mathbf{G}^{ij}_{eh}\frac{\mathbf{h}^j_{n-1/2}-\mathbf{h}^j_{n-3/2}}{\Delta t} \tag{4.179.d}$$

Substituting (4.179.a) into (4.177.a) leads to

$$\begin{aligned}
\mathbf{e}_{n+1}^{i} = \mathbf{e}_{n}^{i} + \Delta t \mathbf{M}_{e}^{i^{-1}} \Bigg(& \mathbf{S}_{eh}^{ii}\mathbf{h}_{n+1/2}^{i} + \mathbf{L}_{eh}^{ii}\mathbf{h}_{n+1/2}^{i} + \sum_{j}\mathbf{L}_{eh}^{ij}\mathbf{h}_{n+1/2}^{j} \\
& +\mathbf{L}_{ee}^{ii}\mathbf{e}_{n}^{i} + \sum_{j}\mathbf{L}_{ee}^{ij}\mathbf{e}_{n}^{j} + \mathbf{j}_{e,n}^{i} \Bigg) \\
& + \mathbf{M}_{e}^{i^{-1}} \Big[\mathbf{G}_{ee}^{ii}\left(\mathbf{e}_{n}^{i} - \mathbf{e}_{n-1}^{i}\right) + \mathbf{G}_{ee}^{ij}\left(\mathbf{e}_{n}^{j} - \mathbf{e}_{n-1}^{j}\right) \\
& +\mathbf{G}_{eh}^{ii}\left(\mathbf{h}_{n+1/2}^{i} - \mathbf{h}_{n-1/2}^{i}\right) + \mathbf{G}_{eh}^{ij}\left(\mathbf{h}_{n+1/2}^{j} - \mathbf{h}_{n-1/2}^{j}\right) \Big]
\end{aligned} \tag{4.180.a}$$

$$\begin{aligned}
\mathbf{h}_{n+1/2}^{i} = \mathbf{h}_{n-1/2}^{i} + \Delta t \mathbf{M}_{h}^{i^{-1}} \Bigg(& \mathbf{S}_{he}^{ii}\mathbf{e}_{n}^{i} + \mathbf{L}_{he}^{ii}\mathbf{e}_{n}^{i} + \sum_{j}\mathbf{L}_{he}^{ij}\mathbf{e}_{n}^{j} + \mathbf{L}_{hh}^{ii}\mathbf{h}_{n-1/2}^{i} \\
& +\sum_{j}\mathbf{L}_{hh}^{ij}\mathbf{h}_{n-1/2}^{j} + \mathbf{j}_{h,n-1/2}^{i} \Bigg) \\
& + \mathbf{M}_{h}^{i^{-1}} \Big[\mathbf{G}_{he}^{ii}\left(\mathbf{e}_{n}^{i} - \mathbf{e}_{n-1}^{i}\right) + \mathbf{G}_{he}^{ij}\left(\mathbf{e}_{n}^{j} - \mathbf{e}_{n-1}^{j}\right) \\
& +\mathbf{G}_{hh}^{ii}\left(\mathbf{h}_{n-1/2}^{i} - \mathbf{h}_{n-3/2}^{i}\right) + \mathbf{G}_{hh}^{ij}\left(\mathbf{h}_{n-1/2}^{j} - \mathbf{h}_{n-3/2}^{j}\right) \Big]
\end{aligned} \tag{4.180.b}$$

Numerical experiments to test the stability using the explicit updating equations in (4.180.a) and the locally implicit updating method can be found in [61].

The time stepping is even more difficult when the susceptibilities contain anisotropy, normal polarization and dispersion. Therefore, from Section 4.1.2, the RK method is used as the time integration scheme, and the reader will directly feel the advantages of its robustness. Before that, the updating equations of DGTD-GSTC for the susceptibility in the form of $\chi = j\omega K$ is presented.

$$\begin{aligned}
\mathbf{e}_{n+1}^{i} = \mathbf{e}_{n}^{i} + \Delta t \mathbf{M}_{e}^{i^{-1}} \Bigg(& \mathbf{S}_{eh}^{ii}\mathbf{h}_{n+1/2}^{i} + \mathbf{L}_{eh}^{ii}\mathbf{h}_{n+1/2}^{i} + \sum_{j}\mathbf{L}_{eh}^{ij}\mathbf{h}_{n+1/2}^{j} \\
& +\mathbf{L}_{ee}^{ii}\mathbf{e}_{n}^{i} + \sum_{j}\mathbf{L}_{ee}^{ij}\mathbf{e}_{n}^{j} + \mathbf{j}_{e,n}^{i} \Bigg) \\
& + \Delta t \mathbf{M}_{e}^{i^{-1}} \Big[K_{ee}^{\tau\tau}\widehat{\mathbf{G}}_{ee}^{ii}\mathbf{e}_{n}^{i} + K_{ee}^{\tau\tau}\widehat{\mathbf{G}}_{ee}^{ij}\mathbf{e}_{n}^{j} + K_{mm}^{\tau\tau}\widehat{\mathbf{G}}_{eh}^{ii}\mathbf{h}_{n+1/2}^{i} + K_{mm}^{\tau\tau}\widehat{\mathbf{G}}_{eh}^{ij}\mathbf{h}_{n+1/2}^{j} \Big]
\end{aligned} \tag{4.181.a}$$

$$\begin{aligned}
\mathbf{h}_{n+1/2}^{i} = \mathbf{h}_{n-1/2}^{i} + \Delta t \mathbf{M}_{h}^{i^{-1}} \Bigg(& \mathbf{S}_{he}^{ii}\mathbf{e}_{n}^{i} + \mathbf{L}_{he}^{ii}\mathbf{e}_{n}^{i} + \sum_{j}\mathbf{L}_{he}^{ij}\mathbf{e}_{n}^{j} + \mathbf{L}_{hh}^{ii}\mathbf{h}_{n-1/2}^{i} \\
& +\sum_{j}\mathbf{L}_{hh}^{ij}\mathbf{h}_{n-1/2}^{j} + \mathbf{j}_{h,n-1/2}^{i} \Bigg) \\
& + \Delta t \mathbf{M}_{h}^{i^{-1}} \Big[K_{ee}^{\tau\tau}\widehat{\mathbf{G}}_{he}^{ii}\mathbf{e}_{n}^{i} + K_{ee}^{\tau\tau}\widehat{\mathbf{G}}_{he}^{ij}\mathbf{e}_{n}^{j} + K_{mm}^{\tau\tau}\widehat{\mathbf{G}}_{hh}^{ii}\mathbf{h}_{n-1/2}^{i} + K_{mm}^{\tau\tau}\widehat{\mathbf{G}}_{hh}^{ij}\mathbf{h}_{n-1/2}^{j} \Big]
\end{aligned} \tag{4.181.b}$$

where the matrix $\widehat{\mathbf{G}}$ is a variation of matrices (4.178.a) that the susceptibility χ is removed and whose elements are

$$\widehat{\mathbf{G}}^{ii}_{ee,n'n} = \frac{\varepsilon_0 Z^j}{2Z^{ij}} \int_{\partial\gamma^i_g} \boldsymbol{\Phi}^i_{n'} \cdot \hat{\mathbf{n}}^i \times \left(\hat{\mathbf{n}}^i \times \boldsymbol{\Phi}^i_n\right) dS \tag{4.182.a}$$

$$\widehat{\mathbf{G}}^{ij}_{ee,n'n} = \frac{\varepsilon_0 Z^j}{2Z^{ij}} \int_{\partial\gamma^i_g} \boldsymbol{\Phi}^i_{n'} \cdot \hat{\mathbf{n}}^i \times \left(\hat{\mathbf{n}}^i \times \boldsymbol{\Phi}^j_n\right) dS \tag{4.182.b}$$

$$\widehat{\mathbf{G}}^{ii}_{eh,n'n} = -\frac{\mu_0}{2Z^{ij}} \int_{\partial\gamma^i_g} \boldsymbol{\Phi}^i_{n'} \cdot \hat{\mathbf{n}}^i \times \boldsymbol{\Psi}^i_n dS \tag{4.182.c}$$

$$\widehat{\mathbf{G}}^{ij}_{eh,n'n} = -\frac{\mu_0}{2Z^{ij}} \int_{\partial\gamma^i_g} \boldsymbol{\Phi}^i_{n'} \cdot \hat{\mathbf{n}}^i \times \boldsymbol{\Psi}^j_n dS \tag{4.182.d}$$

$$\widehat{\mathbf{G}}^{ii}_{he,n'n} = -\frac{\varepsilon_0}{2Z^{ij}} \int_{\partial\gamma^i_g} \boldsymbol{\Psi}^i_{n'} \cdot \hat{\mathbf{n}}^i \times \boldsymbol{\Phi}^i_n dS \tag{4.182.e}$$

$$\widehat{\mathbf{G}}^{ij}_{he,n'n} = -\frac{\varepsilon_0}{2Z^{ij}} \boldsymbol{\Psi} \int_{\partial\gamma^i_g} \boldsymbol{\Phi}^i_{n'} \cdot \hat{\mathbf{n}}^i \times \boldsymbol{\Phi}^j_n dS \tag{4.182.f}$$

$$\widehat{\mathbf{G}}^{ii}_{hh,n'n} = -\frac{\mu_0 Y^j}{2Z^{ij}} \int_{\partial\gamma^i_g} \boldsymbol{\Psi}^i_{n'} \cdot \hat{\mathbf{n}}^i \times \left(\hat{\mathbf{n}}^i \times \boldsymbol{\Phi}^i_n\right) dS \tag{4.182.g}$$

$$\widehat{\mathbf{G}}^{ij}_{hh,n'n} = -\frac{\mu_0 Y^j}{2Z^{ij}} \int_{\partial\gamma^i_g} \boldsymbol{\Psi}^i_{n'} \cdot \hat{\mathbf{n}}^i \times \left(\hat{\mathbf{n}}^i \times \boldsymbol{\Phi}^j_n\right) dS \tag{4.182.h}$$

When simulating a time-varying metasurface, (4.181.a) is a suitable choice, provided that there is no significant phase delay in the fields before and after the meta-atoms (resonance phenomenon), so that the transmission and reflection of each meta-atom can be regarded as time-varying real coefficients $T(t)$ and $R(t)$. Then, the parameter K can be time-dependent

$$K^{\tau\tau}_{ee}(t) = 2c_0 \frac{1 - T(t) - R(t)}{1 + T(t) + R(t)} \tag{4.183.a}$$

$$K^{\tau\tau}_{mm}(t) = 2c_0 \frac{1 - T(t) + R(t)}{1 + T(t) - R(t)} \tag{4.183.b}$$

and (4.181.a) becomes

$$\mathbf{e}^i_{n+1} = \mathbf{e}^i_n + \Delta t \mathbf{M}^{i-1}_e \left(\mathbf{S}^{ii}_{eh}\mathbf{h}^i_{n+1/2} + \mathbf{L}^{ii}_{eh}\mathbf{h}^i_{n+1/2} + \sum_j \mathbf{L}^{ij}_{eh}\mathbf{h}^j_{n+1/2} \right.$$
$$\left. + \mathbf{L}^{ii}_{ee}\mathbf{e}^i_n + \sum_j \mathbf{L}^{ij}_{ee}\mathbf{e}^j_n + \mathbf{j}^i_{e,n} \right)$$

$$+ \Delta t \mathbf{M}_e^{i-1} \left[K_{ee,n}^{\tau\tau} \dot{\mathbf{G}}_{ee}^{ii} \mathbf{e}_n^i + K_{ee,n}^{\tau\tau} \dot{\mathbf{G}}_{ee}^{ij} \mathbf{e}_n^j + K_{mm,n+1/2}^{\tau\tau} \dot{\mathbf{G}}_{eh}^{ii} \mathbf{h}_{n+1/2}^i \right.$$
$$\left. + K_{mm,n+1/2}^{\tau\tau} \dot{\mathbf{G}}_{eh}^{ij} \mathbf{h}_{n+1/2}^j \right] \tag{4.184.a}$$

$$\mathbf{h}_{n+1/2}^i = \mathbf{h}_{n-1/2}^i + \Delta t \mathbf{M}_h^{i-1} \left(\mathbf{S}_{he}^{ii} \mathbf{e}_n^i + \mathbf{L}_{he}^{ii} \mathbf{e}_n^i + \sum_j \mathbb{L}_{he}^{ij} \mathbf{e}_n^j + \mathbf{L}_{hh}^{ii} \mathbf{h}_{n-1/2}^i \right.$$
$$\left. + \sum_j \mathbf{L}_{hh}^{ij} \mathbf{h}_{n-1/2}^j + \mathbf{j}_{h,n-1/2}^i \right)$$
$$+ \Delta t \mathbf{M}_h^{i-1} \left[K_{ee,n}^{\tau\tau} \dot{\mathbf{G}}_{he}^{ii} \mathbf{e}_n^i + K_{ee,n}^{\tau\tau} \dot{\mathbf{G}}_{he}^{ij} \mathbf{e}_n^j + K_{mm,n-1/2}^{\tau\tau} \dot{\mathbf{G}}_{hh}^{ii} \mathbf{h}_{n-1/2}^i \right.$$
$$\left. + K_{mm,n-1/2}^{\tau\tau} \dot{\mathbf{G}}_{hh}^{ij} \mathbf{h}_{n-1/2}^j \right] \tag{4.184.b}$$

where $K_n = K(n\Delta t)$.

4.2.2.2 DGTD-GSTC with Surface Normal Polarization

In this section, the method to consider surface normal polarization with DGTD-GSTC will be introduced. The normal polarization $(P_\perp, M_\perp)$ and tangential polarization $(\mathbf{P}_\parallel, \mathbf{M}_\parallel)$ will be treated as individual unknowns. The normal polarization will be expanded in surface scalar basis function ζ and the tangential one will be expanded in surface vector basis function $\boldsymbol{\gamma}$ as introduced in (4.138.a). The unknown vectors will be updated using RK method according to (4.164).

To incorporate the normal surface polarization into the DGTD-GSTC method, the corresponding numerical flux should be derived first. The GSTC with surface normal polarization can be written as

$$\hat{\mathbf{i}}_\perp \times \Delta \mathbf{H} = \frac{\partial}{\partial t} \mathbf{P}_\parallel - \hat{\mathbf{i}}_\perp \times \nabla_\parallel M_\perp \tag{4.185.a}$$

$$\hat{\mathbf{i}}_\perp \times \Delta \mathbf{E} = \frac{\partial}{\partial t} \mathbf{M}_\parallel - \hat{\mathbf{i}}_\perp \times \frac{1}{\varepsilon_0} \nabla_\parallel P_\perp \tag{4.185.b}$$

A new symbol $\hat{\mathbf{i}}_\perp$ is introduced to represent the normal unit vector of the metasurface. Note that the symbol $\hat{\mathbf{n}}^i$ represents the normal unit vector pointing outside the i-th element, while $\hat{\mathbf{i}}_\perp^a$ represents the normal vector of the a-th GSTC-interface (i.e., the metasurface). When the GSTC is employed to model the metasurface, $\hat{\mathbf{i}}_\perp$ should be strictly defined. Relationship $\hat{\mathbf{i}}_\perp = \hat{\mathbf{n}}$ denotes a GSTC-interface facing the same direction with the element, while $\hat{\mathbf{i}}_\perp = -\hat{\mathbf{n}}$ denotes a GSTC-interface facing the opposite direction with the element, as shown in Figure 4.25.

This difference has an impact on the numerical flux, which depends on both the i-th element and the a-th interface. The direction of $\hat{\mathbf{i}}_\perp$ needs to be accurately identified for metasurfaces with bicharacteristic ($\chi_{em}, \chi_{me} \neq 0$), while it does not influence the simulation of monoisotropic metasurfaces.

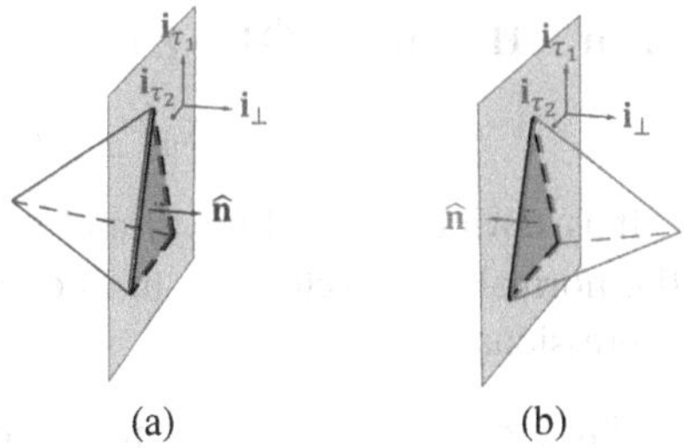

Figure 4.25 The sketch of the GSTC-interface and the element facing (a) the same direction and (b) opposite direction. The gray triangle denotes the GSTC-interface.

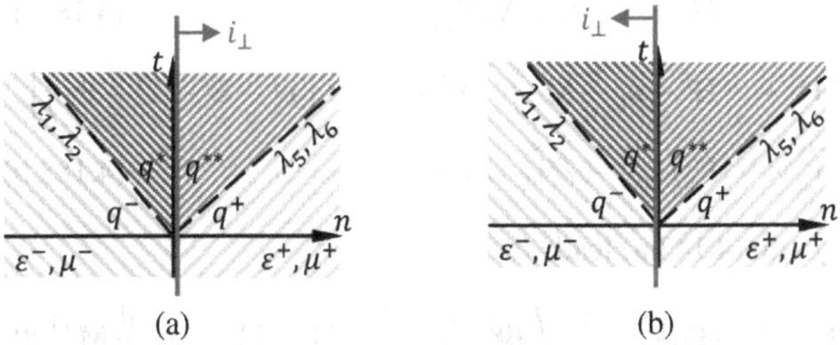

Figure 4.26 The construction of the solution to the interface of the (a) same-facing element and (b) opposite-facing element.

Like the method to derive (4.173.a), combine (4.153), (4.155), (4.155), (4.156), and (4.185.a), and use $\hat{\mathbf{i}}_\perp = \hat{\mathbf{n}}^i$ to obtain a numerical flux with surface normal polarization. For the element as shown in Figure 4.25(a), whose characteristics plot is presented in Figure 4.26(a), its corresponding numerical flux are

$$\begin{aligned} Y^{ij}\hat{\mathbf{n}}^i \times \mathbf{E}^* &= \hat{\mathbf{n}}^i \times \left(Y^i\mathbf{E}^i + Y^j\mathbf{E}^j - \frac{\partial}{\partial t}\mathbf{P}_\parallel \right) + \hat{\mathbf{n}}^i \\ &\quad \times \left[\hat{\mathbf{n}}^i \times \left(\mathbf{H}^j - \mathbf{H}^i - \mu_0 Y^j \frac{\partial}{\partial t}\mathbf{M}_\parallel \right) \right] \\ &\quad + \hat{\mathbf{n}}^i \times (\hat{\mathbf{n}}^i \times \nabla_\parallel M_\perp) + Y^j \hat{\mathbf{n}}^i \times \nabla_\parallel P_\perp / \varepsilon_0 \end{aligned} \tag{4.186.a}$$

$$\begin{aligned} Z^{ij}\hat{\mathbf{n}}^i \times \mathbf{H}^* &= \hat{\mathbf{n}}^i \times \left(Z^i\mathbf{H}^i + Z^j\mathbf{H}^j - \mu_0 \frac{\partial}{\partial t}\mathbf{M}_\parallel \right) - \hat{\mathbf{n}}^i \\ &\quad \times \left[\hat{\mathbf{n}}^i \times \left(\mathbf{E}^j - \mathbf{E}^i - Z^j \frac{\partial}{\partial t}\mathbf{P}_\parallel \right) \right] \\ &\quad - \hat{\mathbf{n}}^i \times (\hat{\mathbf{n}}^i \times \nabla_\parallel P_\perp / \varepsilon_0) + Z^j \hat{\mathbf{n}}^i \times \nabla_\parallel M_\perp \end{aligned} \tag{4.186.b}$$

For the element shown in Figure 4.25(b), whose characteristics plot are presented in Figure 4.26(b), use $\hat{\mathbf{i}}_\perp = -\hat{\mathbf{n}}^i$ to obtain its corresponding numerical flux as

$$\begin{aligned} Y^{ij}\hat{\mathbf{n}}^i \times \mathbf{E}^* &= \hat{\mathbf{n}}^i \times \left(Y^i\mathbf{E}^i + Y^j\mathbf{E}^j - \frac{\partial}{\partial t}\mathbf{P}_\parallel \right) + \hat{\mathbf{n}}^i \times \left[\hat{\mathbf{n}}^i \times \left(\mathbf{H}^j - \mathbf{H}^i - \mu_0 Y^j \frac{\partial}{\partial t}\mathbf{M}_\parallel \right) \right] \\ &\quad - \hat{\mathbf{n}}^i \times (\hat{\mathbf{n}}^i \times \nabla_\parallel M_\perp) - Y^j \hat{\mathbf{n}}^i \times \nabla_\parallel P_\perp / \varepsilon_0 \end{aligned} \tag{4.187.a}$$

$$Z^{ij}\hat{\mathbf{n}}^i \times \mathbf{H}^* = \hat{\mathbf{n}}^i \times \left(Z^i\mathbf{H}^i + Z^j\mathbf{H}^j - \mu_0\frac{\partial}{\partial t}\mathbf{M}_\parallel\right) - \hat{\mathbf{n}}^i \times \left[\hat{\mathbf{n}}^i \times \left(\mathbf{E}^j - \mathbf{E}^i - Z^j\frac{\partial}{\partial t}\mathbf{P}_\parallel\right)\right]$$
$$+ \hat{\mathbf{n}}^i \times (\hat{\mathbf{n}}^i \times \nabla_\parallel P_\perp/\varepsilon_0) - Z^j\hat{\mathbf{n}}^i \times \nabla_\parallel M_\perp \quad (4.187.b)$$

It is obvious that (4.186.a) and (4.187.a) only differ in the last row. Thus, the normal unit vector of a-th interface $(\widehat{\mathbf{i}}_\perp^a)$ can be used to obtain the general expressions [62].

$$Y^{ij}\hat{\mathbf{n}}^i \times \mathbf{E}^* = \hat{\mathbf{n}}^i \times \left(Y^i\mathbf{E}^i + Y^j\mathbf{E}^j - \frac{\partial}{\partial t}\mathbf{P}_\parallel\right) + \hat{\mathbf{n}}^i \times \left[\hat{\mathbf{n}}^i \times \left(\mathbf{H}^j - \mathbf{H}^i - \mu_0 Y^j\frac{\partial}{\partial t}\mathbf{M}_\parallel\right)\right]$$
$$- \hat{\mathbf{n}}^i \times \left(\widehat{\mathbf{i}}_\perp^a \times \nabla_\parallel M_\perp\right) - Y^j\widehat{\mathbf{i}}_\perp^a \times \nabla_\parallel P_\perp/\varepsilon_0 \quad (4.188.a)$$

$$Z^{ij}\hat{\mathbf{n}}^i \times \mathbf{H}^* = \hat{\mathbf{n}}^i \times \left(Z^i\mathbf{H}^i + Z^j\mathbf{H}^j - \mu_0\frac{\partial}{\partial t}\mathbf{M}_\parallel\right) - \hat{\mathbf{n}}^i \times \left[\hat{\mathbf{n}}^i \times \left(\mathbf{E}^j - \mathbf{E}^i - Z^j\frac{\partial}{\partial t}\mathbf{P}_\parallel\right)\right]$$
$$+ \hat{\mathbf{n}}^i \times \left(\widehat{\mathbf{i}}_\perp^a \times \nabla_\parallel P_\perp/\varepsilon_0\right) - Z^j\widehat{\mathbf{i}}_\perp^a \times \nabla_\parallel M_\perp \quad (4.188.b)$$

Substituting (4.188.a) into (4.143.a) leads to

$$\int_{\gamma^i} \mathbf{\Phi}^i \cdot \varepsilon^i\frac{\partial \mathbf{E}^i}{\partial t}dV = \int_{\gamma^i} \mathbf{\Phi}^i \cdot \nabla \times \mathbf{H}^i dV + \frac{1}{Z^{ij}}\int_{\partial\gamma^i} \mathbf{\Phi}^i \cdot \hat{\mathbf{n}}^i \times \left[Z^j(\mathbf{H}^j - \mathbf{H}^i) - \mu_0\frac{\partial}{\partial t}\mathbf{M}_\parallel^a\right] dS$$
$$+ \frac{1}{Z^{ij}}\int_{\partial\gamma^i} \mathbf{\Phi}^i \cdot \hat{\mathbf{n}}^i \times \left[\hat{\mathbf{n}}^i \times \left(\mathbf{E}^i - \mathbf{E}^j + Z^j\frac{\partial}{\partial t}\mathbf{P}_\parallel^a\right)\right] dS - \int_{\gamma^i} \mathbf{\Phi}^i \cdot \mathbf{J}_e^i dV$$
$$- \frac{1}{\varepsilon_0 Z^{ij}}\int_{\partial\gamma^i} \mathbf{\Phi}^i \cdot \hat{\mathbf{n}}^i \times \left(\widehat{\mathbf{i}}_\perp^a \times \nabla_\parallel P_\perp^a\right) dS + \frac{Z^j}{Z^{ij}}\int_{\partial\gamma^i} \mathbf{\Phi}^i \cdot \widehat{\mathbf{i}}_\perp^a \times \nabla_\parallel M_\perp^a dS$$
$$(4.189.a)$$

$$\int_{\gamma^i} \mathbf{\Psi}^i \cdot \mu^i\frac{\partial \mathbf{H}^i}{\partial t}dV = -\int_{\gamma^i} \mathbf{\Psi}^i \cdot \nabla \times \mathbf{E}^i dV + \frac{1}{Y^{ij}}\int_{\partial\gamma^i} \mathbf{\Psi}^i \cdot \hat{\mathbf{n}}^i \times \left[Y^j(\mathbf{E}^i - \mathbf{E}^j) + \frac{\partial}{\partial t}\mathbf{P}_\parallel^a\right] dS$$
$$+ \frac{1}{Y^{ij}}\int_{\partial\gamma^i} \mathbf{\Psi}^i \cdot \hat{\mathbf{n}}^i \times \left[\hat{\mathbf{n}}^i \times \left(\mathbf{H}^i - \mathbf{H}^j - \mu_0 Y^j\frac{\partial}{\partial t}\mathbf{M}_\parallel^a\right)\right] dS - \int_{\gamma^i} \mathbf{\Psi}^i \cdot \mathbf{J}_h^i dV$$
$$- \frac{1}{Y^{ij}}\int_{\partial\gamma^i} \mathbf{\Psi}^i \cdot \hat{\mathbf{n}}^i \times \left(\widehat{\mathbf{i}}_\perp^a \times \nabla_\parallel M_\perp^a\right) dS - \frac{Y^j}{\varepsilon_0 Y^{ij}}\int_{\partial\gamma^i} \mathbf{\Psi}^i \cdot \widehat{\mathbf{i}}_\perp^a \times \nabla_\parallel P_\perp^a dS \quad (4.189.b)$$

where the superscript a denotes the index of a-th GSTC-interface. For these new polarization unknowns based on surface, test their constitutive relation on interface.

$$\int_{\partial\gamma^i} \mathbf{\gamma}^a \cdot \mathbf{P}_\parallel^a dS = -\frac{1}{2}\int_{\partial\gamma^i} \mathbf{\gamma}^a \cdot \hat{\mathbf{i}}_\perp \times \hat{\mathbf{i}}_\perp$$
$$\times \left[\varepsilon_0 \overline{\overline{\chi}}_{ee}^{\tau}\left(\mathbf{E}^{i_1} + \mathbf{E}^{i_2}\right) + \sqrt{\varepsilon_0\mu_0}\,\overline{\overline{\chi}}_{em}^{\tau}\left(\mathbf{H}^{i_1} + \mathbf{H}^{i_2}\right)\right] dS \quad (4.190.a)$$

$$\int_{\partial\gamma^i} \boldsymbol{\gamma}^a \cdot \mathbf{M}_{\parallel}^a dS = -\frac{1}{2}\int_{\partial\gamma^i} \boldsymbol{\gamma}^a \cdot \hat{\mathbf{i}}_{\perp}$$

$$\times \hat{\mathbf{i}}_{\perp} \times \left[\overline{\overline{\chi}}_{me}^{\tau}\left(\mathbf{E}^{i_1}+\mathbf{E}^{i_2}\right) + \sqrt{\varepsilon_0/\mu_0}\overline{\overline{\chi}}_{mm}^{\tau}\left(\mathbf{H}^{i_1}+\mathbf{H}^{i_2}\right)\right] dS \tag{4.190.b}$$

$$\int_{\partial\gamma^i} \zeta^a \cdot P_{\perp}^a dS = \frac{1}{2}\int_{\partial\gamma^i} \zeta^a$$

$$\cdot \left[\varepsilon_0 \chi_{ee}^{\perp\perp}\left(\mathbf{E}^{i_1}+\mathbf{E}^{i_2}\right)\cdot\hat{\mathbf{i}}_{\perp} + \sqrt{\varepsilon_0\mu_0}\chi_{rm}^{\perp\perp}\left(\mathbf{H}^{i_1}+\mathbf{H}^{i_2}\right)\cdot\hat{\mathbf{i}}_{\perp}\right] dS \tag{4.190.c}$$

$$\int_{\partial\gamma^i} \zeta^a \cdot M_{\perp}^a dS = \frac{1}{2}\int_{\partial\gamma^i} \zeta^a$$

$$\cdot \left[\chi_{me}^{\perp\perp}\left(\mathbf{E}^{i_1}+\mathbf{E}^{i_2}\right)\cdot\hat{\mathbf{i}}_{\perp} + \sqrt{\varepsilon_0/\mu_0}\chi_{mm}^{\perp\perp}\left(\mathbf{H}^{i_1}+\mathbf{H}^{i_2}\right)\cdot\hat{\mathbf{i}}_{\perp}\right] dS \tag{4.190.d}$$

where i_1 and i_2 denote the two GSTC-element share the a-th GSTC-interface, and the normal-tangential coupling susceptibilities are ignored as

$$\overline{\overline{\chi}} = \begin{bmatrix} \chi^{\tau_1\tau_2} & \chi^{\tau_1\tau_2} & 0 \\ \chi^{\tau_2\tau_1} & \chi^{\tau_2\tau_2} & 0 \\ 0 & 0 & \chi^{\perp\perp} \end{bmatrix} = \begin{bmatrix} \overline{\overline{\chi}}^{\tau} & \\ & \chi^{\perp\perp} \end{bmatrix} \tag{4.191}$$

The semi-discrete matrix equations of (4.189.a) for the i-th element are

$$\mathbf{M}_e^i \frac{\partial}{\partial t}\mathbf{e}^i - \mathbf{L}_{ep_{\parallel}}^{ia}\frac{\partial}{\partial t}\mathbf{p}_{\parallel}^a - \mathbf{L}_{em_{\parallel}}^{ia}\frac{\partial}{\partial t}\mathbf{m}_{\parallel}^a = \mathbf{S}_{eh}^{ii}\mathbf{h}^i + \mathbf{L}_{eh}^{ii}\mathbf{h}^i$$

$$+ \sum \mathbf{L}_{eh}^{ij}\mathbf{h}^j + \mathbf{L}_{ee}^{ii}\mathbf{e}^i + \sum \mathbf{L}_{ee}^{ij}\mathbf{e}^j + \mathbf{L}_{ep_{\perp}}^{ia}\mathbf{p}_{\perp}^a + \mathbf{L}_{em_{\perp}}^{ia}\mathbf{m}_{\perp}^a + \mathbf{j}_e^i \tag{4.192.a}$$

$$\mathbf{M}_h^i \frac{\partial}{\partial t}\mathbf{h}^i - \mathbf{L}_{hp_{\parallel}}^{ia}\frac{\partial}{\partial t}\mathbf{p}_{\parallel}^a - \mathbf{L}_{hm_{\parallel}}^{ia}\frac{\partial}{\partial t}\mathbf{m}_{\parallel}^a = \mathbf{S}_{he}^{ii}\mathbf{e}^i + \mathbf{L}_{he}^{ii}\mathbf{e}^i$$

$$+ \sum \mathbf{L}_{he}^{ij}\mathbf{e}^j + \mathbf{L}_{hh}^{ii}\mathbf{h}^i + \sum \mathbf{L}_{hh}^{ij}\mathbf{h}^j + \mathbf{L}_{hp_{\perp}}^{ia}\mathbf{p}_{\perp}^a + \mathbf{L}_{hm_{\perp}}^{ia}\mathbf{m}_{\perp}^a + \mathbf{j}_h^i \tag{4.192.b}$$

where the elements of the matrices in (4.159) are updated as

$$\mathbf{L}_{ep_{\parallel},n'n}^{ia} = \frac{Z^j}{Z^{ij}}\int_{\partial\gamma^i} \boldsymbol{\Phi}_{n'}^i \cdot \hat{\mathbf{n}}^i \times \left(\hat{\mathbf{n}}^i \times \boldsymbol{\gamma}_n^a\right) dS \tag{4.193.a}$$

$$\mathbf{L}_{em_{\parallel},n'n}^{ia} = -\frac{\mu_0}{Z^{ij}}\int_{\partial\gamma^i} \boldsymbol{\Phi}_{n'}^i \cdot \hat{\mathbf{n}}^i \times \boldsymbol{\gamma}_n^a dS \tag{4.193.b}$$

$$\mathbf{L}_{hp_{\parallel},n'n}^{ia} = \frac{1}{Y^{ij}}\int_{\partial\gamma^i} \boldsymbol{\Psi}_{n'}^i \cdot \hat{\mathbf{n}}^i \times \boldsymbol{\gamma}_n^a dS \tag{4.193.c}$$

$$\mathbf{L}_{hm_{\parallel},n'n}^{ia} = \frac{\mu_0 Y^j}{Y^{ij}}\int_{\partial\gamma^i} \boldsymbol{\Psi}_{n'}^i \cdot \hat{\mathbf{n}}^i \times \left(\hat{\mathbf{n}}^i \times \boldsymbol{\gamma}_n^a\right) dS \tag{4.193.d}$$

$$\mathbf{L}_{ep_{\perp},n'n}^{ia} = -\frac{1}{\varepsilon_0 Z^{ij}}\int_{\partial\gamma^i} \boldsymbol{\Phi}_{n'}^i \cdot \hat{\mathbf{n}}^i \times \left(\hat{\mathbf{i}}_{\perp} \times \nabla\zeta_n^a\right) dS \tag{4.193.e}$$

$$\mathbf{L}^{ia}_{ep_\perp,n'n} = \frac{Z^j}{\varepsilon_0 Z^{ij}} \int_{\partial\gamma^i} \mathbf{\Phi}^i_{n'} \cdot \widehat{\mathbf{i}}_\perp \times \nabla\zeta^a_n dS \tag{4.193.f}$$

$$\mathbf{L}^{ia}_{hp_\perp,n'n} = -\frac{Y^j}{\varepsilon_0 Y^{ij}} \int_{\partial\gamma^i} \mathbf{\Psi}^i_{n'} \cdot \widehat{\mathbf{i}}_\perp \times \nabla\zeta^a_n dS \tag{4.193.g}$$

$$\mathbf{L}^{ia}_{hm_\perp,n'n} = -\frac{1}{Y^{ij}} \int_{\partial\gamma^i} \mathbf{\Psi}^i_{n'} \cdot \widehat{\mathbf{n}}^i \times \left(\widehat{\mathbf{i}}_\perp \times \nabla\zeta^a_n\right) dS \tag{4.193.h}$$

The semi-discrete matrices equations of (4.190.a) for the a-th GSTC-interface are

$$\mathbf{M}^a_{p_\parallel}\mathbf{p}^a_\parallel = \mathbf{L}^{ai_1}_{p_\parallel e}\mathbf{e}^{i_1} + \mathbf{L}^{ai_2}_{p_\parallel e}\mathbf{e}^{i_2} + \mathbf{L}^{ai_1}_{p_\parallel h}\mathbf{h}^{i_1} + \mathbf{L}^{ai_2}_{p_\parallel h}\mathbf{h}^{i_2} \tag{4.194.a}$$

$$\mathbf{M}^a_{m_\parallel}\mathbf{m}^a_\parallel = \mathbf{L}^{ai_1}_{m_\parallel e}\mathbf{e}^{i_1} + \mathbf{L}^{ai_2}_{m_\parallel e}\mathbf{e}^{i_2} + \mathbf{L}^{ai_1}_{m_\parallel h}\mathbf{h}^{i_1} + \mathbf{L}^{ai_2}_{m_\parallel h}\mathbf{h}^{i_2} \tag{4.194.b}$$

$$\mathbf{M}^a_{p_\perp}\mathbf{p}^a_\perp = \mathbf{L}^{ai_1}_{p\perp e}\mathbf{e}^{i_1} + \mathbf{L}^{ai_2}_{p_\perp e}\mathbf{e}^{i_2} + \mathbf{L}^{ai_1}_{p_\perp h}\mathbf{h}^{i_1} + \mathbf{L}^{ai_2}_{p_\perp h}\mathbf{h}^{i_2} \tag{4.194.c}$$

$$\mathbf{M}^a_{m_\perp}\mathbf{m}^a_\perp = \mathbf{L}^{ai_1}_{m_\perp e}\mathbf{e}^{i_1} + \mathbf{L}^{ai_2}_{m_\perp e}\mathbf{e}^{i_2} + \mathbf{L}^{ai_1}_{m_\perp h}\mathbf{h}^{i_1} + \mathbf{L}^{ai_2}_{m_\perp h}\mathbf{h}^{i_2} \tag{4.194.d}$$

where the elements of the new matrices are

$$\mathbf{M}^a_{p_\parallel,n'n} = \mathbf{M}^a_{m_\parallel,n'n} = \int_{\partial\gamma^i} \boldsymbol{\gamma}^a_{n'} \cdot \boldsymbol{\gamma}^a_n dS \tag{4.195.a}$$

$$\mathbf{L}^{ai}_{p_\parallel e,n'n} = -\frac{\varepsilon_0}{2} \int_{\partial\gamma^i} \boldsymbol{\gamma}^a_{n'} \cdot \widehat{\mathbf{i}}_\perp \times \left[\widehat{\mathbf{i}}_\perp \times \left(\overline{\overline{\chi}}^\tau_{ee}\mathbf{\Phi}^i_n\right)\right] dS \tag{4.195.b}$$

$$\mathbf{L}^{ai}_{p_\parallel h,n'n} = -\frac{\sqrt{\varepsilon_0\mu_0}}{2} \int_{\partial\gamma^i} \boldsymbol{\gamma}^a_{n'} \cdot \widehat{\mathbf{i}}_\perp \times \left[\widehat{\mathbf{i}}_\perp \times \left(\overline{\overline{\chi}}^\tau_{em}\mathbf{\Psi}^i_n\right)\right] dS \tag{4.195.c}$$

$$\mathbf{L}^{ai}_{m_\parallel e,n'n} = -\frac{1}{2} \int_{\partial\gamma^i} \boldsymbol{\gamma}^a_{n'} \cdot \widehat{\mathbf{i}}_\perp \times \left[\widehat{\mathbf{i}}_\perp \times \left(\overline{\overline{\chi}}^\tau_{me}\mathbf{\Phi}^i_n\right)\right] dS \tag{4.195.d}$$

$$\mathbf{L}^{ai}_{m_\parallel h,n'n} = -\frac{\sqrt{\varepsilon_0/\mu_0}}{2} \int_{\partial\gamma^i} \boldsymbol{\gamma}^a_{n'} \cdot \widehat{\mathbf{i}}_\perp \times \left[\widehat{\mathbf{i}}_\perp \times \left(\overline{\overline{\chi}}^\tau_{mm}\mathbf{\Psi}^i_n\right)\right] dS \tag{4.195.e}$$

$$\mathbf{M}^a_{p_\perp,n'n} = \mathbf{M}^a_{m_\perp,n'n} = \int_{\partial\gamma^i} \zeta^a_{n'} \cdot \zeta^a_n dS \tag{4.195.f}$$

$$\mathbf{L}^{ai}_{p_\perp e,n'n} = \frac{\varepsilon_0}{2} \int_{\partial\gamma^i} \zeta^a_{n'} \cdot \chi^{\perp\perp}_{ee} \left(\mathbf{\Phi}^i_n \cdot \widehat{\mathbf{i}}^a_\perp\right) dS \tag{4.195.g}$$

$$\mathbf{L}^{ai}_{p_\perp h,n'n} = \frac{\sqrt{\varepsilon_0\mu_0}}{2} \int_{\partial\gamma^i} \zeta^a_{n'} \cdot \chi^{\perp\perp}_{em} \left(\mathbf{\Psi}^i_n \cdot \widehat{\mathbf{i}}^a_\perp\right) dS \tag{4.195.h}$$

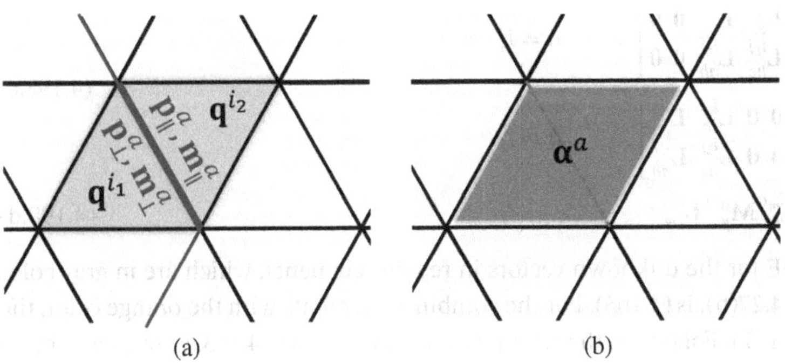

Figure 4.27 The combination of the unknowns over one GSTC-interface. (a) The individual unknowns of i_1-, i_2-elements and a-th surface, and the red lines denote the GSTC-interface. (b) The synthesized unknowns.

$$\mathbf{L}_{m_\perp e,n'n}^{ia} = \frac{1}{2}\int_{\partial\gamma^i} \zeta_{n'}^{a} \cdot \chi_{me}^{\perp\perp}\left(\boldsymbol{\Phi}_n^i \cdot \widehat{\mathbf{i}}_\perp^{a}\right) \tag{4.195.i}$$

$$\mathbf{L}_{m_\perp h,n'n}^{ia} = \frac{\sqrt{\varepsilon_0/\mu_0}}{2}\int_{\partial\gamma^i} \zeta_{n'}^{a} \cdot \chi_{mm}^{\perp\perp}\left(\boldsymbol{\Psi}_n^i \cdot \widehat{\mathbf{i}}_\perp^{a}\right) dS \tag{4.195.j}$$

Equation (4.194.a) is not a PDE, because the susceptibility tensor is a constant, which means that it does not bring extra DoFs.

After combining the unknowns from the i_1-th and the i_2-th elements, and the a-th surface, as shown in Figure 4.27, into a new unknown vector $\boldsymbol{\alpha}^a$,

$$\boldsymbol{\alpha}^a = \left[\mathbf{e}^{i_1}\ \mathbf{h}^{i_1}\ \mathbf{e}^{i_2}\ \mathbf{h}^{i_2}\right]^T \tag{4.196}$$

(4.192.a) and (4.194.a) can be combined as

$$\mathbf{M}_\alpha^a \frac{\partial}{\partial t}\boldsymbol{\alpha}^a = \mathbf{L}_\alpha^a \boldsymbol{\alpha}^a + \sum_j \mathbf{L}_{\alpha q}^{aj}\mathbf{q}^j, \tag{4.197}$$

where the new system matrices $\mathbf{M}_\alpha^a$, $\mathbf{L}_\alpha^a$, and $\mathbf{L}_\alpha^{aj}$ are

$$\mathbf{M}_\alpha = \begin{bmatrix} \mathbf{M}_e^{i_1} - \mathbf{L}_{ep_\parallel e}^{i_1 i_1} & -\mathbf{L}_{em_\parallel h}^{i_1 i_1} & -\mathbf{L}_{ep_\parallel e}^{i_1 i_2} & -\mathbf{L}_{em_\parallel h}^{i_1 i_2} \\ -\mathbf{L}_{hp_\parallel e}^{i_1 i_1} & \mathbf{M}_h^{i_1} - \mathbf{L}_{hm_\parallel h}^{i_1 i_1} & -\mathbf{L}_{hp_\parallel e}^{i_1 i_2} & -\mathbf{L}_{hm_\parallel h}^{i_1 i_2} \\ -\mathbf{L}_{ep_\parallel e}^{i_2 i_1} & -\mathbf{L}_{em_\parallel h}^{i_2 i_1} & \mathbf{M}_e^{i_2} - \mathbf{L}_{ep_\parallel e}^{i_2 i_2} & -\mathbf{L}_{em_\parallel h}^{i_2 i_2} \\ -\mathbf{L}_{hp_\parallel e}^{i_2 i_1} & -\mathbf{L}_{hm_\parallel h}^{i_2 i_1} & -\mathbf{L}_{hp_\parallel e}^{i_2 i_2} & \mathbf{M}_h^{i_2} - \mathbf{L}_{hm_\parallel h}^{i_2 i_2} \end{bmatrix} \tag{4.198.a}$$

$$\mathbf{L}_\alpha = \begin{bmatrix} \mathbf{L}_{ee}^{i_1 i_1} + \mathbf{L}_{ep_\perp e}^{i_1 i_1} & \mathbf{L}_{eh}^{i_1 i_1} + \mathbf{L}_{em_\perp h}^{i_1 i_1} + \mathbf{S}_{eh}^{i_1} & \mathbf{L}_{ee}^{i_1 i_2} + \mathbf{L}_{ep_\perp e}^{i_1 i_2} & \mathbf{L}_{eh}^{i_1 i_2} + \mathbf{L}_{em_\perp h}^{i_1 i_2} \\ \mathbf{L}_{he}^{i_1 i_1} + \mathbf{L}_{hp_\perp e}^{i_1 i_1} + \mathbf{S}_{he}^{i_1} & \mathbf{L}_{hh}^{i_1 i_1} + \mathbf{L}_{hm_\perp h}^{i_1 i_1} & \mathbf{L}_{he}^{i_1 i_2} + \mathbf{L}_{hp_\perp e}^{i_1 i_2} & \mathbf{L}_{hh}^{i_1 i_2} + \mathbf{L}_{hm_\perp h}^{i_1 i_2} \\ \mathbf{L}_{ee}^{i_2 i_1} + \mathbf{L}_{ep_\perp e}^{i_2 i_1} & \mathbf{L}_{eh}^{i_2 i_1} + \mathbf{L}_{ep_\perp h}^{i_2 i_1} & \mathbf{L}_{ee}^{i_2 i_2} + \mathbf{L}_{ep_\perp e}^{i_2 i_2} & \mathbf{L}_{eh}^{i_2 i_2} + \mathbf{L}_{em_\perp h}^{i_2 i_2} + \mathbf{S}_{eh}^{i_2} \\ \mathbf{L}_{he}^{i_2 i_1} + \mathbf{L}_{hp_\perp e}^{i_2 i_1} & \mathbf{L}_{hh}^{i_2 i_1} + \mathbf{L}_{hm_\perp h}^{i_2 i_1} & \mathbf{L}_{he}^{i_2 i_2} + \mathbf{L}_{hp_\perp e}^{i_2 i_2} + \mathbf{S}_{he}^{i_2} & \mathbf{L}_{hh}^{i_2 i_2} + \mathbf{L}_{hm_\perp h}^{i_2 i_2} \end{bmatrix} \tag{4.198.b}$$

$$\mathbf{L}_{\alpha q}^{aj} = \begin{cases} \begin{bmatrix} \mathbf{L}_{ee}^{i_1 j} & \mathbf{L}_{eh}^{i_1 j} & \mathbf{0} & \mathbf{0} \\ \mathbf{L}_{he}^{i_1 j} & \mathbf{L}_{hh}^{i_1 j} & \mathbf{0} & \mathbf{0} \end{bmatrix}^T, & a = i_1 \\ \begin{bmatrix} \mathbf{0} & \mathbf{0} & \mathbf{L}_{ee}^{i_2 j} & \mathbf{L}_{eh}^{i_2 j} \\ \mathbf{0} & \mathbf{0} & \mathbf{L}_{he}^{i_2 j} & \mathbf{L}_{hh}^{i_2 j} \end{bmatrix}^T, & a = i_2 \end{cases} \tag{4.198.c}$$

$$\mathbf{L}_{\psi\gamma\psi}^{i_m i_n} = \mathbf{L}_{\psi\gamma}^{i_m a} \mathbf{M}_{\gamma}^{a^{-1}} \mathbf{L}_{\gamma\psi}^{a i_n} \tag{4.198.d}$$

The PDE for the unknown vectors in regular elements, which are in gray color in Figure 4.27(b), is (4.165). For the combined elements with the orange color, the PDE is (4.197). Following the RK updating equations in (4.163.a), one can obtain the response of the metasurface.

One of the advantages of DGTD is that in addition to the h-refinement (refinement by decreasing the element size h), it can also employ the p-refinement (refinement by increasing polynomial order p) to achieve high accuracy. Readers may be concerned about whether the incorporation of GSTC will destroy the p-refinement capabilities of the DGTD method. The answer is that the tangential surface polarizations $\mathbf{P}_{\parallel}$ and $\mathbf{M}_{\parallel}$ certainly do not break the p-refinement capacity, while the effect on p-refinement from normal components $P_{\perp}$ and $M_{\perp}$ has not been investigated. To illustrate this problem, the TE plane wave incident from a medium with the permittivity ε_1 and the permeability μ_1 upon another medium with the permittivity ε_2 and the permeability μ_2 is considered. Besides, there are scalar tangential surface susceptibilities $\overline{\overline{\chi}}_{ee}^{\tau} = \chi_{ee}$ and $\overline{\overline{\chi}}_{mm}^{\tau} = \chi_{mm}$ on the interface between these two media. The analytical transmission coefficients are:

$$T_{ana}(Z_1, Z_2, \chi_{ee}, \chi_{mm}; \omega, \theta_i) = \frac{4\cos\theta_i Z_2 (2 - j\omega\mu_0\chi_{mm})}{\begin{array}{c} 4j\omega(\varepsilon_0\chi_{ee} Z_1 Z_2 + \mu_0\chi_{mm}\cos\theta_i\cos\theta_t) \\ +(4 - \omega^2\varepsilon_0\mu_0\chi_{ee}\chi_{mm})(\cos\theta_2 Z_1 + \cos\theta_1 Z_2) \end{array}} \tag{4.199}$$

where $Z_1 = \sqrt{\mu_1/\varepsilon_1}$ and $Z_2 = \sqrt{\mu_2/\varepsilon_2}$ are the wave impedances, while θ_i and θ_t are the incident and transmission angles, the relation of which follows the Snell's law as

$$\sin\theta_t = \frac{\sqrt{\mu_1\varepsilon_1}}{\sqrt{\mu_2\varepsilon_2}} \sin\theta_i \tag{4.200}$$

Expand **E** and **H** using the interpolatory basis functions [50] of different orders p and different mesh element numbers K, to obtain the numerical transmission coefficient T_{num}. By plotting the relative error ε between T_{num} and T_{ana} under different p and $\sqrt{K}$ (because of the 2-D DGTD-GSTC solver in this problem, the square root of K can be used as a measure of the average mesh size h).

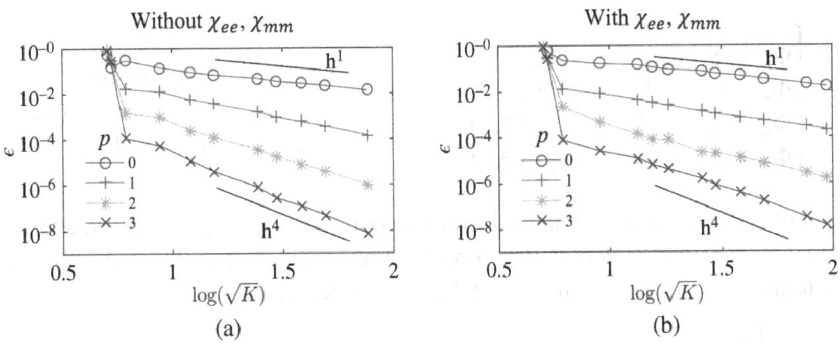

Figure 4.28 The relative error between numerical and analytical transmission coefficients in the 2-D TE plane wave incidence upon interface between two media (a) without surface susceptibilities and (b) with scalar tangential surface susceptibilities.

Figure 4.28(a) shows that for the traditional DGTD method in (4.157.a), as the average mesh size h increases exponentially, the accuracy of different order p also increases exponentially with $(p+1)$ times, which verifies that the convergence rates of the traditional DGTD method is $\mathcal{O}(h^{p+1})$ [46]. As for DGTD-GSTC in (4.189.a) with only tangential surface susceptibilities, Figure 4.28(b) shows that the convergence rates keeps the same as $\mathcal{O}(h^{p+1})$.

Although the effect of normal surface susceptibilities on p-refinement has not been studied, it has been experimentally demonstrated that the introduction of $\chi^{\perp\perp}$ places additional constraints on the basis functions [62]. The normal polarizations, according to (4.190.c) and (4.190.d), depend on the fields, which means that the unknown vectors $\mathbf{p}_{\perp}$ and $\mathbf{m}_{\perp}$ are not only related to the scalar basis functions $\boldsymbol{\gamma}$ but also affected by the vector basis functions $\boldsymbol{\Phi}$ and $\boldsymbol{\Psi}$. Taking P_n and $\mathbf{E}$ as an example, consider the simplest constitutive relation

$$P_n = \mathbf{E} \cdot \hat{\mathbf{n}} \chi_{ee}^{\perp\perp} \tag{4.201}$$

Substituting (4.137.a) and (4.138.c) into (4.201) leads to

$$P_n = \mathbf{p}_{\perp}\zeta = \mathbf{e}(\boldsymbol{\Phi} \cdot \hat{\mathbf{n}})\chi_{ee}^{\perp\perp} \tag{4.202}$$

Equation (4.202) reveals that although P_n is expanded on the basis function ζ, it is also actually expanded on the projection of the basis function $\boldsymbol{\Phi}$ on the surface of the elements. Therefore, the nature of the projection of basis function $\boldsymbol{\Phi}$ will affect the expansion of P_n. While for the tangential polarization, this issue will disappear naturally because the basis functions to expand the fields $\mathbf{E}$ and $\mathbf{H}$ are curl-conforming, which means that their projection on the surface of the elements is known. The following basis function is feasible to expand $\mathbf{E}$ in 2-D TM problem [51]:

$$\begin{cases} \boldsymbol{\Phi}_m = \xi_{m1}\nabla\xi_{m2} - \xi_{m2}\nabla\xi_{m1}, & \\ \boldsymbol{\Phi}_m = (\xi_{m1} - \xi_{m2})(\xi_{m1}\nabla\xi_{m2} - \xi_{m2}\nabla\xi_{m1}), & m = 1,2,\ldots,6, \text{ edge based} \\ \boldsymbol{\Phi}_m = \xi_{m3}(\xi_{m1}\nabla\xi_{m2} - \xi_{m2}\nabla\xi_{m1}) & \\ \boldsymbol{\Phi}_m = \xi_{m2}(\xi_{m3}\nabla\xi_{m1} - \xi_{m1}\nabla\xi_{m3}) & m = 1,2,3,4, \text{ face based} \end{cases} \tag{4.203}$$

where ξ_i denotes the normalized simplex coordinates of the i-th vertex, and the indexes $i1$, $i2$, and $i3$ denote the corresponding vertex of edge (face). Another set of basis functions in the same order, but infeasible, is [63]

$$\begin{cases} \boldsymbol{\Phi}_m = \xi_{m1}\nabla\xi_{m2}, & \\ \boldsymbol{\Phi}_m = \xi_{m2}\nabla\xi_{m1}, & m = 1,2,\ldots,6, \text{ edge based} \\ \boldsymbol{\Phi}_m = \xi_{m3}(\xi_{m1}\nabla\xi_{m2} - \xi_{m2}\nabla\xi_{m1}) & \\ \boldsymbol{\Phi}_m = \xi_{m2}(\xi_{m3}\nabla\xi_{m1} - \xi_{m1}\nabla\xi_{m3}) & m = 1,2,3,4, \text{ face based} \end{cases} \tag{4.204}$$

The reason for its failure lies in the normal component of the edge-based basis functions $\boldsymbol{\Phi}_m$, which is constant along the tangential direction and hence fails to expand $P_\perp$. Besides, $P_\perp$ and $M_\perp$ also require a smaller time step interval, which depends on the value of $\chi_{ee}^{\perp\perp}$ and $\chi_{mm}^{\perp\perp}$. Although there is no rigorous theoretical analysis, the criterion (4.171) can still tell whether the choice of Δt is stable.

4.2.2.3 Methods for Dispersive Susceptibilities

In this section, the method for handling dispersive susceptibilities in DGTD-GSTC method is introduced. With the dispersive susceptibilities, $P_\perp$, $M_\perp$, $\mathbf{P}_\parallel$, and $\mathbf{M}_\parallel$ are now real unknowns with DoFs. Considering the surface susceptibility tensors in the form of (4.191), where the tangential susceptibilities are dispersive and approximated by the sum of Lorentzian functions.

$$\chi(\omega; a_{0,n}, a_{1,n}, a_{2,n}) \approx \sum_{n}^{N} \frac{a_{0,n}}{(j\omega)^2 + a_{1,n}(j\omega) + a_{2,n}} \tag{4.205}$$

Without loss of generality, the following constitutive relation is discussed to illustrate the implementation of dispersive susceptibilities in DGTD-GSTC method:

$$\mathbf{P} = \begin{bmatrix} \dfrac{a_0}{(j\omega)^2 + a_1(j\omega) + a_2} & \dfrac{b_0}{(j\omega)^2 + b_1(j\omega) + b_2} & 0 \\ \dfrac{c_0}{(j\omega)^2 + c_1(j\omega) + c_2} & \dfrac{d_0}{(j\omega)^2 + d_1(j\omega) + d_2} & 0 \\ 0 & 0 & 1 \end{bmatrix} \mathbf{E} \tag{4.206}$$

where the numbers of Lorentzian functions are reduced to 1 because the extension to N is straightforward. The normal susceptibility is constant because the situation of dispersive normal susceptibilities in DGTD-GSTC is very complex and has not been studied enough.

Using the scalar tangential components to rewrite (4.206) leads to

$$P_{\tau_1} = \frac{a_0}{(j\omega)^2 + a_1(j\omega) + a_2} E_{\tau_1} + \frac{b_0}{(j\omega)^2 + b_1(j\omega) + b_2} E_{\tau_2} \tag{4.207.a}$$

$$P_{\tau_2} = \frac{c_0}{(j\omega)^2 + c_1(j\omega) + c_2} E_{\tau_1} + \frac{d_0}{(j\omega)^2 + d_1(j\omega) + d_2} E_{\tau_2} \tag{4.207.b}$$

Multiply the denominator to the left-hand side and assemble the scale expressions into vector form, one will obtain

$$\begin{aligned}(j\omega)^4\mathbf{P}_{\|} &+ (j\omega)^3 \begin{bmatrix} a_1 + b_1 & \\ & c_1 + d_1 \end{bmatrix} \mathbf{P}_{\|} + (j\omega)^2 \begin{bmatrix} a_2 + b_2 + a_1 b_1 & \\ & c_2 + d_2 + c_1 d_1 \end{bmatrix} \mathbf{P}_{\|} \\ &+ (j\omega) \begin{bmatrix} a_1 b_2 + a_2 b_1 & \\ & c_1 d_2 + c_2 d_1 \end{bmatrix} \mathbf{P}_{\|} + \begin{bmatrix} a_2 b_2 & \\ & c_2 d_2 \end{bmatrix} \mathbf{P}_{\|} \\ &= \varepsilon_0 (j\omega)^2 \begin{bmatrix} a_0 & b_0 \\ c_0 & d_0 \end{bmatrix} \mathbf{E}_{\|} + \varepsilon_0 (j\omega) \begin{bmatrix} a_0 b_1 & a_1 b_0 \\ c_0 d_1 & c_1 d_0 \end{bmatrix} \mathbf{E}_{\|} + \varepsilon_0 \begin{bmatrix} a_0 b_2 & a_2 b_0 \\ c_0 d_2 & c_2 d_0 \end{bmatrix} \mathbf{E}_{\|}\end{aligned} \tag{4.208}$$

Using the inverse Fourier transform, the corresponding time-domain expressions of (4.208) are

$$\begin{aligned}\frac{\partial^4}{\partial t^4}\mathbf{P}_{\|} &+ \frac{\partial^3}{\partial t^3} \begin{bmatrix} a_1 + b_1 & \\ & c_1 + d_1 \end{bmatrix} \mathbf{P}_{\|} + \frac{\partial^2}{\partial t^2} \begin{bmatrix} a_2 + b_2 + a_1 b_1 & \\ & c_2 + d_2 + c_1 d_1 \end{bmatrix} \mathbf{P}_{\|} \\ &+ \frac{\partial}{\partial t} \begin{bmatrix} a_1 b_2 + a_2 b_1 & \\ & c_1 d_2 + c_2 d_1 \end{bmatrix} \mathbf{P}_{\|} + \begin{bmatrix} a_2 b_2 & \\ & c_2 d_2 \end{bmatrix} \mathbf{P}_{\|} \\ &= \varepsilon_0 \frac{\partial^2}{\partial t^2} \begin{bmatrix} a_0 & b_0 \\ c_0 & d_0 \end{bmatrix} \mathbf{E}_{\|} + \varepsilon_0 \frac{\partial}{\partial t} \begin{bmatrix} a_0 b_1 & a_1 b_0 \\ c_0 d_1 & c_1 d_0 \end{bmatrix} \mathbf{E}_{\|} + \varepsilon_0 \begin{bmatrix} a_0 b_2 & a_2 b_0 \\ c_0 d_2 & c_2 d_0 \end{bmatrix} \mathbf{E}_{\|}\end{aligned} \tag{4.209}$$

Employing the basis functions to expand the corresponding components on the a-th GSTC-interface and using the Galerkin's method to test them will lead to

$$\begin{aligned}\mathbf{M}^a_{p_\|} \frac{\partial}{\partial t}\mathbf{p}^{(3),a}_{\|} &= \mathbf{L}^a_{p^3_\|}\mathbf{p}^{(3),a}_{\|} + \mathbf{L}^a_{p^2_\|}\mathbf{p}^{(2),a}_{\|} + \mathbf{L}^a_{p^1_\|}\mathbf{p}^{(1),a}_{\|} + \mathbf{L}^a_{p_\|}\mathbf{p}^a_{\|} + \mathbf{L}^{ai_1}_{p_\| e^2}\mathbf{e}^{(2),i_1} \\ &+ \mathbf{L}^{ai_2}_{p_\| e^2}\mathbf{e}^{(2),i_2} + \mathbf{L}^{ai_1}_{p_\| e^1}\mathbf{e}^{(1),i_1} + \mathbf{L}^{ai_2}_{p_\| e^1}\mathbf{e}^{(1),i_2} + \mathbf{L}^{ai_1}_{p_\| e}\mathbf{e}^{i_1} + \mathbf{L}^{ai_2}_{p_\| e}\mathbf{e}^{i_2}\end{aligned} \tag{4.210.a}$$

$$\mathbf{M}^a_{p_\|} \frac{\partial}{\partial t}\mathbf{p}^{(2),a}_{\|} = \mathbf{M}^a_{p_\|}\mathbf{p}^{(3),a}_{\|} \tag{4.210.b}$$

$$\mathbf{M}^a_{p_\|} \frac{\partial}{\partial t}\mathbf{p}^{(1),a}_{\|} = \mathbf{M}^a_{p_\|}\mathbf{p}^{(2),a}_{\|} \tag{4.210.c}$$

$$\mathbf{M}^a_{p_\|} \frac{\partial}{\partial t}\mathbf{p}^{a}_{\|} = \mathbf{M}^a_{p_\|}\mathbf{p}^{(1),a}_{\|} \tag{4.210.d}$$

$$\mathbf{M}_e^i \frac{\partial}{\partial t} \mathbf{e}^{(1),i} = \mathbf{M}_e^i \mathbf{e}^{(2),i} \tag{4.210.e}$$

$$\mathbf{M}_e^i \frac{\partial}{\partial t} \mathbf{e}^i = \mathbf{M}_e^i \mathbf{e}^{(1),i} \tag{4.210.f}$$

where the new matrices are

$$\mathbf{L}^a_{p_\parallel^3,n'n} = \int_{\partial\gamma^i} \boldsymbol{\gamma}^a_{n'} \cdot \left[\begin{bmatrix} a_1 + b_1 & \\ & c_1 + d_1 \end{bmatrix} \cdot \hat{\mathbf{i}}_\perp \times \left(\hat{\mathbf{i}}_\perp \times \boldsymbol{\gamma}^a_n\right)\right] dS \tag{4.211.a}$$

$$\mathbf{L}^a_{p_\parallel^2,n'n} = \int_{\partial\gamma^i} \boldsymbol{\gamma}^a_{n'} \cdot \left[\begin{bmatrix} a_2 + b_2 + a_1 b_1 & \\ & c_2 + d_2 + c_1 d_1 \end{bmatrix} \cdot \hat{\mathbf{i}}_\perp \times \left(\hat{\mathbf{i}}_\perp \times \boldsymbol{\gamma}^a_n\right)\right] dS \tag{4.211.b}$$

$$\mathbf{L}^a_{p_\parallel^1,n'n} = \int_{\partial\gamma^i} \boldsymbol{\gamma}^a_{n'} \cdot \left[\begin{bmatrix} a_1 b_2 + a_2 b_1 & \\ & c_1 d_2 + c_2 d_1 \end{bmatrix} \cdot \hat{\mathbf{i}}_\perp \times \left(\hat{\mathbf{i}}_\perp \times \boldsymbol{\gamma}^a_n\right)\right] dS \tag{4.211.c}$$

$$\mathbf{L}^a_{p_\parallel,n'n} = \int_{\partial\gamma^i} \boldsymbol{\gamma}^a_{n'} \cdot \left[\begin{bmatrix} a_2 b_2 & \\ & c_2 d_2 \end{bmatrix} \cdot \hat{\mathbf{i}}_\perp \times \left(\hat{\mathbf{i}}_\perp \times \boldsymbol{\gamma}^a_n\right)\right] dS \tag{4.211.d}$$

$$\mathbf{L}^{ai}_{p_\parallel e^1,n'n} = -\frac{\varepsilon_0}{2} \int_{\partial\gamma^i} \boldsymbol{\gamma}^a_{n'} \cdot \left[\begin{bmatrix} a_0 b_1 & a_1 b_0 \\ c_0 d_1 & c_1 d_0 \end{bmatrix} \cdot \hat{\mathbf{i}}_\perp \times \left(\hat{\mathbf{i}}_\perp \times \boldsymbol{\Phi}^i_n\right)\right] dS \tag{4.211.e}$$

$$\mathbf{L}^{ai}_{p_\parallel e^2,n'n} = -\frac{\varepsilon_0}{2} \int_{\partial\gamma^i} \boldsymbol{\gamma}^a_{n'} \cdot \left[\begin{bmatrix} a_0 & b_0 \\ c_0 & d_0 \end{bmatrix} \cdot \hat{\mathbf{i}}_\perp \times \left(\hat{\mathbf{i}}_\perp \times \boldsymbol{\Phi}^i_n\right)\right] dS \tag{4.211.f}$$

$$\mathbf{L}^{ai}_{p_\parallel e,n'n} = -\frac{\varepsilon_0}{2} \int_{\partial\gamma^i} \boldsymbol{\gamma}^a_{n'} \cdot \left[\begin{bmatrix} a_0 b_2 & a_2 b_0 \\ c_0 d_2 & c_2 d_0 \end{bmatrix} \cdot \hat{\mathbf{i}}_\perp \times \left(\hat{\mathbf{i}}_\perp \times \boldsymbol{\Phi}^i_n\right)\right] dS \tag{4.211.g}$$

Now, for the unknown vector $\boldsymbol{\alpha}^a$ based on the a-th GSTC-interface, there are seven new unknown components $\mathbf{p}_\parallel^{(3),a}$, $\mathbf{p}_\parallel^{(2),a}$, $\mathbf{p}_\parallel^{(1),a}$, $\mathbf{e}^{(2),i_1}$, $\mathbf{e}^{(1),i_1}$, $\mathbf{e}^{(2),i_2}$, and $\mathbf{e}^{(1),i_2}$. Then, the new combined unknown vector is

$$\boldsymbol{\alpha}^a = \begin{bmatrix} \mathbf{e}^{i_1} & \mathbf{h}^{i_1} & \mathbf{e}^{i_2} & \mathbf{h}^{i_2} & \mathbf{p}^a_\parallel & \mathbf{e}^{(2),i_1} & \mathbf{e}^{(1),i_1} & \mathbf{e}^{(2),i_2} & \mathbf{e}^{(1),i_2} & \mathbf{p}_\parallel^{(3),a} & \mathbf{p}_\parallel^{(2),a} & \mathbf{p}_\parallel^{(1),a} \end{bmatrix}^T \tag{4.212}$$

Meanwhile, there are also 12 systems of PDEs synthesizing the system equation of $\boldsymbol{\alpha}^a$, including two groups of Maxwell's equations in (4.192.a), one group of equations from (4.210.a)–(4.210.d), and two groups of equations from (4.210.e)–(4.210.f). So the problem is still well-posed, and the only difference is the larger DoFs in $\boldsymbol{\alpha}^a$. The time integration is also the RK method by simply replacing the vector in (4.196) with the new one in (4.212). This shows the advantage of treating the polarizability as individual unknowns and using the RK method. Regardless of the dispersive models, it incorporates the surface susceptibilities model like a plug-in. The code only needs to adjust $\boldsymbol{\alpha}$ on the related GSTC-elements and to increase DoFs for the new PDEs without additional manipulation.

4.2.3 Simulation Examples

According to the introduction of the DGTD-GSTC method, three corresponding examples are presented, which mainly focus on (1) the metasurface with time-varying characteristics that are difficult for the frequency-domain method, (2) the dispersive and anisotropic meta-atoms that is more difficult in finite-element based method than in finite-difference method, and (3) the meta-atoms arranged in a curved array that DGTD-GSTC performs better than FDTD-GSTC, which is one of the main reasons the DGTD-GSTC method is developed.

4.2.3.1 The Time-Varying Metasurface

Our initial analysis focuses on an amplitude-modulating metasurface capable of converting a normally incident monochromatic carrier into a dual-sideband modulated signal. This dynamic metasurface exhibits time-modulated transmission properties, with its complex transmission coefficient expressed as $T = \frac{1}{2}$ $[\sin(\omega_s t) + 1]$, where ω_s is the angular frequency of the signal. This kind of metasurface with a specified transmission and reflection coefficient in a certain frequency band has a special form of susceptibilities as shown in (4.85). In DGTD-GSTC method, it can be solved according to the equations in (4.181.a). Here, the time-varying parameters are

$$K_{ee}(t) = K_{mm}(t) = 2c_0 \frac{1 - \sin \omega_s t}{3 + \sin \omega_s t} \tag{4.213}$$

In the present case, the spectral component of the impinging carrier wave is a Gaussian beam with a sinusoidal signal whose parameters are $f_0 = 300$ MHz and $\omega_s = 2\pi \cdot 180 \times 10^6$ rad/s. The results, obtained through (4.184.a), including electric field distribution at a single moment and the frequency spectrums, are presented in Figure 4.29.

As illustrated, the DGTD-GSTC method can simulate time-varying metasurface and accurately reproduce the expected modulation function of the metasurface.

4.2.3.2 Dispersive and Anisotropic Cross-I Meta-Atom

In this example, an infinite array composed of metal cross-I patch is simulated, as shown in Figure 4.30. This meta-atom has I-shapes of different lengths in the two tangential directions, which result in different frequencies of resonance when excited by incident waves of different polarizations. This example, although simple, is a good choice to test the ability of the algorithm to describe the anisotropy and dispersion.

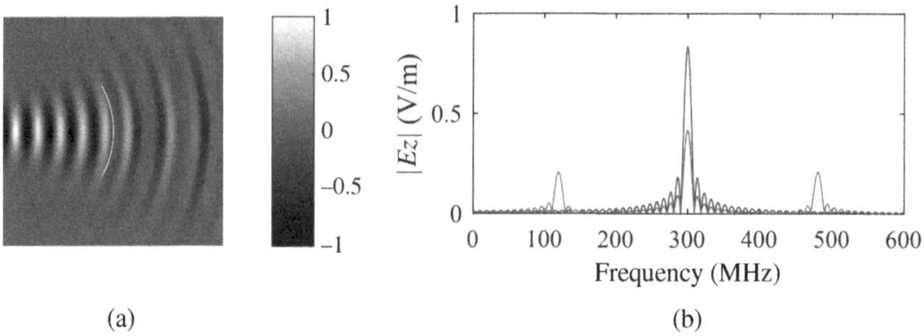

Figure 4.29 The simulation result of the modulation metasurface. (a) The z-direction electric field distribution at one moment. The white curve denotes the metasurface. (b) The frequency spectrums of the incident and transmitted waves.

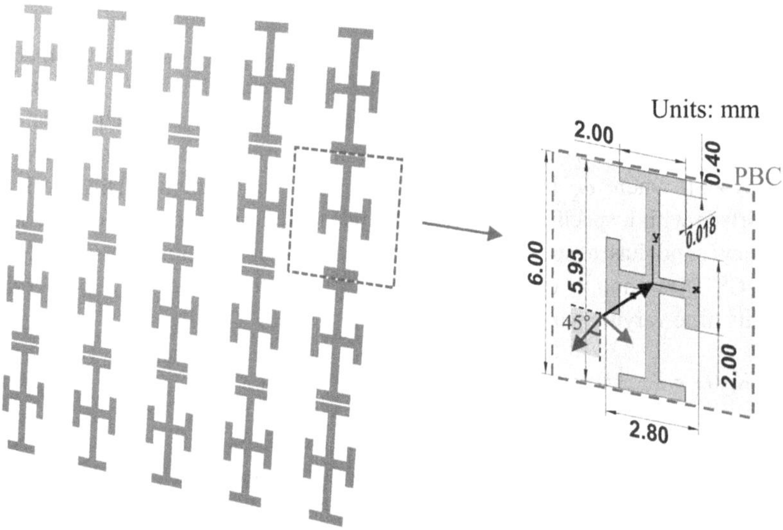

Figure 4.30 The sketch of a dispersive and anisotropic cross-I atom.

The equivalent surface susceptibility tensors, preset with 16 tangential elements and derived according to the method described in Section 2.3.1, have two dominant elements as

$$\chi_{ee}^{xx} = \frac{1.39 \times 10^{19}}{(j\omega)^2 + 3.02 \times 10^8 (j\omega) + 3.77 \times 10^{22}} \tag{4.214.a}$$

$$\chi_{ee}^{yy} = \frac{4.42 \times 10^{19}}{(j\omega)^2 + 5.92 \times 10^8 (j\omega) + 5.85 \times 10^{21}} \tag{4.214.b}$$

The resonance occurs at approximately 30.9 GHz in the x-direction and about 12.2 GHz in the y-direction. A normal incident plane wave polarized at 45° with the

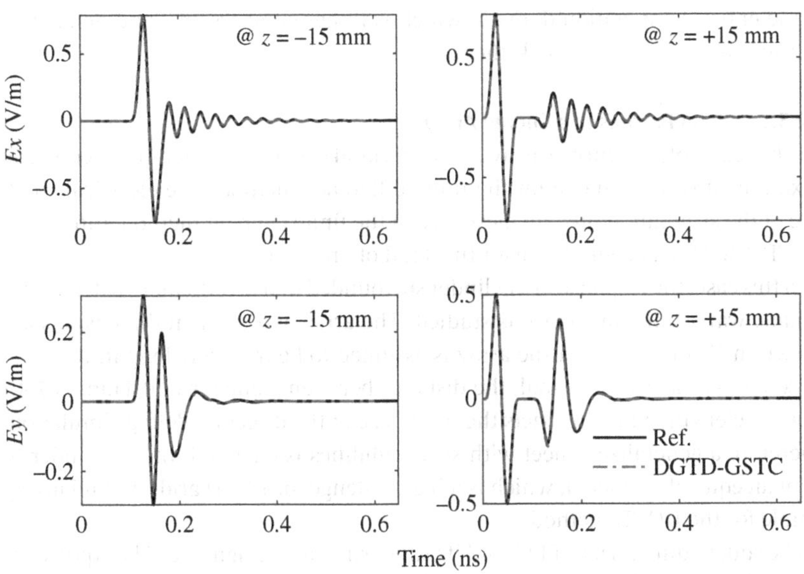

Figure 4.31 The result of the metal cross-I patch in the infinite array.

x-axis is set as excitation, and the pulse is a BHW1d pulse with the characteristic frequency of 22 GHz. The reference result is from the CST time-domain solver with the actual structure. The temporal results of electric fields before and behind the array are shown in Figure 4.31.

Besides, the criterion in (4.171) is used for stability analysis and it calculates the maximum eigenvalue $\lambda = \text{eig}\left(\mathbf{M}_q^{-1}\mathbf{L}_q\right)$ corresponding to the regular elements in (4.165), and $\lambda_\alpha = \text{eig}\left(\mathbf{M}_\alpha^{-1}\mathbf{L}_\alpha\right)$ corresponding to the GSTC-elements in (4.212). Selecting different sizes of time interval Δt from 10^{-14} s to 10^{-10} s, the values of $f(z)$ in (4.171) are shown in Figure 4.32.

As $f(z) < 1$ is the stable region, it can be seen from Figure 4.32 that the maximal stable step size of the regular element is less than that from the GSTC-element. Therefore, one can take any value less than the regular maximal time step interval, and the introduction of the GSTC-equivalence will not lead to numerical divergence. This method is only a necessary criterion, the necessary and sufficient criterion should examine the eigenvalue of the large system matrix corresponding

Figure 4.32 The stability function of regular elements and GSTC-elements with different time step intervals.

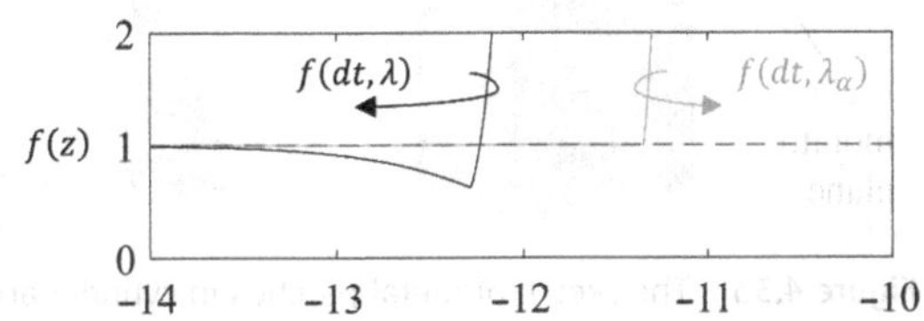

to the entire computation domain, which is rather cumbersome, and hence this method can be used as a quick check.

4.2.3.3 Metal Patches in a Curved Array

The last example is a curved array. Since this kind of simulation requires geometric flexibility, it is not suitable for the finite-difference method. The electrical small size of the structure limits the mesh size of the finite element method, thus using DGTD-GSTC is a better choice for this kind of problem.

In this case, the response of a cylinder surrounded by a metal square patch under a plane wave radial incidence is studied. The array and element dimensions are shown in Figure 4.33, and the array is assumed to be infinitely long in the axial direction. For a 10 GHz signal, the distance between elements of 0.4 mm is 75th of a wavelength, which restricts the mesh size of the direct modeling simulation. Therefore a generalized sheet with susceptibilities is enclosed into a cylinder to form an equivalent model, which is a big challenge for FDTD grids, but relatively simple for the DGTD method.

The square unit is set as PEC, and its supporting base is ignored. The equivalent susceptibilities of this meta-atom is [39]

$$\chi_{ee}^{\tau_1\tau_1} = \chi_{ee}^{\tau_2\tau_2} = 3.095 \times 10^{-3}, \quad \chi_{mm}^{\perp\perp} = -1.367 \times 10^{-3} \tag{4.215}$$

where the normal polarization must be considered because the plane wave is an oblique incidence for the units on both flanks. The field monitor plane is set perpendicular to the z-axis. The excitation is a z-polarized plane wave, and the pulse is BHW1d with the characteristic frequency of 10 GHz. The reference is from the CST time-domain solver with the actual structure, with PBC set on $\pm z$ boundary.

The numerical results and the reference are shown in Figure 4.34. The DGTD-GSTC method reproduces the macro field distribution well compared to the reference.

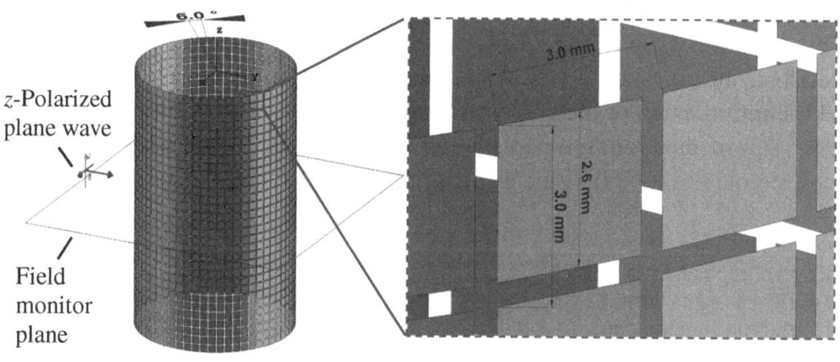

Figure 4.33 The sketch of metal patches in cylinder arrays.

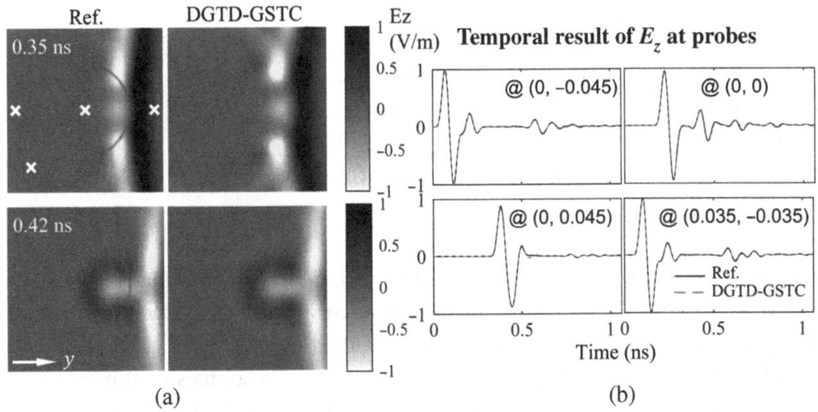

Figure 4.34 The result of metal patches in cylinder arrays. (a) The electric field distributions on the field monitor plane at different moments, and the symbol "x" denotes the probes whose data are shown in (b). (b) The temporal results of the electric fields at different probes.

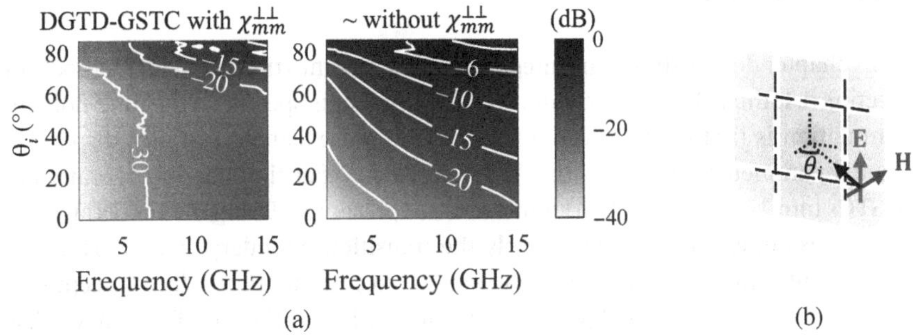

Figure 4.35 (a) The comparison of the relative error of the amplitude of transmission coefficient T of the metal square array between the DGTD-GSTC method with and without the normal susceptibilities for different incident angles and frequencies. The white lines represent the equi-error contour lines. (b) The schematic of extraction of T of the actual periodic metal square patch array, where the black dotted line represents PBC.

To quantitatively analyze the error introduced by the equivalence under different incident angle, the relative error is presented as a function of incident angles, as shown in Figure 4.35(a). For a 12 GHz plane wave at 65° oblique incidence, the relative error between the transmission coefficients T reproduced by (4.215) and by the actual structure is −20 dB (1%), while at 75° the error is about −13 dB (5%). The relative errors between susceptibility-T and actual-T of this metal square patch are shown in Figure 4.35(a), in which there is also a result obtained by DGTD-GSTC without normal surface susceptibilities. The relative

Table 4.3 Computing resource of Example 4.2.3.3.

	Time	Memory
DGTD-GSTC	110 s	1.6 GB
CST	33 min, 46 s	18.4 GB

error could tell the importance of normal polarization in homogenization. In [62], the DGTD-GSTC simulation results of an array with the same unit but only arranged in a 150° arc are shown, whose reproduction is better.

Table 4.3 shows the computing time and memory resources of this example, used by the DGTD-GSTC method (MATLAB 2016a) and CST (2018), both on the same workstation with an Intel Core i7-9700 CPU at 3.00 GHz and with 32-GB RAM.

4.3 Conclusion

This chapter demonstrates the integration of GSTCs into time-domain methods. In Section 4.1, the FDTD method and some key concepts specific to the time-domain simulation is firstly introduced, including the pulse signals, stability, dispersion model parameter fitting, and the ADE approach. After that, the incorporation of GSTCs into 2-D and 3-D FDTD frameworks is presented. Using the IBC as an illustrative example, the method to apply the transition boundary conditions in Yee grids is introduced. Furthermore, the method to consider the surface susceptibilities model's unique normal polarization effects within FDTD-GSTC framework is presented. Finally, the ADE method to deal with the dispersive susceptibilities is introduced. To verify the accuracy and efficiency of the FDTD-GSTC, two numerical examples are provided. The first example addresses a synthesis problem, where surface susceptibility distributions are synthesized based on predefined incident and transmitted fields. The FDTD-GSTC method is then employed to solve the response of these distributions under plane wave excitation. Despite the complexity of susceptibility dispersion, the FDTD-GSTC method demonstrates robust performance, effectively validating its capabilities. The second example focuses on practical metasurface elements. The result shows that FDTD-GSTC method not only accurately reproduces the response of the resonant metasurface but also outperforms commercial software in terms of computational speed and memory efficiency.

In Section 4.2, the DGTD method is introduced, focusing on the concepts relevant to GSTC, including the basis functions for space discretization, the numerical flux, and the time integration schemes. Next, the combination of GSTC and DGTD is presented through the numerical flux and the use of combined unknowns to describe the dispersive susceptibilities. It is also verified that tangential polarization does not affect the p-refinement capability of DGTD. To verify the algorithm, three examples are provided, including a time-varying metasurface arranged in a curved array, an anisotropic classical cross-I meta-atom arranged within the PBC, and densely arranged conformal metal square patches on cylinder surfaces. The integration of DGTD with GSTCs demonstrates significant advantages in time-domain simulations of curved and conformal metasurfaces. By leveraging the flexibility of DGTD in handling complex geometries and the accuracy of GSTCs in modeling metasurface, this combined approach provides a powerful tool for analyzing advanced metasurface designs, particularly those involving non-planar and conformal configurations.

References

1. K. Yee, "Numerical solution of initial boundary value problems involving Maxwell's equations in isotropic media," *IEEE Trans. Antennas Propag.*, vol. 14, no. 3, pp. 302–307, 1966.
2. A. Taflove and M. E. Brodwin, "Numerical-solution of steady-state electromagnetic scattering problems using time-dependent Maxwell's equations," *IEEE Trans. Microwave Theory Techn.*, vol. 23, pp. 623–630, 1975.
3. A. Taflove and M. E. Brodwin, "Computation of electromagnetic-fields and induced temperatures within a model of a microwave-irradiated human eye," *IEEE Trans. Microwave Theory Tech.*, vol. 23, pp. 888–896, 1975.
4. G. Mur, "Absorbing boundary-conditions for the finite-difference approximation of the time-domain electromagnetic-field equations," *IEEE Trans. Electromagn. Compat.*, vol. 23, pp. 377–382, 1981.
5. K. Umashankar and A. Taflove, "A novel method to analyze electromagnetic scattering of complex objects," *IEEE Trans. Electromagn. Compat.*, vol. 24, pp. 397–405, 1982.
6. Wikipedia contributors, Thomas Weiland. 2024. https://en.wikipedia.org/wiki/Thomas_Weiland. Sep. 23, 2024.
7. J. P. Berenger, "A perfectly matched layer for the absorption of electromagnetic waves," *J. Comput. Phys.*, vol. 114, pp. 195–200, 1994.
8. B. Engquist and A. Majda, "Absorbing boundary-conditions for numerical simulation of waves," *Math. Comput.*, vol. 31, pp. 629–651, 1977.
9. A. F. Peterson, S. L. Ray, and R. Mittra, *Computational Methods for Electromagnetics*. New York: IEEE Press, 1998.

10 S. D. Gedney, "An anisotropic perfectly matched layer-absorbing medium for the truncation of FDTD lattices," *IEEE Trans. Antennas Propag.*, vol. 44, pp. 1630–1639, 1996.

11 J. P. Berenger, *Perfect Matched Layer (PML) for Computational Electromagnetics*. San Rafael, CA: Morgan and Claypool Publishers, 2007.

12 W. C. Chew and W. H. Weedon, "A 3D perfectly matched medium from modified Maxwell's equations with stretched coordinates," *Microw. Opt. Technol. Lett.*, vol. 7, pp. 599–604, 1994.

13 J. A. Roden and S. D. Gedney, "Convolutional PML (CPML): An efficient FDTD implementation of the CFS-PML for arbitrary media," *Microw. Opt. Technol. Lett.*, vol. 27, pp. 334–339, 2000.

14 S. D. Gedney, *Introduction to the Finite-Difference Time-Domain (FDTD) Method for Electromagnetics*. San Rafael, CA: Morgan & Claypool Publishers, 2011.

15 F. J. Harris, "Use of windows for harmonic-analysis with discrete Fourier-transform," *Proc. IEEE*, vol. 66, pp. 51–83, 1978.

16 D. E. Merewether, R. Fisher, and F. W. Smith, "On implementing a numeric Huygen's source scheme in a finite difference program to illuminate scattering bodies," *IEEE Trans. Nucl. Sci.*, vol. 27, pp. 1829–1833, 1980.

17 R. Courant, K. O. Friedrichs, and H. Lewy, "On the partial differential equations of mathematical physics (translated)," *IBM J. Res. Dev.*, vol. 11, pp. 215–234, 1967.

18 S. D. Gedney and J. A. Roden, "Numerical stability of nonorthogonal FDTD methods," *IEEE Trans. Antennas Propagat.*, vol. 48, no. 2, pp. 231–239, 2000.

19 H. Choi, J.-W. Baek, and K.-Y. Jung, "Comprehensive study on numerical aspects of modified Lorentz model-based dispersive FDTD formulations," *IEEE Trans. Antennas Propagat.*, vol. 67, no. 12, pp. 7643–7648, 2019.

20 P. Triverio, "System- and data-driven methods and algorithms: 8 vector fitting," in *System- and Data-Driven Methods and Algorithms*. De Gruyter, 2021, pp. 275–310.

21 E. Levy, "Complex-curve fitting," *IRE Trans. Autom. Control.*, vol. AC–4, no. 1, pp. 37–43, 1959.

22 C. Sanathanan and J. Koerner, "Transfer function synthesis as a ratio of two complex polynomials," *IEEE Trans. Autom. Control*, vol. 8, no. 1, pp. 56–58, 1963.

23 G. H. Golub and C. F. V. Loan, *Matrix Computations*. Johns Hopkins University Press, 1996.

24 B. Gustavsen and A. Semlyen, "Rational approximation of frequency domain responses by vector fitting," *IEEE Trans. Power Deliv.*, vol. 14, no. 3, pp. 1052–1061, 1999.

25 B. Gustavsen, "Improving the pole relocating properties of vector fitting," *IEEE Trans. Power Deliv.*, vol. 21, no. 3, pp. 1587–1592, 2006.

26 Sintef, Vector Fitting 2026. https://www.sintef.no/en/software/vector-fitting/ Apr. 14, 2026.

27 MathWorks, Curve Fitting Toolbox 2026. https://ww2.mathworks.cn/products/curvefitting.html Apr. 14, 2026.

28 X. Jia, X. Wang, and Y. Vahabzadeh, "Fast computation of resonant metasurfaces in FDTD scheme using dispersive surface susceptibility model," *IEEE Trans. Antennas Propagat.*, vol. 71, no. 1, pp. 713–722, 2022.

29 K. Hosseini and Z. Atlasbaf, "PLRC-FDTD modeling of general GSTC-based dispersive bianisotropic metasurfaces," *IEEE Trans. Antennas Propagat.*, vol. 66, no. 1, pp. 262–270, 2018.

30 K. Wu and Q. Ren, "A Vector Fitting Enhanced GSTC-FDTD Method for Broadband Analysis of Dispersive Metasurfaces," in 2023 International Applied Computational Electromagnetics Society Symposium (ACES-China), Aug. 2023.

31 V. Nayyeri, M. Soleimani, and O. M. Ramahi, "Modeling graphene in the finite-difference time-domain method using a surface boundary condition," *IEEE Trans. Antennas Propagat.*, vol. 61, no. 8, pp. 4176–4182, 2013.

32 Y. Vahabzadeh, N. Chamanara, and C. Caloz, "Generalized sheet transition condition FDTD simulation of metasurface," *IEEE Trans. Antennas Propagat.*, vol. 66, no. 1, pp. 271–280, 2018.

33 T. J. Smy, S. Stewart, and S. Gupta, "Integrated Generalized Sheet Transition Conditions (GSTCs) in a Yee-Cell based Finite-Difference Time-Domain (FDTD) Simulation of Electromagnetic Metasurfaces," Jun. 14, 2017, arXiv: arXiv:1706.10136. doi:10.48550/arXiv.1706.10136.

34 K. Achouri, M. A. Salem, and C. Caloz, "General metasurface synthesis based on susceptibility tensors," *IEEE Trans. Antennas Propagat.*, vol. 63, no. 7, pp. 2977–2991, 2015.

35 X. Jia, Y. Vahabzadeh, C. Caloz, and F. Yang, "Synthesis of spherical metasurfaces based on susceptibility tensor GSTCs," *IEEE Trans. Antennas Propagat.*, vol. 67, no. 4, pp. 2542–2554, 2019.

36 A. Taflove, *Computational Electrodynamics: The Finite Difference Time Domain.* Boston: Artech House, 2005.

37 K. Wu, P. Coquet, Q. J. Wang, and P. Genevet, "Modelling of free-form conformal metasurfaces," *Nat. Commun.*, vol. 9, no. 1, Art. no. 1, 2018.

38 J. Yi, L. J. Wong, C. Molardi, and P. Genevet, "Controlling electromagnetic fields at boundaries of arbitrary geometries," *Phys. Rev.*, vol. 94, no. 2, 023820, 2016.

39 X. Jia, F. Yang, X. Liu, M. Li, and S. Xu, "Fast nonuniform metasurface analysis in FDTD using surface susceptibility model," *IEEE Trans. Antennas Propagat.*, vol. 68, no. 10, pp. 7121–7130, 2020.

40 F. Yang, J. Chen, R. Qiang, and A. Elsherbeni, "A simple and efficient FDTD/PBC algorithm for scattering analysis of periodic structures," *Radio Sci.*, vol. 42, no. 4, pp. 1–9, 2007.

41 D. R. Smith, W. J. Padilla, D. C. Vier, S. C. Nemat-Nasser, and S. Schultz, "Composite medium with simultaneously negative permeability and permittivity," *Phys. Rev. Lett.*, vol. 84, no. 18, pp. 4184–4187, 2000.

42 Y. Wang, J. Su, Z. Li, Q. Guo, and J. Song, "A prismatic conformal metasurface for radar cross-sectional reduction," *IEEE Antennas Wirel. Propag. Lett.*, vol. 19, no. 4, pp. 631–635, 2020, doi: 10.1109/LAWP.2020.2974018.

43 Z. H. Jiang, L. Kang, and D. H. Werner, "Conformal metasurface-coated dielectric waveguides for highly confined broadband optical activity with simultaneous low-visibility and reduced crosstalk," *Nat. Commun.*, vol. 8, no. 1, p. 356, 2017.

44 W. H. Reed and T. R. Hill, Triangular mesh methods for the neutron transport equation, Los Alamos Scientific Laboratory Report, LA-UR-73-479, Oct. 1973.

45 J. S. Hesthaven and T. Warburton, "Nodal high-order methods on unstructured grids: I. Time-domain solution of Maxwell's equations," *J. Comput. Phys.*, vol. 181, no. 1, pp. 186–221, 2002.

46 J. S. Hesthaven and T. Warburton, *Nodal Discontinuous Galerkin Methods, Texts in Applied Mathematics*, vol. 54. Springer New York: New York, NY, 2008.

47 J. Álvarez González, "A Discontinuous Galerkin Finite Element Method for the Time-Domain Solution of Maxwell Equations," Ph.D. dissertation, Universidad de Granada, 2014.

48 J. Chen and Q. H. Liu, "Discontinuous Galerkin time-domain methods for multiscale electromagnetic simulations: A review," *Proc. IEEE*, vol. 101, no. 2, pp. 242–254, 2013.

49 M. Dong, L. Chen, P. Li, L. Jiang, and H. Bagci, "Discontinuous Galerkin time-domain method in electromagnetics: From nanostructure simulations to multiphysics implementations," in *Advances in Time-Domain Computational Electromagnetic Methods*. IEEE, 2023, pp. 135–198.

50 R. D. Graglia, D. R. Wilton, and A. F. Peterson, "Higher order interpolatory vector bases for computational electromagnetics," *IEEE Trans. Antennas Propagat.*, vol. 45, no. 3, pp. 329–342, 1997, doi: 10.1109/8.558649.

51 L. S. Andersen and J. L. Volakis, "Hierarchical tangential vector finite elements for tetrahedra," in IEEE Antennas and Propagation Society International Symposium, vol. 1, No. 98, Jun. 1998, pp. 240–243.

52 V. Shankar, A. H. Mohammadian, and W. F. Hall, "A time-domain, finite-volume treatment for the Maxwell equations," *Electromagnetics*, vol. 10, no. 1–2, pp. 127–145, 1990.

53 R. J. LeVeque, *Finite volume methods for hyperbolic problems.* New York: Cambridge University Press, 2002.

54 A. Catella, V. Dolean, and S. Lanteri, "An unconditionally stable discontinuous Galerkin method for solving the 2-D time-domain Maxwell equations on unstructured triangular meshes," *IEEE Trans. Magnet.*, vol. 44, no. 6, pp. 1250–1253, 2008.

55 J. C. Butcher, *Numerical Methods for Ordinary Differential Equations*, 2nd ed. Chichester; Hoboken, NJ: Wiley, 2008.

56 R. Diehl, K. Busch, and J. Niegemann, "Comparison of low-storage RK schemes for discontinuous Galerkin time-domain simulations of Maxwell's equations," *J. Comput. Theor. Nanosci.*, vol. 7, no. 8, pp. 1572–1580, 2010, doi: 10.1166/jctn.2010.1521.

57 L. Fezoui, S. Lanteri, S. Lohrengel, and S. Piperno, "Convergence and stability of a discontinuous Galerkin time-domain method for the 3D heterogeneous Maxwell equations on unstructured meshes," *ESAIM: M2AN*, vol. 39, no. 6, pp. 1149–1176, 2005.

58 K. Wu, S. Tian, and Q. Ren, "A Vectorial Discontinuous Galerkin Time-Domain Method Incorporating Generalized Sheet Transition Conditions," *IEEE Trans. Microwave Theory Tech.*, vol. 70, no. 8, pp. 3765–3775, 2022.

59 S. Tian, K. Wu, and Q. Ren, "Modeling of metasurfaces using discontinuous Galerkin time domain method based on generalized sheet transition conditions," *IEEE Trans. Antennas Propagat.*, vol. 70, no. 8, pp. 6905–6917, 2022.

60 K. Wu and Q. Ren, "A New Numerical Flux for the DGTD-GSTC Method in Simulation of Metasurfaces," in 2022 International Applied Computational Electromagnetics Society Symposium (ACES-China), Dec. 2022.

61 L. Chen, M. B. Ozakin, R. Zhao, and H. Bagci, "A locally-implicit discontinuous Galerkin time-domain method to simulate metasurfaces using generalized sheet transition conditions," *IEEE Trans. Antennas Propagat.*, vol. 71, no. 1, pp. 869–881, 2023.

62 K. Wu and Q. Ren, "An improved DGTD-GSTC method equipped with normal surface polarizabilities for oblique incidence," *IEEE Trans. Antennas Propagat.*, vol. 72, no. 12, pp. 9317–9328, 2024.

63 J. S. Savage and A. F. Peterson, "Higher-order vector finite elements for tetrahedral cells," *IEEE Trans. Microwave Theory Tech.*, vol. 44, no. 6, pp. 874–879, 1996.

5
Conclusions and Remarks

The main goal of research and development of metasurfaces is to enrich the ability to shape and manipulate electromagnetic waves. With the rapid development of computing technology, numerical simulation has become an indispensable part of metasurface research. Efficient simulation for designing metasurfaces is still a challenging task. In this chapter, we summarize the content of the book and discuss the future challenges of metasurface modeling and related numerical algorithms.

5.1 Conclusions on the GSTC and Surface Susceptibility Model

Beginning with the asymptotic analysis, the Maxwell's equations in the region around the periodic structure are rigorously transformed into a set of PDEs, in which the electric and magnetic fields are decoupled into periodicity-structure-dependent and -independent fields. When the periodic structures are electrically small, the periodicity-structure-dependent fields vary rapidly compared to the wavelength. Hence, the fields can be expanded according to the ratio between the periodicity and the working wavelength. When we want to replace some periodic and extremely small structures with equivalent parameters, the periodicity-structure-independent field, called the effective field, should be the same as the effective field before the replacement. In other words, the electrically small periodic structure determines the real field distribution, and the equivalent model only needs to reproduce the effective field. This is the theoretical basis of GSTC and other homogenized models.

The GSTC is a surface-based homogenization, and it uses the surface susceptibility model to replace the metasurface. Compared with the bulk-based homogenized parameters $\bar{\bar{\varepsilon}}$, $\bar{\bar{\mu}}$, $\bar{\bar{\xi}}$, and $\bar{\bar{\zeta}}$, the surface susceptibility tensors $\bar{\bar{\chi}}$ reduce one dimension of the vertical spatial characterization of metasurfaces. Just like the engineers often treat 0.018-mm-thick metal patches as PEC boundaries with zero thickness in simulation software, modeling thin metasurfaces as GSTC boundaries with zero-thickness is very efficient in simulation. Compared with other conventional boundary conditions, the advantage of the GSTC and surface susceptibility model is that it provides more DoFs for describing the properties of the meta-atoms, especially the capacity of expressing the normal polarization, which provides a wider effective range of incident angles in the equivalent models.

Determining the surface susceptibility tensors is a typical inverse problem. According to the uniqueness theorem, when the source and the tangential components of the electric or magnetic fields are specified on the entire boundary surface of a given region, the field distribution within that region is unique. Therefore, we mainly use the first and second formulas of GSTC to obtain the surface susceptibility tensors, and the last two divergence formulas may be redundant. The tangential surface susceptibilities are relatively simple to determine because they are not dependent on spatial derivatives and can form a closed solution in many cases. In contrast, the normal surface susceptibility is more complicated because it involves the spatial derivative. By using the wave number instead of the spatial derivative, the problem has a closed solution at the cost of being only valid within a certain wavenumber range. Although there is a range limit, it is more accurate to include the normal components of surface susceptibility than to be without them, especially when the angular response of the metasurface is obvious.

5.2 Conclusions on Numerical Algorithms

In this book, we have introduced a total of six algorithms and their incorporation with GSTC, among which SEM in Section 3.3 is a special version of FEM in Section 3.2. Each of these methods has its own pros and cons. In terms of the maturity of combination with GSTC, IE-GSTC in Section 3.5 and FDTD-GSTC in Section 4.1 have been developed relatively well, while FEM-GSTC in Section 3.2 and DGTD-GSTC in Section 4.2 are still in a fast-developing stage.

The FDFD-GSTC described in Section 3.1 is easy to implement, and FDFD is the most suitable algorithm for flat metasurfaces with no other complex scatterers. In addition, when there is no analytical solution or reliable numerical solution with the actual structures, the solution of the FDFD-GSTC method can also be used as a reference for other equivalent algorithms with GSTC. Similarly, FDTD-GSTC in Section 4.1 also has the advantage of being easy to implement. As a time-domain method, FDTD-GSTC needs to consider the fitting process of dispersive surface susceptibilities. The accuracy of the dispersion fitting described in Section 4.1.2 will directly affect the accuracy of FDTD-GSTC.

In Section 4.1.3, we introduce the coupling methods of GSTC and FDTD, and the auxiliary differential equation method for dispersive surface susceptibilities. FDFD-GSTC and FDTD-GSTC methods are easier to deal with the normal surface susceptibilities and anisotropic tangential surface susceptibilities because the field quantities are just the unknowns. When we want to investigate curved metasurfaces, FEM-GSTC, IE-GSTC, and DGTD-GSTC are all good choices. In Section 3.2, we introduce the basic theory of FEM-GSTC with derivation and simulation. The basis functions and boundary conditions of FEM are the key points to incorporate GSTC. We then use the FEM to derive the stiffness matrix, mass matrix, and the matrix originating from GSTC. In Section 3.4, we introduce the only integral-equation-based algorithm, IE-GSTC. The advantage of IE-GSTC lies in its ability to handle curved metasurfaces and fewer DoFs compared to the PDE-based methods. As the last algorithm introduced in this book, the DGTD method appeared later than other algorithms in computational electromagnetics. Because the basic logic of the DGTD method accepts the discontinuity of the field between adjacent grids, it is very suitable for incorporating GSTC. From the perspective of modifying the existing codes, the DGTD method requires the least effort. As a variation of the finite element method, DGTD-GSTC is also geometrically flexible and suitable for curved metasurfaces. With the help of GSTC, each method will be much faster than directly modeling the actual structures. According to the characteristics of the metasurface, one can choose which algorithm is the best fit. In addition, the GSTC coupling techniques introduced in this book are all based on the basic versions of these algorithms. They may conflict with some advanced techniques indeed; however, for most of the techniques, GSTC is friendly in cooperation. For example, GSTC is compatible with the message passing interface (MPI) and GPU parallelization in the DGTD method [1].

In terms of efficiency, GSTC-incorporated methods are undoubtedly faster than the brute-force method. Comparing the three methods in Chapter 3, the speed of solving the system of linear equations $\mathbf{Ax} = \mathbf{b}$ is similar for them. For the example in Section 3.4.3.4, under the same mesh size, the solving time of FEM-GSTC is also about 100 s, similar to that of IE-GSTC. Although the matrix $\mathbf{A}$ of IE is dense and the matrix $\mathbf{A}$ of FEM is sparse, the size of the matrix $\mathbf{A}$ of the IE method is much smaller than that of FEM because it only needs to place unknowns on the GSTC interface. As the size of the problem increases, it is necessary to solve the system equation by an iterative method. Usually, FDFD and FEM methods employ the Krylov subspace method [2], while the IE method uses the FMM [3]. Comparing the FDTD and DGTD methods in Chapter 4, we use a set of quantitative comparisons to illustrate the efficiency. We simulate the problem mentioned in Section 4.1.2.2 using FDTD-GSTC and DGTD-GSTC, respectively. The calculation time and related information are shown in Table 5.1.

In Table 5.1, p, dx, N_{elem}, Δt, and N_T denote the order of the basis function, length of the mesh grids, number of the grids, time interval, and number of time steps. The term "Time" denotes the CPU time used in the time-stepping process, and "Error" represents the error between numerical and analytical transmission coefficients. Figure 5.1 presents the mesh grids of these three simulations.

The FDTD-GSTC method has second-order accuracy in both spatial and time discretization. It consumes the least simulation time. The DGTD-GSTC with first-order basis function has second-order spatial accuracy and has second-order temporal accuracy by using the second-order RK method. There are two reasons why DGTD-GSTC ($p=1$) takes more time than FDTD-GSTC with the same order of accuracy. First, the Δt of DGTD is smaller than that of FDTD, so the DGTD method has more time steps than FDTD. Second, each DoF of DGTD also needs to communicate more with other DoFs in the time-stepping process than FDTD. The advantage of the DGTD is its p-refinement ability, as shown by the DGTD-GSTC ($p=2$), which has third-order accuracy and hence is easier to achieve higher precision.

Table 5.1 The detailed information of the simulation of TE wave incidence on the interface between two different media using FDTD-GSTC and DGTD-GSTC methods.

	p	dx (mm)	N_{elem}	DoFs	Δt (ps)	N_T	Time	Error (dB)
FDTD-GSTC	—	5	10,201	30,603	4.16	3100	20.0	−23.68
DGTD-GSTC	1	20	1460	16,060	1.95	9200	34.4	−23.52
	2	50	240	5040	2.92	2200	24.9	−26.85

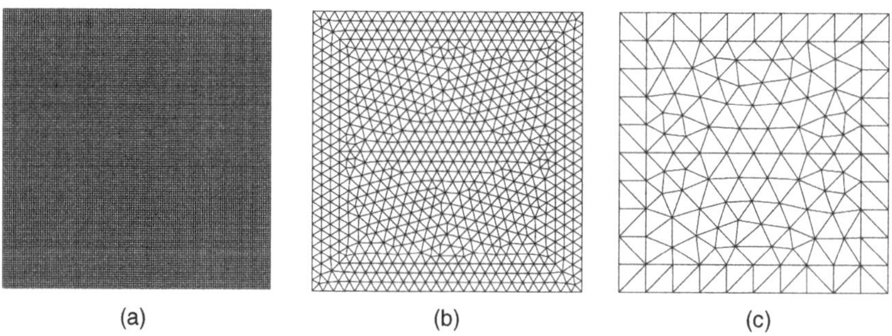

Figure 5.1 The mesh grids of the TE wave simulations in (a) FDTD-GSTC, (b) DGTD-GSTC with $p=1$, and (c) DGTD-GSTC with $p=2$.

5.3 Future Challenges

This book summarizes the GSTC theory and its incorporation with several numerical algorithms to accelerate metasurface simulations. In this section, we present some future challenges to the GSTC theory and fast simulation of the metasurface.

5.3.1 Artificial Intelligence in Susceptibility Retrieval

By the time this book was finished, the pioneers in artificial intelligence (AI) had won the Nobel Prize in both physics and chemistry. AI, deep learning, and neural networks, the hottest words in the scientific community, are not new things in the field of CEM [4–6]. Many researchers have made good achievements by using deep learning instead of traditional numerical methods to solve partial differential equations. On the other hand, certain progress has been made by using AI-assisted design of metasurfaces and metamaterials [7]. In any case, many aspects of AI in electromagnetic computing demonstrate the speed and efficiency of subverting traditional methods. For this book, the surface susceptibilities retrieval described in Section 2.3.4.3 is a very suitable process of AI.

For an actual metasurface element, the retrieval of its tangential surface susceptibilities only requires one PBC simulation; however, it is not rigorous to confirm the normal surface susceptibilities with only one oblique incidence. The rigorous approach should sweep the angle and then determine the normal surface susceptibilities according to the tolerance of the error between the approximated and actual S-parameters. Because sweeping the angles involves multiple full-wave simulations, it is very time-consuming. But this task is very suitable for AI; the geometry of meta-atoms is not a very complex input for AI. Moreover, the characteristic information of typical meta-atoms can also be used as input for AI training, such as the radius of the split ring, the periodicity of the SRRs, and the unit length resistance of the ring of the SRRs. The surface susceptibilities extracted by full-wave simulation can be used as the output for training the neural network. After enough training, we can use the neural network to perform susceptibility retrieval. Given AI's recent success in predicting permittivity and permeability, the susceptibility retrieval is a good fit for AI.

5.3.2 Rationality of the Surface Susceptibilities Model

Compared with surface impedance and admittance, the most prominent feature of the surface susceptibility model is its normal component, which makes it have better wide-angle effectiveness. However, with the deepening of research on metasurfaces, especially spatial dispersive metasurfaces, the surface susceptibilities are incident-angle dependent [8, 9], and the stronger the angle dependence is, the narrower the effective angular domain of the equivalent model is. From the perspective of exploration, the less dependent the parameters of the equivalent model are on excitation, the more intrinsic the model becomes. Whether the surface susceptibility model can be modified to improve its intrinsic degree or whether there will be a more suitable model is one of the challenges in the future. Considering that the normal-tangential coupling elements $\chi^{\tau\perp}$ and $\chi^{\perp\tau}$ have not been fully studied, the authors believe that the surface susceptibility model still has room for optimization.

Besides, surface susceptibilities also have the non-causal problem [10–12], which has a greater impact on the time-domain simulation. While this is not a unique problem of the surface susceptibility model, bulk-based equivalent parameters also have this problem [13]. According to the practice of researchers in bulk-based equivalence theory, non-local constitutive relations are used to explain non-causality [14], and this idea can also be adopted for the surface susceptibility model [14]. Another idea to handle non-causal susceptibilities is to use an enlarged formulation of the GSTC used in all-dielectric metasurfaces, in which the layer is excluded from the computational domain [12].

5.3.3 Calculation Rules of Surface Susceptibilities

Most well-trained circuit engineers can easily use the basic components of the circuit theory, resistance R, inductor L, and capacitor C, to build a circuit that meets very complex needs. However, designing an electromagnetic metasurface that meets specific requirements seems to be a challenging task. In fact, the difficulty of metasurface design is that there is no corresponding calculation rule for the surface susceptibilities, like series and parallel connections in circuit theory.

As shown in Figure 5.2, with the known surface susceptibilities of several elements $\bar{\bar{\chi}}_1, \bar{\bar{\chi}}_2, \ldots, \bar{\bar{\chi}}_n$, how can we obtain the surface susceptibilities of arranging them sequentially in the normal or tangential direction? There are two options. The first is treating them as a new element and re-extracting the surface susceptibilities [15], and the second is converting susceptibilities into circuit parameters, using the series and parallel connection in circuit theory to obtain the total resistance, and then converting them back [16]. So it is important to explore the convenient calculation rules f_s and f_p for the surface susceptibilities, just like what we do for circuits.

5.3.4 GSTC in the High-Frequency Method

This book introduces the application of GSTC in full-wave methods. In fact, for the high-frequency approximation method using impedance boundary conditions,

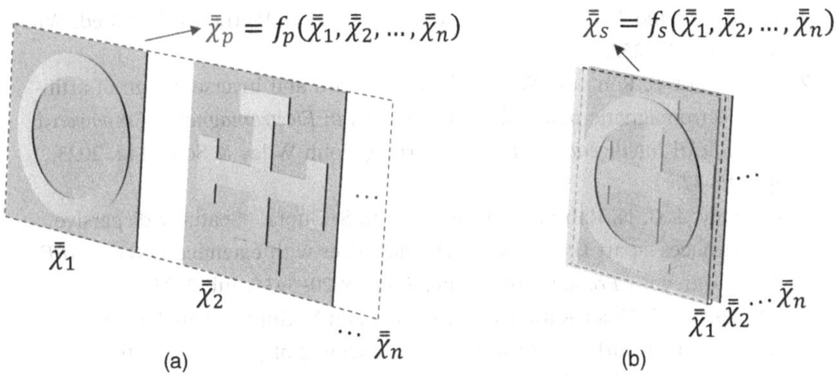

Figure 5.2 The meta-atoms with known surface susceptibilities arranged in (a) tangential or (b) normal direction.

GSTC can be a better substitution for impedance boundary conditions. The theoretical base of surface impedance and surface susceptibilities are both the equivalent source distribution, while the latter provides additional DoFs in the normal direction. These two equivalent models can be converted to each other for single-angle or narrow-angle validity cases, but for wide-angle validity cases, surface susceptibilities provide a richer capability.

References

1 L. Chen, M. B. Ozakin, R. Zhao, and H. Bagci, "A locally-implicit discontinuous Galerkin time-domain method to simulate metasurfaces using generalized sheet transition conditions," *IEEE Trans. Antennas Propag.*, vol. 71, no. 1, pp. 869–881, 2023.

2 G. H. Golub and C. F. van Loan, *Matrix Computations*, 3rd ed. Baltimore: Johns Hopkins University Press, 1996.

3 R. Coifman, V. Rokhlin, and S. Wandzura, "The fast multipole method for the wave equation: a pedestrian prescription," *IEEE Antennas Propag. Mag.*, vol. 35, no. 3, pp. 7–12, 1993.

4 Q. Ren, Y. Wang, Y. Li, *et al.*, *Sophisticated Electromagnetic Forward Scattering Solver Via Deep Learning*. Singapore: Springer Singapore, 2022.

5 T. G. Grossmann, U. J. Komorowska, J. Latz, and C.-B. Schönlieb, "Can physics-informed neural networks beat the finite element method?," *IMA J. Appl. Math.*, vol. 89, no. 1, pp. 143–174, Jan. 2024.

6 R. Lupoiu and J. A. Fan, "Machine learning advances in computational electromagnetics," in *Advances in Electromagnetics Empowered by Artificial*

Intelligence and Deep Learning, S. Campbell and D. Werner, Eds, 1st ed. Wiley, 2023, pp. 225–252.

7 J. M. Malof, S. Ren, and W. J. Padilla, "Forward and inverse design of artificial electromagnetic materials," in *Advances in Electromagnetics Empowered by Artificial Intelligence and Deep Learning*. John Wiley & Sons, Ltd, 2023, pp. 345–370.

8 T. J. Smy, J. G. N. Rahmeier, J. Dugan, and S. Gupta, "Spatially dispersive metasurfaces—part II: IE-GSTC-SD field solver with extended GSTCs," *IEEE Trans. Antennas Propag.*, vol. 71, no. 7, pp. 5920–5934, Jul. 2023.

9 J. Dugan, J. G. Nizer Rahmeier, T. J. Smy, and S. Gupta, "Part III – Spatially dispersive metasurfaces: Zero thickness modeling of periodic and finite non-uniform surfaces," *IEEE Trans. Antennas Propag.*, vol. 71, no. 7, pp. 5935–5945, 2023.

10 T. Lepetit and B. Kanté, "Cramér–Rao bounds for determination of electric and magnetic susceptibilities in metasurfaces," *Opt. Express*, vol. 23, no. 3, p. 3460, 2015.

11 M. Goodarzi and T. Pakizeh, "Retrieving effective surface susceptibilities of high-index metasurfaces based on dipole approximation," *Opt. Commun.*, vol. 483, 126659, Mar. 2021.

12 N. Lebbe, K. Pham, and A. Maurel, "Stable GSTC formulation for Maxwell's equations," *IEEE Trans. Antennas Propag.*, vol. 70, no. 8, pp. 6825–6840, Aug. 2022.

13 T. Koschny, P. Markoš, D. R. Smith, and C. M. Soukoulis, "Resonant and antiresonant frequency dependence of the effective parameters of metamaterials," *Phys. Rev. E*, vol. 68, no. 6, Dec. 2003.

14 K. Mnasri, A. Khrabustovskyi, M. Plum, and C. Rockstuhl, "Retrieving effective material parameters of metamaterials characterized by nonlocal constitutive relations," *Phys. Rev. B*, vol. 99, no. 3, Jan. 2019.

15 V. Tiukuvaara, T. J. Smy, K. Achouri, and S. Gupta, "Surface susceptibilities as characteristic models of reflective metasurfaces," *IEEE Trans. Antennas Propag.*, vol. 70, no. 11, pp. 10583–10591, Nov. 2022.

16 X. Jia, X. Dang, X. Pan, and M. Wang, "Metasurfaces design using primary elements and characteristic models: Connection between SSM/SPM and circuit quantities," *IEEE Trans. Antennas Propag.*, vol. 71, no. 3, pp. 2837–2842, 2023.

Index

Printed in the USA/Agawam, MA
August 25, 2026

909557.016